W0082785

The Annotated
OLD FOURLEGS
The updated Story of the Coelacanth

Mike Bruton

University Press of Florida
Gainesville/Tallahassee/Tampa/Boca Raton
Pensacola/Orlando/Miami/Jacksonville/Ft. Myers/Sarasota

First published in 2017 by Struik Nature, an imprint of
Penguin Random House South Africa (Pty) Ltd

North American edition published by the University Press of Florida. All rights reserved.
University Press of Florida, 15 Northwest 15th Street, Gainesville, FL 32611-2079
http://upress.ufl.edu

The University Press of Florida is the scholarly publishing agency for the State University
System of Florida, comprising Florida A&M University, Florida Atlantic University, Florida
Gulf Coast University, Florida International University, Florida State University, New College
of Florida, University of Central Florida, University of Florida, University of North Florida,
University of South Florida, and University of West Florida.

1 3 5 7 9 10 8 6 4 2

Copyright © in original *Old Fourlegs* text, 1957: Longmans, Green and Co Ltd
(Readers Union edition), London
Copyright © in text, 2017: Mike Bruton
Copyright © in photographs and illustrations, 2017: See picture credits alongside images
Copyright © in maps, 2017: Penguin Random House South Africa (Pty) Ltd
Copyright © in published edition, 2017: Penguin Random House South Africa (Pty) Ltd

Publisher: Pippa Parker
Editor: Helen de Villiers
Designer: Janice Evans
Typesetter: Deirdré Geldenhuys
Picture researcher: Colette Stott
Proofreader: Emsie du Plessis

Reproduction by Resolution Colour (Pty) Ltd
Printed and bound in China by RR Donnelley Asia Printing Solutions Ltd, Hong Kong

All rights reserved. No part of this publication may be reproduced,
stored in a retrieval system, or transmitted, in any form or by any means,
electronic, mechanical, photocopying, recording or otherwise,
without the prior written permission of the copyright owner(s).

While every effort has been made to contact photographers whose images have been used
in this book, we have, in a few instances – and despite our best efforts – been unable to
contact all copyright holders. We apologize for this and undertake to update image credits
in subsequent editions if the missing information is provided.

9780813064642

Front cover: Coelacanth photographed in the iSimangaliso Wetland Park (Peter Timm);
JLB Smith identifies the second coelacanth, found in 1952; Front cover of the original title,
Old Fourlegs – The Story of the Coelacanth
Title page: Illustration of a coelacanth (Elaine Heemstra)
Contents page: Collage of an unborn coelacanth by Aidon Westcott (Aidon Westcott)
Back cover: Jacket of the original title, *Old Fourlegs – The Story of the Coelacanth*

CONTENTS

Acknowledgements   4

Preface   5

Foreword   6

Introduction to the story of the coelacanth   8

The annotated Old Fourlegs   34

Coelacanth discoveries   298

The significance of the coelacanth   311

Postscript   329

Glossary of terms   330

Acronyms used in the text   332

References and Further reading   332

ACKNOWLEDGEMENTS

I am very grateful to Dr Angus Paterson, CEO of SAIAB, for access to the Institute's archives and library and for the use of photographs, drawings and information in the Institute's collections. Sincere thanks are also due to Sally Schramm and Maditaba Metlaf in the SAIAB library for helping me to access material from the SAIAB library and archives as well as to Kylie van Zyl for facilitating access to the Cory Library at Rhodes University.

My sincere thanks to William Smith for his Foreword and to William and Jenny Smith for supporting this unique book about their parents and the fish that they pursued. Special thanks also go to Jean Pote, who worked with the Smiths and me, and to Rik Nulens, the ultimate coelacanth archivist.

I am very grateful to the following for the use of their photographs: SAIAB, ACEP, Rhodes University, East London Museum, Albany Museum, University of KwaZulu-Natal, South African Air Force (SAAF), JASEC, Muséum National d'Histoire Naturelle de Paris (MNHM), Tellumat, Peter Barnett, Hans Fricke, Jürgen Schauer, Karen Hissmann, Irene McCulloch, Rob Gess, Rik Nulens, Mark Erdmann, Chris Fallows, the late Eugene Balon, Christine Flegler-

Gregorydavid at the English language Wikipedia

Brass plate at Latimer's Landing, East London

http://www.cryptozoonews.com/top-coel-2013/

Coelacanth display in the International Cryptozoology Museum in Portland, USA

Balon, Robin Stobbs, Jean-Louis Geraud, Wolfgang Grulke, the late Augusto Cabral, and Eve Marshall, fiancée of the late Peter Timm.

I would like to thank Struik Nature for agreeing to publish this rather unusual book and especially Pippa Parker for commissioning this title and guiding me through the publishing labyrinth. I thank Helen de Villiers of Struik Nature for her meticulous editing and attention to detail – it was a pleasure working with her. I also thank Janice Evans who not only designed the book with flair but was also instrumental in assessing, researching and finding additional images; and Colette Stott of the Penguin Random House SA Image Library for her untiring efforts to locate suitable images.

I am also grateful to the many people who provided new information or checked facts, including Nancy Tietz, Ofer Gon, Rob Gess, John and Lizzie Rennie, Phil and Elaine Heemstra, Allan Heydorn, Martin Davies, Paul Murray, Fleur Way-Jones, Penny Haworth, Brian and Sue Allanson, Geraldine Morcom, Willie 'Bomber' Burger, Keith Hunt and Graeme Murray.

Finally, I am grateful to my wife, Carolynn, for assisting me with the preparation of this book and for giving me the space to pursue my own passion for the coelacanth.

PREFACE

One of the more fascinating books on my bookshelf is *The Annotated Alice – Lewis Carroll*, published in 1965 by Penguin and edited by Martin Gardner. A poem on the back cover sets the scene, 'Said Gardner to Carroll, come, let us not quarrel 'bout Wonderland logic or Looking Glass lore. I'm a man without malice: I'll annotate Alice .. I'll garnish and season your rhyme with my reason. And we two'll give Alice a new party dress.'

Lewis Carroll was, of course, the well-kept pen name of the Oxford University mathematician, poet and author of brilliantly absurd fiction, the Reverend Charles Lutwidge Dodgson (1832–1898). Besides *Alice's Adventures in Wonderland* (1865) and *Through the Looking Glass, and What Alice Found There* (1871), Carroll also wrote nonsensical poems and parodies, including *Phantasmagoria and Other Poems* (1869), *The Hunting of the Snark – An Agony in Eight Fits* (1876) and the novel *Sylvie and Bruno* (1889).

He penned a curious and complicated, even abstruse, form of nonsense, for another century, and the reader needs to know a great many things that are not part of the text if he or she wishes to capture its full wit and flavour. His books have been analysed by linguists, psychologists, moralists, chess grand masters and even political commentators but probably most profitably by mathematicians and cosmologists, as they are riddled with mathematical, scientific and logical puzzles as well as jokes, puns, brain teasers ('knots') and symbolisms.

Martin Gardner's 'Annotated Alice', published in 1965

JLB Smith's book, *Old Fourlegs – The Story of the Coelacanth*, is also richly permeated with deep emotions, subtle undertones, political and social commentary, international rivalries, and descriptions of the intimate thoughts of a practising scientist, and reveals a great deal about his character. It is useful, when reading the book, to know some background in order to fully understand the importance and impact of what he has had to say.

It struck me, therefore, that an excellent way to bring *Old Fourlegs*, and JLB Smith himself, back to life again would be to publish an annotated version, with the original text fully intact but with extensive notes in the margins that comment on Smith's writing and thinking; and, at the same time, bringing the remarkable story of the coelacanth up to date. In this way I could help to set the scene for JLB's dramatic memoir, which tells of events that took place nearly 80 years ago in a world that was very different from ours today. I also felt that this approach would be the best way to commemorate the contributions of this great South African scientist and to place the discovery of the first and second coelacanths in their proper perspective.

Of course, there is little similarity between the books of Dodgson and Smith. The former are whimsical tales of fantasy whereas the latter is a hard, factual account of a pursuit by an obsessed scientist. I do, however, detect one similarity between the two authors – they both originally trained and became competent in one discipline but excelled in another.

Mike Bruton
July 2016, Cape Town

FOREWORD

When asked to write this Foreword, I was so pleased that my Dad and Mom would be remembered and acknowledged again. I, too, am keen to have *Old Fourlegs* shared with new generations and to take the story to the next level. I am delighted that Mike Bruton has taken on this task and I know of no-one better positioned, with the necessary scientific and literary experience, to make this story come to life again.

We arrived at almost the same time – the coelacanth in December 1938 and I, six months later. I was told that during the months preceding my arrival it was all hands to 'the fish' so when I was born there were no baby clothes (thank heaven for a granny). Fortunately I was not born with scales as some had predicted!

My earliest recollections were of two parents who did nothing but work, both at university and at home (which I assumed all parents did). This, no doubt, was one of the reasons for their great success. And this shaped me so that when I chose a life partner, it had to be someone with whom I could share life and work and I was lucky enough to find Jenny.

It was only some years later that I realized that JLB and Margaret were not like other parents. Dad was incredibly bright (considered at the time to be one of the three greatest minds South Africa had ever produced) and, as a result, he had only select friends. He hated parties and small-talk, and people were scared of him. As a youngster growing up it was impossible for me to win at anything – he always came out on top, which was tough but I had to learn to cope. This made me the person I am today.

When I discovered JLB's lack of interest and disdain for film and television, I knew I had found my own passion. When I produced my first movie as a student he was convinced I would become a criminal. Fortunately he was wrong or at least I never got 'caught' but, despite his being perplexed, I think he would have been proud. Later I successfully developed mathematics and science TV teaching through the Star Schools programme. At its height

the live, phone-in interactive broadcasts reached over 100 million viewers daily in 27 countries.

The realization that the East London find was a coelacanth is an excellent example of JLB's exceptional mind. This fish was only known to the scientific world through fossil remains as a lake (fresh water) dweller, 30 cm long, in the northern hemisphere, and had become extinct some 65 million years ago. At the time it must have been quite something for an unknown South African to claim that he had in his possession a 140 cm, southern-hemisphere, bright blue sea creature which was a real, live coelacanth.

Imagine the world's consternation at this South African 'bloke', who was nothing but a fisherman in their eyes, making ridiculous claims from a relatively obscure country, in scientific terms! There was much scepticism about a chemist talking about ichthyology and even the British Museum mocked JLB, suggesting that his specimen must have been some other sea creature mutilated by a ship's propeller.

It was Margaret, on the other hand, who supplied the warmth and love and kept everything together. It was her ability, not just as a scientist and artist, but also with people, that made my parents a formidable team. She was dearly loved by everyone: once, when her car was stolen, it was taken to the informal settlement near Grahamstown where it was recognized as belonging to Margaret Smith by the community, who forced the thief to return it. She found it parked the next day in the exact spot that it had been taken from, washed and cleaned!

I was too young to remember much about the years between the two coelacanths. I know it was an intense time while the search was on for their home. JLB predicted, using his knowledge of the ocean currents and having studied the fish, that they must occur along the East African coast.

The thrill of the find in the Comoros, and getting that fish out from under the noses of the French and back to South Africa in a military Dakota thanks to the then Prime Minister, Dr Malan, is well documented.

It was amazing that I, as a 14 year-old, could be part of that flight if only from Grahamstown to Cape Town. While Dad was showing the fish to the Prime Minister, I got a much better deal by being taken to the Air Force base – to a teenage boy, fighter planes were much more exciting than an 'old fish'.

During my 10th year of schooling I accompanied my folks on a four-month expedition to the Seychelles. The whole trip was carried out with military precision. Sixteen-hour work days, every day, avoiding cyclones and reefs, and I watched in horror as the boat carrying both my parents was blown into the air by an underwater explosion when the current unexpectedly changed direction. There was 20 kg of unexploded gelignite in that boat! That was the day I nearly became an orphan. *Fishes of Seychelles* was published the following year.

During my Masters at Natal University a group of students felt it unfair that the screening of a film on the birth of a baby was for females only. I made a stink bomb, which cleared the theatre in a very short time so no-one got to see the movie. When I was hauled up in front of the entire Senate, and the Vice-Chancellor (Ernst Malherbe) demanded to know what chemical I had used, my reply was well prepared, 'The same stuff that you and my father used at Stellenbosch University as students to break up a political rally'. I think my answer probably increased my punishment but the effect was well worth it!

William Smith with a photograph of the Knysna Heads in the background (image extracted from a video)

With the royalties from *Old Fourlegs* JLB purchased the Western Head of Knysna for no other reason but to make sure that he could always reach his favourite fishing spot, which is still known today as 'JLB's Rock'. Today the Western Head has been carefully preserved and has become known across the world as the 'Featherbed Nature Reserve', visited by thousands of tourists annually. It is one of the few self-sustaining nature reserves in South Africa and is a flagship in the Western Cape, a pioneer in eco-tourism.

Having proved his theories on coelacanths, JLB felt that he had achieved what he set out to do. It was time to make way for younger ichthyologists. It was sad to see him getting old – although I still was not able to win against such a giant intellect. The brain that had carried him so successfully through life was failing. Not wanting to be a burden, and to allow my mother to continue his work unencumbered, in January 1968 he took his own life, using his chemical knowledge to do so.

I was very privileged to be part of the family and very proud too. I have often wished that I could have inherited my Dad's great mental ability as well as my Mom's people skills.

William Smith
Perth, Western Australia, January 2016

William Smith teaching kids

INTRODUCTION TO THE STORY OF THE COELACANTH

'One friend who kindly read the manuscript asked me if I realised how it revealed myself. I do not mind. No man is a god.'

JLB Smith, in his Foreword to the first edition of *Old Fourlegs*

I first read *Old Fourlegs* as a teenager living in East London, where the first coelacanth was caught in December 1938. By the time I was 10 years old I had gazed in wonder at the first specimen on display in the East London Museum and also met the extraordinary Curator of the museum, Miss (later Dr) Marjorie Courtenay-Latimer.

My first meeting with the formidable Professor Smith took place in the old Department of Ichthyology building at Rhodes University in Grahamstown in 1966, when I was a first-year student and he was 69 years old. He was a stern man who led a spartan life with all the trivialities trimmed away so that he could focus his energy on fishes. When I first managed to penetrate the formidable barrier of women 'protectors', led by his wife Margaret, I asked him a number of questions about fishes that he answered politely, and I then took my leave. Not once during our discussion did he look up from the fish that he was examining under the microscope.

During our second 'meeting' he whacked me over the knuckles with a steel ruler for touching a scale on a coelacanth specimen. Later, we became more familiar and I even went on country walks with him. Then, many years later, after completing my PhD in ichthyology and studying at the Natural History Museum in London, I joined the staff of

Drawing of the old Department of Ichthyology building in Grahamstown

the then JLB Smith Institute of Ichthyology as Senior Lecturer in Ichthyology. Margaret, JLB's widow, was the Director at the time and we worked closely together to develop the teaching of ichthyology in South Africa.

When Margaret retired in 1982 I took over as the second Director of the Ichthyology Institute and continued to work with her to develop the full potential of her 'baby'. I also became fascinated by the life stories of the two 'Fishy Smiths' and, inevitably, found myself irrevocably entwined in the coelacanth story.

JLB Smith (or simply JLB, as he was widely known), was born in Graaff-Reinet, South Africa, on 26th September 1897. He had the same birthday as his wife, and their son, William, grew up thinking that all parents were born on the same day of the year! Smith was educated at schools in Noupoort, De Aar, Aliwal North and, from 1912 to 1914, at Diocesan College (Bishops), Cape Town. At Bishops he excelled in his academic studies and was an active member of the cadets.

Other important scientists and technologists who were graduates of Diocesan College include James Greathead (1844–1896), inventor of the 'Greathead Shield' that was used to drill the tunnels for the first London Underground trains, and Alexander Logie du

Toit (1878–1948), a geologist who was an early supporter of the theory of continental drift.

After matriculating, JLB entered Victoria College (the precursor to Stellenbosch University) in 1915 and, at the end of the year, passed with top marks in his subjects in South Africa and was awarded several bursaries and exhibitions. He then enlisted with the 12th South African Infantry battalion as a machine gunner in the East Africa campaign. Along with many other young soldiers, he soon contracted several tropical diseases, including malaria, dysentery and Malta fever, and was discharged as medically unfit. He retained his sense of humour, though, and, according to his friend Ernst Malherbe, he regaled incredulous ladies who visited him in hospital with tall tales of giant mosquitoes, 'We had to fight them with our bayonets – they were huge creatures almost like little aeroplanes.' The aftermath of these diseases, however, affected him for the rest of his life.

In his account of life at Stellenbosch University JLB does not mention the mischievous pranks that he perpetrated with the 'Heavenly Quartet', probably because he had a pact with the other three that only the last survivor would reveal the truth. The 'quartet' comprised four men who would all achieve fame in their professions: JLB Smith (ichthyology), Ernst Malherbe (Vice-Chancellor of the University of Natal), Frikkie Meyer (Head of Iscor) and W Kupferburger (pioneering mining geologist).

Malherbe out-survived the rest and, in his memorable 1981 autobiography, *Never a Dull Moment*, describes some of their outrageous pranks. One involved turning all the clocks at the university and on the church tower forward by 30 minutes. This caused chaos on the campus and in town as people did not carry wrist watches in those days. The 'quartet' was able to carry out this prank as they were all expert climbers.

James Greathead

Smith returned to Victoria College in the third term of 1916, quickly made up for lost time, and in 1917 graduated as a Bachelor of Arts of the University of the Cape of Good Hope, achieving the highest marks for chemistry in the Union of South Africa. By the end of 1918 he had completed his MSc degree (with distinction) in chemistry, one of the most brilliant chemistry students in the early history of South Africa. He also had a keen interest in classical literature and was apparently an authority on George Bernard Shaw and William Shakespeare.

After a brief stint as a university staff member in Stellenbosch, during which he also ran a paint factory, JLB entered Selwyn College at Cambridge University in England in 1919, where he studied mustard gases under Sir William Pope. While abroad he travelled widely in Britain and on the continent, learnt to speak German, and made many valuable scientific contacts.

He received his PhD from Cambridge University in 1922 and returned to Grahamstown in 1923, where he taught chemistry for 23 years and conducted pioneering research on photo-sensitizing dye-stuffs, the essential oils of indigenous plants, and other organic chemistry topics. He tutored many students who went on to become South Africa's leading organic chemists.

NRF-SAIAB

The 'Heavenly Quartet' in 1917. Standing: JLB Smith and W Kupferburger; sitting: Frikkie Meyer and Ernst Malherbe

While at Stellenbosch University he often went camping, cycling and mountaineering with friends. He also played golf, rugby and tennis but his favourite pastime, after bee-keeping, was angling. Fishes became his new passion and he travelled to the coast by train, ox-wagon and bicycle to catch fishes, once cycling from Stellenbosch to Cape Point and back on unpaved, sandy roads.

JLB Smith had an inquisitive and active mind and was frustrated that he could not identify the fishes that he caught. Later, at Rhodes University College in the 1920s and 1930s, he found the available books on fishes to be inadequate and, using the analytical approach that he had developed as a chemist, he soon developed unique identification keys for South African fishes based on counts of their scales and fin spines and rays, which are still used today. While he was still practising chemistry he started to publish revisionary papers on fishes, with the first paper in 1932 followed by a steady stream of increasingly authoritative articles. His work attracted attention locally and abroad and he was soon appointed as the Honorary Curator of

NRF-SAIAB

JLB Smith as a student in Stellenbosch, ca 1918

Fishes at the museums in Grahamstown, East London and Port Elizabeth.

In the 1940s he was encouraged by his medical doctors to spend more time outdoors for health reasons. Typically, he embraced this challenge with relish; it is estimated that he walked the equivalent of twice around the world during his last 25 years. He also withstood the rigours of several arduous fish-collecting expeditions up the East African coast and, at the age of 70 years, was still working long hours and walking an average of 50 km per week.

In April 1938 JLB married Margaret Mary Macdonald, who had studied and worked under him in the Department of Chemistry. On p. 141 of *Old Fourlegs* he states, 'When I asked my wife to marry me, I said that I did not know if I could bring her happiness, but I could at least promise that she would never be bored.' He lived up to his promise.

Margaret was from sturdy Scots-Afrikaner stock. Her many times great grandmother was Sarah Uys who, as a girl of 13, had loaded her father's guns during the Battle of Blood River. Her father was a Scots medical doctor and her mother had been the first woman mayor in the Cape Province.

Eight months after the wedding the first coelacanth was caught off East London and taken to the local museum, and their lives changed forever. As the only practising ichthyologist in the Eastern Cape, JLB was summoned to identify the specimen. It was hailed as the greatest biological discovery of the 20th century and JLB and Margaret became entwined in one of the most dramatic tales in modern natural history.

At this stage another remarkable person enters the fray – Miss (later Dr) Marjorie Courtenay-Latimer, Curator of the East London Museum. Marjorie was

NRF-SAIAB

The 'Heavenly Quartet' in about 1937 – from left, JLB Smith, Frikkie Meyer, Ernst Malherbe and W Kupferburger

the first museologist in South Africa to develop three-dimensional ecological dioramas to display wild plants and animals in a natural setting, particularly with young visitors in mind. She single-handedly developed the museum's first array of public displays. She and a Xhosa assistant, Enoch Thwate, were the only employees of the museum for the first 15 years, and Marjorie served as its Director for a phenomenal 42 years.

The museum building in which Marjorie worked in the 1930s and 1940s (and the one that JLB mentions throughout the narrative on the first coelacanth) is not the same building as that to which the East London Museum moved in 1951, in Upper Oxford Street, where it is still located. The original building, situated nearby, is now the Buffalo City Public Further Education and Training College.

The capture of the 1938 coelacanth was a classic example, in the field of science, of a 'black swan' event, something unpredictable that came as a total surprise and then had major 'downstream' consequences. If Captain Hendrik Goosen had not shot his trawl off the Chalumna River mouth on that fateful day in December 1938 and the coelacanth had not been caught, JLB would probably have remained an amateur ichthyologist, the Ichthyology Department and Institute in Grahamstown would probably not have been established, and ichthyology would have followed a very different course in South Africa, almost certainly based in Cape Town or Durban.

JLB and Margaret Smith formed a very effective team, with her handling the administration and logistics so that he could be free to pursue his research. She later commented, 'A wife can be independent or indispensable, not both. I chose to be indispensable'. She later also became a competent fish illustrator and scientist in her own right.

JLB was not an easy companion. He was lean and hard and, in later life, had a maniacal focus on being efficient and functional, with no frills or luxuries. He cared little about his clothes or appearance, often

NRF-SAIAB

Margaret and JLB Smith with their dog 'Marlin' at Rhodes University in Grahamstown in about 1946.

East London Museum

Dr Marjorie Courtenay-Latimer in the 1940s (left) and in 2004 in her doctoral robes on the cover of the Journal of the Border Historical Society

East London Museum

The new East London Museum, which was opened to the public in 1951.

wearing a pith helmet, a khaki shirt, shorts and sandals (or a flannel khaki suit), and sporting a severe, crew-cut hairstyle. He also followed a strict though bizarre diet, alternating on different days between fish and fruit and nuts. On expeditions he ate fish prepared in special ways, bananas and cheese. Margaret said that he needed a tailor-made diet because of 'the whiff of gas that he had caught in World War 1'.

He was also moody and intense and did not suffer fools gladly. Peter Barnett, who accompanied the Smiths on a 1951 fish-collecting expedition to Mozambique, described him as follows, 'he had a patriarchal visage, lean cheeks, and deep hollows at the temples; he seemed to me the epitome of perennial youth … he gave the impression of great superiority over the uncertainties of smaller men'. His eyes were 'blue, penetrating, and coldly speculative. They gave the impression of immense force of character, and single-minded purpose.'

According to his wife, Margaret, and others, JLB's field expeditions were characterized by 'discomfort, hardship, danger, and unending hard labour' as he was a harsh

NRF-SAIAB

JLB Smith in 1939

task master who expected everyone to share his maniacal devotion to duty. Although he was apparently a jovial companion who thrived on practical jokes as a student, and enjoyed sports during his early career at Victoria College and Rhodes University College, he became a stern and demanding man.

When the story of old fourlegs was 'dragged from my reluctant pen by the unflagging determination of my wife …', as JLB Smith wrote in his August 1955 Foreword to the first edition published in 1956, I doubt that he anticipated the impact his memoir would have on the world. *Old Fourlegs* became an international best seller and one of the most popular books of science non-fiction in the world at the time.

It has been published in six English editions and in nine other language editions (see table opposite).

I doubt that any other book written by a South African scientist has appeared in so many different language editions.

The translations can be divided into two groups based on their titles (although the body of the text remains the same). The English, Afrikaans, Russian, Latvian and Japanese editions have titles that are variations on *Old Fourlegs – The Story of the Coelacanth*, the original title, whereas the American, German, French, Dutch, Estonian, Slovak and Czech editions use different titles that place the emphasis on the 'search for' or 'discovery of' the coelacanth. The Czech and Slovak titles start with 'cesta za', which means 'in pursuit of' or 'in search of'. The Dutch title, *Vis op die Loop* ('Fish on the Walk') is innovative but perhaps takes the 'old four legs' analogy too far.

One of the major difficulties that translators encountered was to find a word for the term 'old four legs' (a misnomer anyway as the coelacanth does not 'walk' on its lobed fins!). The Slovak and Czech editions use the word *štvornožcom*, which means 'quadruped', so 'old four legs' becomes 'ancestor quadruped', but this is overly technical and would discourage buyers. For instance, one translation of the Latvian title that I have been sent is 'Ancient Quadrupeds. How we Discovered Coelacanths'. Another, lighter translation of 'old four legs' is the 'four-legged animal', but this is misleading as many readers do not regard fishes as animals. In addition, not all the foreign language editions were translated from English – the Latvian and Estonian editions were translated from Russian.

The Russian edition of 1962 (one of the first books by a South African to be translated into Russian) had an extraordinary first print run of 100,000. Jean Pote has pointed out to me that Margaret Smith first heard about the Estonian edition when she received a letter dated 4th September 1968 from a 14 year-old

PUBLICATION HISTORY

Year	Language	Title	English translation of title	Translator	Publisher
1956	English	Old Fourlegs – The Story of the Coelacanth			Longmans, Green & Co., London, UK
1956	English	The Search Beneath the Sea			Henry Holt and Co, Inc., New York , USA
1957	German	Vergangenheit steigt aus dem Meer	The Past Ascends from the Sea	Kurt Lamerdia	Günther Verlag, Stuttgart, Germany
1957	English	Old Fourlegs – The Story of the Coelacanth			Reader's Union, London, UK
1958	English	Old Fourlegs – The Story of the Coelacanth			Pan Books, London, UK
1960	French	À la Poursuite du Coelacanthe	The Search for the Coelacanth	D Meunier	Librairie Plon, Paris, France
1962	Russian	СТАРЦНА ЧТВРОНОГ	Old Fourlegs	PP Gdarova	State Publisher of Geographical Literature, Moscow, Russia
1964	Estonian	Kuidas Avastati Latimeeria	Old Fourlegs – The Story of the Coelacanth	I Veldre	Eesti Riiklik Kirjastus, Tallinn, Estonia
1965	Afrikaans	Ou Vierpoot – Die Verhaal van die Selakant	Old Fourlegs – The Story of the Coelacanth	PC du Plessis	Tafelberg Uitgewers, Cape Town, South Africa
1970	Slovak	Cesta za štvornožcom	A Journey to the Fourleg	Alžbeta Kubišová	Mladé Letá, Bratislava, Slovakia
1973	Dutch	Vis op de Loop	Fish on the Walk	J Mart Duiven	BV Uitgeverij Nijgh & Van Ditmar's – Gravenhage, Netherlands
1977	Latvian	Sencis Četrkājis: Kā Tika Atklāts Celakants	Ancient Fourlegs: How the Coelacanth was Discovered	A Lauzis	Apvārsnis, Riga, Latvia
1981	Japanese	生きた化石：シーラカンス発見物語	Old Fourlegs – The Story of the Coelacanth	Yoshihisa Kajitani	Tuttle Mori, Tokyo, Japan
1993	English audio-book Old Fourlegs – The Story of the Coelacanth				South African Library for the Blind, Grahamstown, South Africa
2012	English	The Search Beneath the Sea – The Story of the Coelacanth			Literary Licensing, Whitefish, MS, USA

CO-EDITIONS OF JLB SMITH'S BOOK OLD FOURLEGS – THE STORY OF THE COELACANTH

1956 English edition

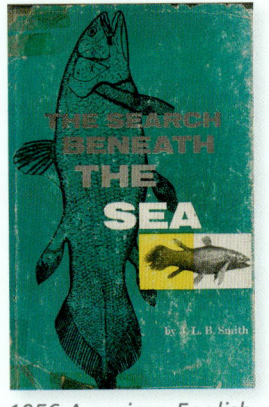

1956 American English edition

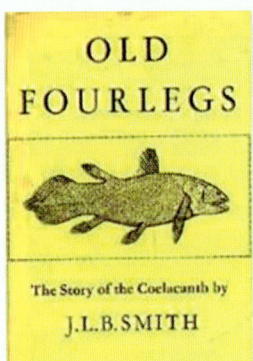

Reprint of the 1957 Reader's Union edition

1957 German edition

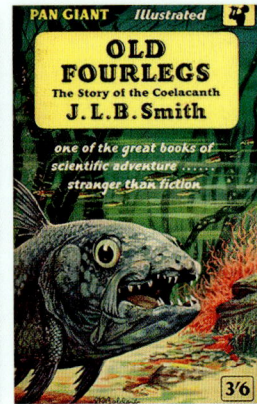

1958 Pan Books English edition

1960 French edition

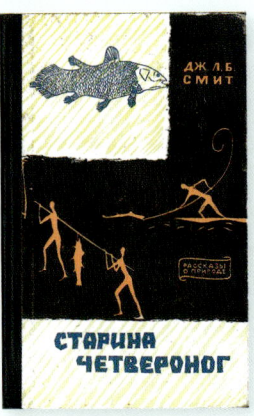

1962 Russian edition

1964 Estonian edition

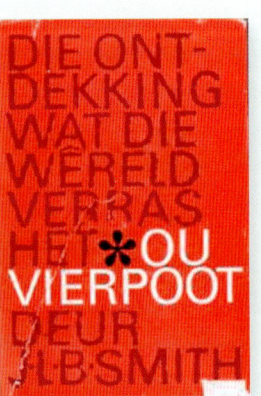

1965 Afrikaans edition

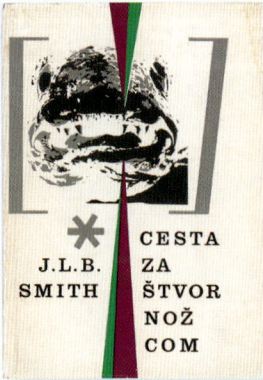

1970 Slovak edition

1973 Dutch edition

1977 Latvian edition

1981 Japanese edition

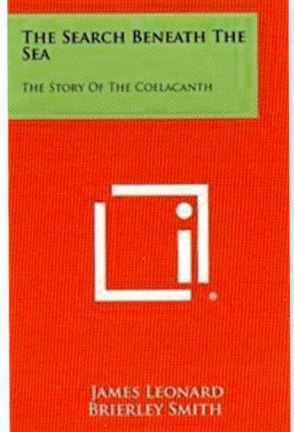

2012 English Literary Licensing edition

schoolboy, Jaan Elken, from Tallinn on the Baltic Sea coast of Estonia. In an accompanying letter he wrote, 'By the way, the impression of the book was 30,000. The number of Estonians is about 900,000. As the whole impression is sold already, it comes out that every 30th Estonian has got your husband's book!'

Similar estimates of the popularity of the book (and the optimism of its publishers) can be made for some other countries. In 1977 Latvia had about 2,485,000 inhabitants; their print run of 65,000 would provide one book for every 38 Latvians. In Russia, where the print run was 100,000, the population in 1962 was about 122,600,000 and there was therefore one book for every 1,226 Russians!

A scan of the entire text of *Old Fourlegs*, with imperfections, is now available through: https://archive.org/details/searchbeneathsea00smit. Furthermore, the online company, Amazon, in Germany also offers a scan of the book on their website: https://www.amazon.de/Search-Beneath-Sea-Story-Coelacanth/dp/1258280833, with the following comment: 'This is a reproduction of a book published before 1923 [sic]. This book may have occasional imperfections such as missing or blurred pages, poor pictures, errant marks, etc. that were either part of the original artefact, or were introduced by the scanning process. We believe this work is culturally important, and despite the imperfections, have elected to bring it back into print as part of our continuing commitment to the preservation of printed works worldwide. We appreciate your understanding of the imperfections in the preservation process, and hope you enjoy this valuable book.'

Old Fourlegs is a riveting account of the exploits of a driven scientist, but its significance is far greater than that alone. At a time when science was not uppermost in the minds of South Africans, it brought science into the living rooms of thousands. Then, through the numerous editions and translations of the book, and radio broadcasts by the SABC and the BBC, it raised the profile of South African science internationally. This was at a time when South Africa, like many other countries, had learned from the Second World War

that there was an urgent need to invest in science and technology. Research became a national priority and science a matter of national prestige.

The Council for Scientific & Industrial Research (CSIR) had been formed in 1946 but there was little interface between scientists and the general public at the time. JLB was one of the first South African scientists to bridge that gap; Eugène Marais, Jan Smuts, James Stevenson-Hamilton, Cecil von Bonde and Sydney Skaife were other exceptional early examples.

JLB Smith's success as a scientist, and as a publicist for ichthyology, also paved the way for the establishment in 1977 of the JLB Smith Institute of Ichthyology at Rhodes University in Grahamstown after he died. This Institute, and the teaching Department of Ichthyology & Fisheries Science that it spawned in 1981, has thrived and is now the leading fish research institute in Africa. In 2001 the Institute's research and curation mandate was broadened to include not only fishes but all aquatic animals, and it is now known as the South African Institute for Aquatic Biodiversity (SAIAB), an agency of the National Research Foundation. Hundreds of students have benefited from the establishment of the Institute and Department, and our knowledge of African marine and freshwater fishes, and our efforts to manage their use sustainably, have greatly improved.

Front entrance of the South African Institute for Aquatic Biodiversity, previously the JLB Smith Institute of Ichthyology

Mike Bruton

Part of the massive fish collection in the South African Institute for Aquatic Biodiversity in Grahamstown

By April 2015 the Institute's collections included over 992,400 preserved fishes as well as more than 27,000 tissue samples, skeletons, otoliths (ear bones) and jaws as well as an archive of over 23,700 ink drawings, colour paintings, colour slides, digital photographs and X-rays of fishes. This is by far the largest fish collection in Africa and one of the best in the world. The fish collection comprises 7,456 different species including 5,162 marine, 497 estuarine and 2,143 freshwater species (some occur in more than one realm), which originate from 133 countries and all the major oceans of the world. They are contained in 507 fibreglass tanks and over 95,000 bottles, all of which are meticulously recorded and labelled. In keeping with its broadened mandate, the collection now also includes diatoms, sea squirts, frogs and other aquatic organisms.

Privately, JLB also benefited from the publication of *Old Fourlegs*, as the book and its translations were very profitable. With the proceeds he bought one of the prime pieces of real estate in South Africa – the Western Heads at Knysna, which included his favourite fishing spot, now known as 'JLB's Rock', in the 'Narrows' (the mouth of the lagoon). For many years JLB and his dog Marlin were a familiar sight as they sped across Knysna Lagoon in his aluminium boat 'Blikkie' or as he paddled his innovative, fold-up canvas boat.

JLB Smith was, of course, an expert angler who combined his knowledge of fishes, aquatic habitats and organic chemistry to outwit his prey. He never let his line, hook or bait wallow in the bilges (if he was fishing from a boat), where they could acquire 'foreign tastes'. He was aware that the pheromones (hormones released into the outside environment) of humans might repel fish if they were detected on the bait, and even postulated that the relative success of women anglers might be because their pheromones contain more fish attractants! He was also renowned for his ability to predict what he would catch on a given day, and usually lived up to his prognosis. He filled his boat on days when other anglers caught nothing and seemed to be able to land a given prey at will.

However, as I state in my autobiography, *When I was a Fish – Tales of an Ichthyologist*, 'It is almost sacrilegious to say so, as JLB Smith is a demi-god in South African ichthyology circles, but he also caught, with rod and line, far more fishes than necessary. Judging by the groaning ropes of fishes that he hooked off the southern and eastern Cape coasts, portrayed in photographs in the Ichthyology archives, he was one of the most insatiable anglers of his day. He could not have eaten all the fishes that he caught, nor did he need so many replicates for his scientific collections.'

Although JLB did catch prodigious numbers of fishes, he was also a 'fair' angler who tried to level the playing fields between predator and prey. He used light tackle and small hooks, and made optimal use of his knowledge of fish behaviour, habitat and bait preferences and activity patterns. He was also very conscious of the factors that scared fishes away, such as noise in the boat or human scent on the bait or line. He rubbed his hands and fishing line in fish guts before casting and insisted on complete silence while fishing. He also released fish that had been hooked in the mouth only (as they could recover) but kept those that had been hooked in the throat or stomach. He quickly despatched those fishes that he kept, and broke off one or two gill arches to allow the fish to bleed, which improved its flavour.

Of course, catching fishes is only part of the reason for angling. It also brings serenity of mind and provides a rare opportunity for philosophical meditation, which JLB would have savoured. It is more a mind-set than a sport and one often hears anglers say that the fishing was good although the catch was poor. There is little doubt that JLB would have contemplated and probably solved many difficult conundrums in fish taxonomy while angling.

Angling is now the most popular form of hunting in which modern humans engage and it is becoming more popular every year. However, if the fish stocks decline to levels below which they cannot sustain themselves, even the best anglers will come away empty handed.

It is interesting that there is little mention of fish, or ocean, conservation in *Old Fourlegs*, although JLB must have been aware of the depletion of inshore fish stocks. He was, however, vitally concerned about the conservation of the coelacanth. In 1956 he wrote

The young angler with his catch near Hamburg, Eastern Cape, in the 1920s

'JLB's Rock' in the narrows at Knysna

a caustic letter to *The Times of London* suggesting that the French had sufficient specimens and that their present policy is 'debasing a once important scientific quest to the level of senseless slaughter of one of our most precious heritages in biology'. In 1963 he wrote a dramatic article entitled 'The Atomic Bomb and the Coelacanth' in which he stated, 'In my view there is a very real danger that this priceless heritage from the past may suffer extermination unless steps are taken to prevent it'.

Also, in 1963, on the 25th anniversary of the first capture, he gave a rousing speech during which he appealed for the formation of an international society to conserve the coelacanth; unfortunately, nothing came of this proposal at the time. In April 1987, during one of our coelacanth research expeditions to the Comoros, we founded the Coelacanth Conservation Council/Conseil pour la Conservation du Coelacanthe (CCC) in a grubby café in Moroni. The CCC subsequently played a major role in cataloguing coelacanth catches and promoting and co-ordinating coelacanth research and conservation internationally.

Old Fourlegs was written some time before the environmental movement had really caught on. Although the origins of the modern 'green' movement can be traced to the Industrial Revolution, the primary environmental cause that was addressed until the 1950s was the mitigation of air pollution and improvements in the quality of life of humans.

The publication in 1962 of *Silent Spring* by Rachel Carson focused the world's attention on living resource conservation, especially the impact of pesticides on wildlife, particularly birds. This book spurred a reversal in national pesticide policies, led to bans on DDT usage, and, with the support of books by Paul Ehrlich, Barry Commoner and the Club of Rome and of various international conservation initiatives (and the first views of our fragile planet from outer space), inspired a global movement that led to the creation of environmental protection agencies in most countries.

This was after JLB's time but it is nevertheless alarming to read in *Old Fourlegs* about the vast numbers of fishes that were collected and the methods used. Today 'blast fishing' is regarded as one of the most harmful and senseless ways of harvesting fishes, yet JLB made extensive use of explosives (including hand grenades) during his fish-collecting expeditions. Thousands of fishes were blasted out of the water, and he even hoped to catch coelacanths in this way (p. 250).

Admittedly Smith had a solid scientific understanding of explosives and their harmful effects and carried out his collecting under strictly controlled conditions. He also needed to make optimal use of his limited time over remote tropical reefs, and to winkle fishes out of deep crevices and coral heads efficiently. He did this with great success, and laid the foundation for the world-class fish collection that we have in the Institute in Grahamstown – although we don't use the blast-fishing method today!

The use of rewards to encourage fishermen to catch coelacanths was also novel and somewhat controversial, but it did serve to raise awareness about the fish, as JLB's £100 reward led directly to his securing the second specimen. His example was soon followed by others as the French doubled the reward to £200 in 1953 for a live coelacanth and the American government offered $5,000 for a specimen, a heft bounty for any fish. In 1975 the Steinhart Aquarium even offered a reward of a two-week round trip to Mecca for a lucky Muslim coelacanth fisherman!

The racial overtones of some of JLB's prose, written in another time and place, are obvious and embarrassing today, and we do not in any way support or condone them. He states, for instance, that 'Comoran natives are not distinguished by great energy; indeed, in that respect they fall below the average, already low' (p. 155) and that 'this surprisingly intelligent behaviour on the part of a native was due to him being of a higher class' (p. 157). He also refers to the 'bestiality of the Kikuyu' (p. 225) and opines 'thank God for native indolence', referring to the fact that the second coelacanth was not gutted (p. 252).

JLB's pro-British and occasional anti-French stance is obvious throughout the book, as is his support for the Nationalist Prime Minister, DF Malan, who would later become one of the main architects of apartheid, rather than for the more liberal internationalist, General Jan Smuts (see, for instance, pp. 98 and 99). He based his opinions on the direct evidence of how the two leaders had treated him during his search for the coelacanth, so we cannot really question his judgement in this respect.

After the publication of *Old Fourlegs* in 1956 Smith continued to study fishes and publish his research in books and scientific journals. He mounted his last major fish-collecting expedition, to Pinda in northern Mozambique, in 1956, accompanied by his wife, Margaret, and their son, William, then 17 years old. In 1963, he and Margaret published *Fishes of the Seychelles*, which he described as a 'comprehensive treatise covering fishes of the tropical zone of the Western Indian Ocean'.

JLB was occasionally criticized for being too hasty in describing new fish species. He described 375 species as 'new' but most of them (about 70%) are now junior synonyms as they belong to species that had already been described by other scientists. From 1948, when Keppel Barnard of the South African Museum in Cape Town completed his marine fish work, JLB was the only marine fish taxonomist working in South Africa until the 1960s, and he had limited contact with overseas scientists. Furthermore, much of his research was conducted under adverse conditions with relatively poor library and laboratory facilities, although he did have excellent fish specimen reference collections, which he had mainly made himself.

Moreover, JLB did tend to publish, after 1946, largely in the in-house journals of his Ichthyology Department, which were not subject to strict peer review. Also, because of his failing health, he was in a hurry to complete his work. Despite these shortcomings, there is no doubt that he made very significant contributions to African fish taxonomy and zoogeography.

By the end of his life, JLB Smith had published nine scientific books (three in chemistry and six in ichthyology), 14 scientific papers in chemistry and over 200 scientific papers in ichthyology. His books included the monumental *Sea Fishes of Southern Africa*, first published in 1949 and re-published in various editions in 1950, 1953, 1961, 1965, 1970 and 1977, with completely revised editions published in 1986, 1995 and 2003. The 1977 edition, published by Valiant, was titled *Smith's Sea Fishes*, with JLB as the sole author. The completely revised 1986 edition, called *Smiths' Sea Fishes*, was published by Macmillan and co-edited by Margaret Smith and Phil Heemstra, with 76 authors of the 270 fish family accounts. Heemstra authored or co-authored 74 of the family accounts whereas Margaret authored or co-authored 43 family accounts, three of them based on the late JLB's contributions. The apostrophe after the 's' of 'Smiths' indicates that the book commemorates the contributions of both JLB (as author of the original edition) and Margaret (as illustrator).

The latest editions of *Smiths' Sea Fishes* were published by Southern Book Publishers (1995) and Struik (2003). In 1986 Springer-Verlag in Germany re-published *Smiths' Sea Fishes* in soft cover (for sale outside southern Africa only) and also produced an e-book edition in December 2012.

JLB shared his knowledge of fishes with the public through radio broadcasts in English, Afrikaans and Portuguese, hundreds of informal articles in newspapers and magazines, and a series of popular books in English and Afrikaans. These included *Old Fourlegs* (*Ou Vierpoot*) as well as *Fishes of the Tsitsikama Coastal National Park* (*Visse van die Tsitsikama Seekus Nasionale Park*, 1966), the latter title with Margaret. Shortly after his death in January 1968 two popular books on fishes written by JLB Smith were also published, *High Tide* and *Our Fishes* (*Ons Visse*), both based on radio interviews that he had done in Afrikaans for the SABC programme 'Uit die Natuur' ('Out of Nature').

Smith was a great intellect who strode the South African and international ichthyology scene with authority and respect. He was also a brilliant and much-loved teacher who inspired several generations of students and made a lasting impact on chemistry and ichthyology in Africa and beyond. Despite his frail body he was a man of incredible energy, drive and enthusiasm who lived several life times in one. He was intolerant of laziness and inefficiency and set very high standards for himself and his colleagues.

When his mental powers and eyesight began to fail him in his early 70s he dreaded becoming bedridden, or suffering a stroke and becoming a 'useless hulk'.

Then, like Eugène Marais before him, he took matters into his own hands and committed suicide on 8th January 1968. In a written statement he indicated that he had felt for some time that his mental powers were deteriorating, and that he took his own life '... probably only a brief anticipation of nature'. He was to have received the degree of Doctor of Science (*honoris causa*) from Rhodes University in April 1968. It is regrettable that this well-deserved honour was not conferred on him earlier, as he richly deserved it. Honorary doctorates have been conferred by Rhodes University on several other key players who have helped to develop ichthyology in Grahamstown and the Eastern Cape, including Rex Jubb (1970, DSc), Marjorie Courtenay-Latimer (1971, DPhil), Margaret Smith (1986, DLaws), Humphry Greenwood (1992, DSc), and the author (2014, DSc).

At the time of JLB's death I was doing postgraduate research on the freshwater fishes of northern Zululand through the Institute for Freshwater Studies of Rhodes University at Lake Sibaya Research Station. JLB and Margaret were due to join us for an expedition to the coastal lakes of southern Mozambique, but they never arrived.

JLB's passing left a huge gap in South African ichthyology but his legacy has made it possible for new generations of scientists to be trained and practise as fish researchers, fisheries scientists and aquatic resource managers. Ichthyology and fisheries science in South Africa have never been stronger.

Several popular books have been published on the coelacanth since Smith's pioneering tome, including Jean Anthony's *Opération Coelacanthe* (1976, in French). Peter Barnett, who accompanied JLB and Margaret (and William) on a fish-collecting expedition to Mozambique in 1951, wrote an interesting account of his experiences in *Sea Safari with Professor Smith* (privately published in 1953 and re-published by Richford Enterprises in 1979). Shirley Bell, previously editor of *Field & Tide*, wrote a children's book entitled *Old Man Coelacanth* (1969, Voortrekkerpers) based on *Old Fourlegs*. Keith Thomson, CEO of the Academy of Natural Sciences

RELATED PUBLICATIONS

High Tide, *1968*

Our Fishes, *1968*

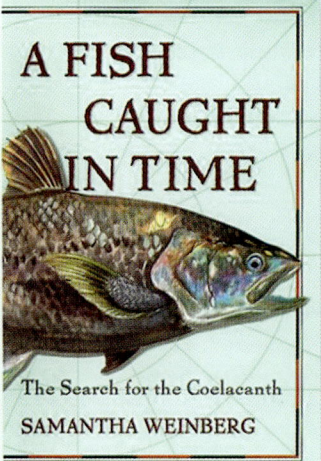

Fish Caught in Time *by Samantha Weinberg*

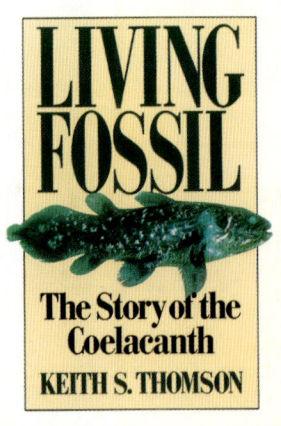

Living Fossil *by Keith Thomson*

in Philadelphia, wrote *Living Fossil – The Story of the Coelacanth* (1991, Norton & Co.; re-published in English, German, Swedish, Japanese and Chinese) and Samantha Weinberg produced *A Fish Caught in Time* (1999, Fourth Estate Ltd; re-published in English, Italian, German and Japanese).

Keith Thomson's book includes an excellent review of the anatomy and physiology of the coelacanth but is unfortunately riddled with factual errors, especially on the early history. This is despite the fact that he bemoans the inaccuracy of other early accounts of coelacanth research and, in particular, that 'a romantic overlay has been applied, especially to the role of Professor Smith' (p. 9)!

Hans Fricke published his autobiographical account, *Die Jagd nach dem Quastenflosser, Der Fisch, der Aus der Urzeit Kam* ('The Search for the Coelacanth, The Fish that Came from the Past') in 2007 through Verlag CH Beck in Germany; this book was re-published with the title reversed by DTV-Verlag in 2010. My autobiography, *When I was a Fish – Tales of an Ichthyologist*, which includes a detailed review of coelacanth research and my involvement in it, was published in 2015 by Jacana Media in Cape Town.

More scientific but still readable books on 'old fourlegs' have also been written or edited by John McCosker and Michael Lagios (*The Biology and Physiology of the Living Coelacanth*, 1979, California Academy of Sciences), the Royal Society of London (*Advances in Coelacanth Research*, 1980), Jack Musick, Mike Bruton and Eugene Balon (*The Biology of* Latimeria *and Evolution of Coelacanths*, 1991, Kluwer) and Peter Forey (*History of the Coelacanth Fishes*, 1998, Chapman & Hall). Frequent reference is also made to coelacanth research in the golden anniversary volume, *The JLB Smith Institute of Ichthyology – 50 Years of Ichthyology*, edited by Paul Skelton and Johann Lutjeharms and produced in 1997, and a dedicated edition of the *South African Journal of Science* (volume 102, 9/10) was published in 2006, on the proceedings of a conference of the 'African Coelacanth Ecosystem Programme' (ACEP). Further literature on the coelacanth is listed in the References and Further reading section at the end of this book.

Hans Fricke's 2007 book

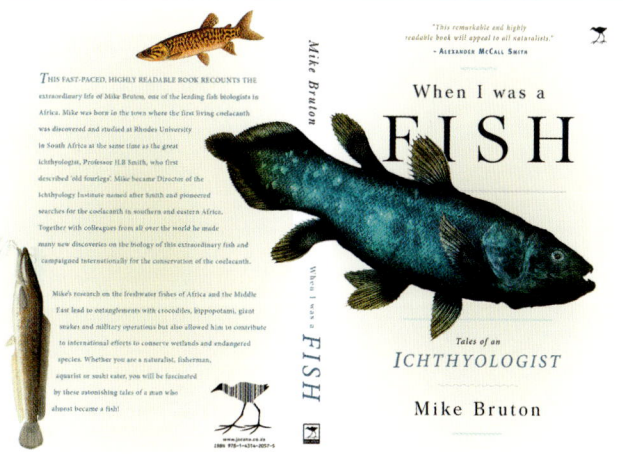

When I was a Fish. Tales of an Ichthyologist, *the author's autobiography*

Peter Forey's authoritative book on the coelacanth

OLD FOURLEGS

NOTE: Some of the blank pages that occur in the original printed version of *Old Fourlegs* have been left out of this facsimile edition.

JLB Smith describes this historic moment in December 1952 as follows, 'Eventually we got away, and there on the deck, swathed in cotton-wool, was the fish. I could not bring myself to touch it and I asked them to open it, and they did and I knelt down to look, and I'm not ashamed to say, I wept … for it was true … it was a Coelacanth – and what was more wonderful, a species different from that of 1938 – another coelacanth. It was more than worth while all that long strain.'

A few moments after the critical identification as a Coelacanth, on Hunt's vessel at Pamanzi, Comores, 29th December 1952.
On left, Capt. E. E. Hunt; *on right*, holding fin, Monsieur P. Coudert, Governor of the Comores; *on extreme right*, one of his staff. *Behind, left to right*, Lt. D. M. Raîston, Capt. P. Letley, Lt. W. J. Bergh, Comd. J. P. D. Blaauw, Cpl. F. Brink.

Old Fourlegs

THE STORY OF
THE COELACANTH

BY

J. L. B. SMITH

LONGMANS, GREEN AND CO
LONDON · NEW YORK · TORONTO

The nickname for the coelacanth, 'old four legs', was coined by journalists after the taxidermist mounted the lobed fins pointing straight downwards. We now know that the coelacanth does not 'walk' on its fins but hovers above the bottom with the paired fins performing exquisite sculling movements.

LONGMANS, GREEN AND CO LTD
6 & 7 CLIFFORD STREET LONDON W 1
BOSTON HOUSE STRAND STREET CAPE TOWN
531 LITTLE COLLINS STREET MELBOURNE

LONGMANS, GREEN AND CO INC
55 FIFTH AVENUE NEW YORK 3

LONGMANS, GREEN AND CO
20 CRANFIELD ROAD TORONTO 16

ORIENT LONGMANS PRIVATE LTD
CALCUTTA BOMBAY MADRAS
DELHI VIJAYAWADA DACCA

First Published 1956

PRINTED AND BOUND IN GREAT BRITAIN BY
HAZELL WATSON AND VINEY LTD
AYLESBURY AND LONDON

This book is dedicated to

MISS M. COURTENAY-LATIMER

one of South Africa's

most able women

[signatures]

JLB Smith always gave credit to Marjorie Courtenay-Latimer for the pivotal role that she had played in the discovery of the first coelacanth and for her broader contributions to museology in South Africa.

This copy of the book was signed by three members of the crew of the *Flying Fishcart*, the military Dakota that took JLB Smith to the Comoros in December 1952 to fetch the second coelacanth. They are DM Ralston, JWJ 'Vanski' van Niekerk and WJ Bergh.

FOREWORD

THIS story has been dragged from my reluctant pen by the unflagging determination of my wife, consciously aided and abetted by numerous friends and unwittingly by publishers and literary agents from several countries.

In succumbing, for the sake of historical record, it has been my aim to present this extraordinary event as accurately as possible.

This has involved the mention of many different persons who played their part in the creation and course of this story. I have spared nobody, least of all myself, which is the extenuation I offer to those inclined to find my descriptive words harsh.

The general public is apt to regard people like leading scientists or cabinet ministers as almost superhuman and beyond or above ordinary human emotions. They are not, emphatically not, and to scale the heights a man must be prepared to wage an unending, bitter battle with those persistent fundamental weaknesses that constantly plague us all. One friend who kindly read the manuscript asked me if I realised how it revealed myself. I do not mind. No man is a god.

J. L. B. SMITH

Grahamstown
August 1955

During his lifetime Smith was one of the few scientists who took the trouble to write popular articles and books, and give radio interviews and lectures in English, Afrikaans and Portuguese, to the lay public on his research.

JLB makes the point that scientists and cabinet ministers are not superhuman. He praises DF Malan, the conservative politician who had little appreciation for the value of the coelacanth but did arrange a military airplane to poach the second coelacanth from French territory, yet he criticizes Jan Smuts, a more enlightened and science-orientated politician who did not support his work.

The unfolding text of *Old Fourlegs* does, indeed, confirm the opinion of his friend: that it reveals a great deal about JLB Smith himself. He wrote the book as he lived his life – with brutal honesty and without compromise.

NRF-SAIAB

JLB Smith as a student at Cambridge University in England in 1920

JLB's mention of 'devastating criticism' of the manuscript by his wife is surprising as he was very much the dominant partner in their scientific relationship, although she became much more assertive after he died.

ACKNOWLEDGEMENTS

I wish to acknowledge permission to reproduce matter, granted by: Miss M. Courtenay-Latimer; Dr. E. I. Nielsen; Dr. J. Millot; Captain E. E. Hunt; S.A. Broadcasting Co.; *Die Burger*, Cape Town; *Daily News*, Durban; *The Times*, London; *Nature*, London; *Evening Post*, Jersey; Jersey Electricity Co.

I am indebted to Dr. H. J. van Eck, Advocate Adrian A. Roberts (formerly Law Adviser to the Union Government), and Professor W. E. G. Louw for helpful criticism of the manuscript, also to Messrs. A. S. Wheeldon and H. Rushmere for opinions on certain points.

To my wife I am indebted for her constant support, for valuable if initially devastating criticism and for numerous illustrations.

This book stands as a tribute to the foresight of the South African Council for Scientific and Industrial Research, whose continued generous support enabled me to pursue this long quest.

CONTENTS

Book I

THE PAST SURGES FROM THE SEA

1 THE STAGE IS SET	*page*	3
2 THIRTY MILLION GENERATIONS		11
3 CINDERELLA		22
4 STRANGER THAN FICTION		27
5 JEKYLL AND HYDE		44

Book II

TROUGH AND CREST

6 NO DEEP-SEA REFUGEE	59
7 OBSESSION	67
8 DUNNOTTAR DILEMMA	80
9 HIS OWN SHEEP	91
10 STARTS AND STOPS	100
11 I MUST SPEAK TO HIM	116
12 DAKOTA DASH	125
13 DZAOUDZI DRAMA	140
14 UP IN THE CLOUDS	152
15 MALAN AND MALANIA	168

Book III

THE WAVE RECEDES

16 FLOTSAM AND JETSAM   *page* 179

17 FALLING THROUGH   193

18 PORTCULLIS AND DRAWBRIDGE   210

19 MARCHAND DE BONHEUR   217

APPENDICES

A CHARACTERISTICS OF COELACANTHS AND HOW
     THEY DIFFER FROM MODERN FISHES   229

B WHY THE DISCOVERY OF THE COELACANTH
     AROUSED SUCH WIDE INTEREST   233

C THE LATEST POSITION ABOUT COELACANTHS   237

D ARTICLE FROM *THE TIMES*   243

E THE COELACANTH BROADCAST FROM DURBAN   247

     INDEX   255

PLATES

A FEW MOMENTS AFTER THE CRITICAL IDEN-
TIFICATION AS A COELACANTH, ON HUNT'S
VESSEL AT PAMANZI, 29TH DECEMBER 1952

frontispiece

1 MISS M. COURTENAY-LATIMER   *facing*
  MISS LATIMER'S SKETCH AND NOTES   *page*
  A COELACANTH FOSSIL, WHICH DATES BACK 170
    MILLION YEARS, FOUND BY DR. E. NIELSEN
    IN GREENLAND   32

2 THE FIRST THREE COELACANTHS   33

3 THE FAMOUS COELACANTH LEAFLET   112

4 MONSIEUR P. COUDERT, GOVERNOR OF THE
  COMORES, NEAR THE WHARF AT DZAOUDZI,
  29TH DECEMBER 1952
  'HUNT'S TRIM VESSEL,' AT THE WHARF,
  DZAOUDZI, SHOWING THE COELACANTH BOX
  ON THE LEFT   113

5 THE LANDING-STRIP AT PAMANZI, COMORES
  'THE BELOVED ISLE' OF MOZAMBIQUE, P.E.A.   128

6 IN THE LATTER PART OF AN UNENDING DAY,
  29TH DECEMBER 1952, TELEPHONING BRIGA-
  DIER MELVILLE IN PRETORIA
  DR. D. F. MALAN EXAMINES THE COELACANTH
  AT THE STRAND, 30TH DECEMBER 1952
  THE COELACANTH QUARTET, NAIROBI, 24TH
  OCTOBER 1953   129

BOOK I

THE PAST SURGES FROM THE SEA

Chapter One

THE STAGE IS SET

THESE are wonderful times, and it is thrilling to be living now, though it would thrill me even more to know that I could still be here a hundred or a thousand years hence, for this immediate future promises to be of intense interest, even excitement, certainly to the scientist.

With a mind constantly reaching towards the potential marvels of the future, it has been my quite fantastic privilege to reveal to the world a living part of the utterly remote past, covering a span of time so great as to be almost beyond the grasp of the ordinary mind. In this process an obscure scientific name, Coelacanth (pronounced 'seelakanth'), jumped into prominence and into a permanent place in the common speech of mankind.

Such things do not happen easily. The appearance of the Coelacanth was like a gigantic tidal wave which washed me violently from my path, held me in its grip, carried me along, and set my feet on a quest that dominated some of the best years of my life. It caused me to lead an unusual life, of which many people came to acquire an attractive but distorted picture, seeing in me a scientist who dashed off on eventful expeditions to romantic tropical islands where wonderful fishes new to science were just waiting to jump into my net. They read of me as having almost casually telephoned a Prime Minister to ask for an aeroplane in which to make a sensational flight to fetch an incredible fish that attracted world-wide attention.

Whenever I return to civilisation, people want to know something about this apparently fascinating life, so I have been virtually compelled to give many lectures, over the radio and in person. I do not conceal the discomfort, hardship, danger, and unending hard labour that our work involves, but these do not obscure the glamour, and a constant stream of eager young folk, men and women, come to me with the same query. 'My present work is

Although JLB was a highly focused scientist, people who had personal contact with him (the author included) found that he had a holistic view of the world and the place of science in it. He was passionate about science and about the central role that it should play in society.

A retiring and private person, at least during the last four decades of his life, he was nevertheless an excellent publicist for science, and especially for ichthyology. While he professed to hate publicity he also courted it and was a master at manipulating the media.

According to Denys Davis, an illustrator who worked with the Smiths in the 1940s, 'He cut music out of his life completely as it stirred emotions, and this was wasteful'. This must have been hard on Margaret, who loved music and singing.

Drostdy tower at the entrance to Rhodes University

Rhodes University crest. Rhodes University College became an autonomous university in 1951.

Fortunately, job opportunities for ichthyologists and other aquatic scientists are now more widely available, largely due to the increased commercial importance of fish and heightened awareness of the vital role of biodiversity conservation.

JLB Smith Institute of Ichthyology *South African Institute for Aquatic Biodiversity*

In 1968, JLB Smith's widow, Margaret Mary Smith, persuaded Rhodes University and the Council for Scientific and Industrial Research (CSIR) to establish the JLB Smith Institute of Ichthyology. Margaret became the Institute's first Director.

JLB states that his attitude 'did not always create the most cordial relations'. He had the sort of personality that always made an impact on people, either good or bad; no-one was neutral about him. Decades after he died his ex-colleagues and students, such as Doug Rivett and Keith Hunt, would become very animated when they spoke about him.

Sir George Edward Cory

Another chemistry academic at Rhodes University who became better known for his late-career achievements, was Sir George Cory (1862–1935), Head of the Department of Chemistry while JLB was a Senior Lecturer. He had an avid interest in history and wrote the multi-volume work, *The Rise of South Africa (1910–1939)*.

dull. How can I become an ichthyologist?' So I tell them. First you must get a University Master's degree in Biology, better still a Doctorate, which means a minimum of five years of University life. Then for ten years at least you must be prepared to do laborious donkey-work, almost certainly poorly paid, as an assistant to some expert in that line, much of it dull, monotonous routine, like counting scales on hundreds and thousands of small fish, probably more deadly than counting pennies in a bank, and those at least don't smell. Even then you may not be good enough to get anywhere, and there are few positions where you will be paid a good salary as an ichthyologist. Most turn sadly away, but the few takers have made good!

This book is to tell the almost incredible story of the Coelacanth, but as this is inextricably bound up with my own personality, it would be as well to tell you something of how that was shaped.

My life has throughout been a series of contrasts and changes, many due to the peculiar circumstances of South Africa. Of English parents, I was born in 1897 at the inland Karoo town of Graaff Reinet. In the midst of the stress and bitterness of the Boer War my early years were spent in an atmosphere of deification of all that was British, and hatred and scorn of the 'Boers', and indeed of anything South African as distinct from British, including the country itself.

It has always been my uncomfortable instinct not to accept uncritically the opinions of others, and while this has ultimately been an asset in my scientific career, it did not always create the most cordial relations at home or at school.

My early education was at several small Karoo village mixed schools, and later at the abruptly different atmosphere of 'Bishops', modelled on an English Public School. The next violent contrast was the Victoria College at Stellenbosch, predominantly Afrikaans and reputedly steeped in Nationalism and Politics, but I encountered a peaceful tolerance towards my firm political views. There I gave my heart to Chemistry.

When the Great War came, in company with thousands of others of like age, on the 7th August 1914, I was called up from school and put into khaki and barracks at Wynberg, then into the tender care of a Regular British regiment for training. The enforced close company of this strange unnatural substratum of society was a bewildering experience. For example, venereal disease changed abruptly from the remote subject of schoolboy jokes to stark reality, for in the lives of these men it appeared to occupy the status of the common cold in mine, a curse but inevitable. After about a month some of us were returned to school as too young for campaign, and I went on to University life at Stellenbosch. As I was set on taking part in the war, I arranged at once to go to England to join the Royal Flying Corps after my 'Intermediate' Examination at the end of the year (1915). However, General Smuts, at that time almost a god to me, appealed to everyone to enlist for German East Africa first, so instead of learning to soar through the skies, I became an earth-bound, foot-slogging infantry-man instead. Thousands of half-trained men of all ages were jammed into a transport at Durban, and fed mainly on bread, tinned rabbit, and tea. While most others gambled I counted heads and life-boats and was appalled at the quotient, but we got safely to Mombasa, and thence to the badly mismanaged campaign that followed.

After sundry misadventures, including contracting malaria, dysentery, and the acute rheumatic enlargement of several major joints, I spent some months in military hospitals, first in Kenya, where I nearly died, then was shipped, helpless, back to the Union, and to hospital at Wynberg. Eventually I returned, virtually a physical wreck, to University life at Stellenbosch, where again even those students most strongly opposed to my convictions respected them, perhaps even more than before. Still racked with fever, and more often ill than well, I continued my studies until the end of 1918. Then came another abrupt change from Afrikaner Stellenbosch to Cambridge in England, where I carried out research work in chemistry. University life there was in many ways different from what I had known. Some of the students occasionally indulged in destructive riots, and the cost of the damage to public and University property, sometimes thousands of pounds, was covered by levies imposed by the University, which had to be paid equally by all students, the innocent majority as well.

In some of my vacations I travelled and tramped various countries on the Continent, learning to speak German and some Italian, and a good deal besides. I travelled widely and saw a good

If JLB had joined the Royal Flying Corps (where the survival rate was very low), or had succumbed to one of several tropical diseases, ichthyology would have followed a very different course in South Africa.

UNIVERSITY OF CAMBRIDGE

Crests of Stellenbosch University and the University of Cambridge

JLB does not mention that he was one of the most brilliant chemistry students in the early history of South Africa.

He was a prodigious catcher of fish as evidenced by the photographs in the Ichthyology Institute's archives of the loads of fish he caught off the Eastern Cape coast, in the Knysna Lagoon or in East Africa. One wonders how much he contributed to the local demise of some species, such as galjoen, in the Eastern Cape.

NRF-SAIAB

JLB Smith the angler, ca 1930

deal of the people and the country of my origin. Despite my undiluted English blood and early upbringing, I found myself resentful of criticism of South Africa, especially of comments on Smuts I heard in quite high circles. I became conscious for the first time of being a 'South African', and those from my own country I met over there were no longer 'English' or 'Afrikaans', but my own people. The childhood-fostered gap between 'Briton' and 'Boer' in my mind just closed up.

On my return to South Africa in 1923 I took an appointment at Rhodes University College, where I taught chemistry, a subject I loved as much as ever, and managed to find time for research, publishing a series of papers.

My father was fond of angling, and as a very small boy, with some of his cast-off gear, I vividly remember catching my first 'Dassie', a Bream-like fish, at Knysna. This wonderful shining thing I had pulled up from the unknown world below the water had a terrific effect on me, probably more than anything ever since. From then on angling has been a passion, a madness, sometimes even a reproach. In South Africa in my young days 'fishing', sea fishing, was rather frowned upon as a pastime for a member of a University staff. It is strange to look back on it now, when even the greatest are proud to display their catches. I had soon got to know all the common kinds of fishes, but as I attained manhood wanted to know more and found great difficulty in identifying strange types, and there were many. No one could help me, and the only books available were beyond easy use. The 'keys' were intelligible only to those already so expert as not to need them, so that it was a fearful job trying to identify unknown fishes.

I struggled on alone, baffled, but eventually worked out a numerical system for identifying fishes. It took all my spare time for more than a year, and its compilation involved the writing of more than a million figures, but it worked. It enabled me to track in a few minutes quite unknown fishes, and even to identify them from mere fragments. This was a tremendous step forward, and gave me power that normally comes only from much longer experience.

Having mastered the fishes that came from angling, I went on to collect others systematically on the Eastern Province coast, discovering to my astonishment that I was the first to do this.

Almost every tide there was something rare or new to South Africa, or even new to science. I made contact with the Albany Museum in Grahamstown, and was encouraged by John Hewitt, the Director, in my first timid entry into the scientific field of ichthyology. In 1931 I published my first short paper in the annals of that museum, with my own illustrations that seemed satisfactory. But an acquaintance of Cambridge days, a zoologist, wrote to say that he was surprised to see a chemist publishing a paper on fishes; the text was reasonable but the illustrations were terrible. That was my first step towards appreciation of the importance of good illustration in biology, which has become a feature of my work, in more recent times thanks largely to the skill of my wife.

My long and thorough training in the mathematical sciences, not generally part of the equipment of systematists, assisted me at every turn. My progress was indeed so rapid that it was not long before anglers and others came to me for information, and an increasing number of fishes were brought and sent for my opinion and identification. Correspondence on this matter steadily increased, and all phases of this work grew so rapidly that there were times when I was almost overwhelmed. Everywhere I turned there were new and fascinating things all round, my time was so fully occupied that one by one the ordinary pastimes of life fell away.

Chemistry covers a vast field, and is the basis of an enormous part of general life and of industry. The subject is continually changing, almost like a moving picture, and to keep abreast of developments is more than a full-time job.

With two such different and full fields to occupy my time, I was in a difficult situation. During working hours in term time I conscientiously did nothing else but chemistry, even when I was bursting to get on with a new fish. My free time and vacations were devoted to fish. I had papers on fishes and on chemistry published at the same time, and even managed to produce three text-books in chemistry.

In South Africa the character of the Universities has been influenced by the Scottish educational system, in which the emphasis is on a high standard of teaching. Their development has also been moulded by having to train young men and women

John Hewitt (1880–1961) was a British-born zoologist (who studied mainly reptiles) and archaeologist. The Albany Museum would later develop a first-rate reputation in freshwater ichthyology through the work of Rex Jubb, Paul Skelton and Jim Cambray.

After JLB's poor attempt to illustrate his first paper in 1932 (see below), he relied on an armada of artists (mainly women) to illustrate his work. They included Margaret Smith, Patricia Parkin, Denys Davis, Hester Locke, Valerie de la Harpe and, later, Jean Michelle Vincent, Virginia McCrostie, Anne Levebre and Moira Lambert.

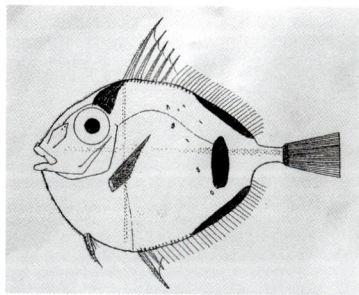

JLB Smith's diagrammatic first (and last) fish illustration

A spin-off of his work was that a strong international reputation emerged in South Africa for fish illustration. In later years the Ichthyology Institute was served by an excellent team of artists, including Liz Tarr (trained by Margaret), Dave Voorvelt and Elaine Heemstra.

Throughout JLB's writing the reader is aware of his analytical approach to life. He is constantly counting, classifying and evaluating events and objects around him. The chemistry text-books that he wrote provided keys to the identification of unknown chemical substances using progressive tests with known reagents. Whether he was studying chemistry or ichthyology, his inclination was towards addressing problems of identification and classification.

JLB makes it clear that, to a large extent, he did not conform to the norms of society at the time. He was an academic who was keen on fishing, and a chemist who studied fishes. He also achieved a much higher public profile than most of his academic colleagues and had a self-centred lifestyle; these factors resulted in some resentment developing against him. His manic work ethic allowed him to pursue his varied interests – and sustain his elevated status – without compromising standards.

NRF-SAIAB

JLB Smith in 1939

JLB's wry sense of humour surfaces occasionally in the book, as in his remark about his 'partial dismemberment' by medical doctors.

In 1902 Dr JDF Gilchrist established a marine research laboratory and the first public aquarium in sub-Saharan Africa in a magnificent stone building at St James on the shores of False Bay near Cape Town. Sadly, this historic building was demolished in 1948.

Keppel Barnard (1887–1964) typified the dying breed of polymath scientist of the 19th and early 20th centuries. He researched and published in fields as diverse as marine and freshwater fishes, cave animals and insect and crustacean classification and biology, and was Director of the South African Museum (now Iziko Museums of South Africa) from 1946 to 1956.

for specific occupations essential to the welfare of a rapidly developing economy.

Research in Universities in South Africa has occupied a subordinate, and in some ways uneasy, position. University staffs are normally appointed and paid for teaching, and while research is officially encouraged, anyone who devotes more than normal time to such work runs the risk of being regarded as not giving proper attention to the teaching for which he is paid. It is certainly looked on as peculiar and possibly even as reprehensible to teach in one subject and to do research work in another. At the time of the first Coelacanth I was told that it is competent for the head of a University to order a member of the staff to desist from doing research work, even in his spare time, if in the opinion of the head it may be prejudicing the efficiency of his teaching work. All this is fundamentally sound. In general, no man can serve two masters; at least, not for long.

For many years the aftermath of the East African campaign led to continued ill-health, the precise origin of which baffled those I consulted. In succession they took away my teeth, my tonsils, and my appendix; but I have no harsh feelings towards those who assisted at my partial dismemberment, and am rather grateful that they did not focus their attention on any other organs as well. In desperation, my wife and I came to seek health in our food, and in a few years I achieved a new lease of life, which made possible the strenuous expeditions in tropical waters that ultimately led to the second Coelacanth.

The most important collection of South African fishes up to 1930 was in the South African Museum at Cape Town, collected and partly worked on by the late J. D. F. Gilchrist.* They had been the basis of a large monograph by K. H. Barnard,† Assistant Director of the South African Museum, at that time the leading authority on South African fishes. In the Eastern Cape there were provincial museums at Grahamstown, Port Elizabeth, East London, and Kingwilliamstown, each with a staff of only a

* Professor J. D. F. Gilchrist, a scientific pioneer, a small man of great heart and great mind, whose ability, energy, and endurance laid the foundations of South African ichthyology and of the great fisheries industry of today.

† One of the most able, versatile, and industrious biologists ever to settle in South Africa. His researches in several widely divergent fields are most valuable contributions to the advancement of scientific knowledge in South Africa.

Curator or Director, who was exhibition officer, scientist, as well as consultant on everything else. They were pleased to have my services as Honorary Curator of Fishes for their museums, which I visited regularly, and they kept or sent their fish rarities for my investigation.

I tried to get trawler crews to hunt and keep unusual specimens from their catches and especially from the 'rubbish', but found them indifferent, and came to realise that more direct contact was necessary. So I endured the miseries of small trawlers on South Africa's stormy seas, often so seasick as barely able to crawl along the slippery heaving decks to scratch among the slimy rubbish shoved aside.

To the crews I was no longer a remote scientist who expected them to do his dirty work while he stayed in a comfortable museum ashore, and they changed from indifference to interest and sometimes to enthusiasm.

I went out with small line-boats and lived with the coastal trek-netters. I walked to remote lighthouses, and to coastal farms and stores, always talking fish, fish, fish. All this took time and effort but paid handsome dividends, and a steady stream of treasures came rolling in.

The study of fishes is very much a full-time occupation even when not complicated by any other duty, but in my early days a few glimpses into a work on fossil fishes set me to find odd moments to explore this fascinating new, or rather ancient, world. I acquired a general knowledge of the types that had lived and died before our time, and found this perhaps the most absorbing of all scientific fields; but my life was already so desperately full that I dared not indulge that desire very far. Nevertheless, those weird creatures of bygone days were constantly flitting in and out of my consciousness, constantly filling me with almost an agony that they had gone for ever and could never be seen again. Fossil fishes are comparative rarities in most parts of South Africa. If it had been otherwise, I have often wondered if they would not have pulled me right away.

And so, by 1938, as all this shows, it was just as if the stage had been set for the Coelacanth. I was in close contact with the various museums, had by constant visits and voyages established cordial personal relations with trawler crews and with the firms

It was through this network of collaboration with local museums that the first coelacanth was brought to Smith's attention in early 1939. Marjorie Courtenay-Latimer, the young Curator of the East London Museum, followed Smith's example and secured the help of trawler crews to catch fishes for her displays.

It was typical of JLB that he climbed down from his ivory tower at the university and mucked about on trawlers and line-boats on stormy seas in order to secure fish specimens. In the 1920s he even used an ox-wagon to trek from Trappes Valley near the Great Fish River to reach remote angling spots.

NRF-SAIAB

JLB Smith in the 1920s with the ox-wagon that he used to go fishing

In fact, we now know that South Africa has a rich fossil fish fauna, with many specimens known from the Eastern Cape. Recently, Robert Gess of the Albany Museum discovered fossils of juvenile coelacanths, *Serenichthys kowiensis*, in road cuttings near Grahamstown. They are over 360 million years old and represent the oldest find of a nursery of prehistoric coelacanths. One of the sites, at Waterloo Farm, is under the same hills that I hiked with JLB Smith!

Albany Museum

Robert Gess

Albany Museum

Serenichthys kowiensis

The Ichthyology Institute in Grahamstown has maintained the tradition of developing strong contacts with anglers, fish farmers and aquarists. Margaret published a list of the common and scientific names of marine fishes, and a useful booklet, *Sea and Shore Dangers*, in 1975. In 1982 we launched ICHTHOS, the Society of Friends of the JLB Smith Institute of Ichthyology, which published a regular newsletter that reached its 76th edition in August 2005, when it was discontinued.

Members of the society also delivered popular talks on fishes, published fish recipe books, held regular fish suppers, organized colloquia on fishes and fish conservation, and mounted exhibitions on fishes in art. Perhaps their most significant initiative was to establish an Angling Museum in Knysna in 1994.

Mike Bruton

Coelacanth model on display in the Knysna Angling Museum

that ran them, had widespread close contact with anglers, partly because I was one myself, and my brain held not only a rapidly increasing and almost comprehensive knowledge of the fishes living in our waters, but also a sketchy panorama of the long line of fascinating fishy creatures that from remote ages past had come and gone. After all, one of them was my own remote ancestor.

Chapter Two

THIRTY MILLION GENERATIONS

WHEN it was said that Coelacanths had been thought to be extinct for 50 million years, many people found it fantastic that scientists should even be prepared to make statements of that type. Such a period of time is of course enormous, but it is short compared with the time that covers the full history of our earth. Before we can show where the Coelacanth fits, it would be advantageous to make a rapid survey of what scientists now believe lies behind us.

Although fossils have been known for quite a long time, it is astonishing that their true significance has been realised only in comparatively recent times. One of the earliest fossils to be described was an almost perfect skeleton of a large salamander, found in rock strata in Germany, and it was regarded as the remains of 'a poor sinner overwhelmed by the [Biblical] flood'.

The science of 'Palaeontology' (knowledge of old life) is in one sense quite new, and in the last half-century it has developed in a manner that the first workers could scarcely have foreseen. In less than a century of intensive work some of the most remarkable intellects of all time have, from often only fragmentary remains, been able to unravel much of the history of life from the most remote past until today, and to present an almost complete picture of the main forms of life that have inhabited the earth. With this has come a rapidly increasing perception of the vast ages that lie behind us, and methods have been developed by which it has become possible to construct a scale to measure past time in a manner undreamt of not so long ago. The methods by which this is done are highly technical, and still newer and finer techniques are continually being developed.

Many people are curious about this. Here is one method by which the approximate age of a rock may be found. Uranium

The concept of dating fossils or fossil-bearing rocks is now accepted by the lay public in most Western countries but not in many Muslim countries (or in parts of Europe and the USA where creationists abound).

When I taught science in Bahrain from 2012 to 2015 I found that most people (scientists and non-scientists alike) did not believe (or were not allowed to believe) that we could accurately date fossils from the past and establish a chronological sequence of their existence.

The discoverer of the first coelacanth fossil, Dr Louis Agassiz, who described *Coelacanthus granulatus* from England in 1836, was a well-respected palaeontologist who later established the Museum of Comparative Zoology at Harvard University in the USA. Despite his stature as a scientist he never accepted

Louis Agassiz

Darwin's theory of evolution by natural selection and believed that species were 'ideas in the mind of God'.

While living in Bahrain in 2012 I found a coelacanth fossil, discovered by oil prospectors, on display in the Oil Museum in the Sakhir Desert. Coelacanth fossils are now known from all continents, including Antarctica.

Fish fossil, labelled as a coelacanth, in the Oil Museum in Bahrain

Mike Bruton

Smith's knowledge of chemistry and geology enabled him to understand the intricacies of uranium-lead dating, carbon isotopes and radiocarbon dating better than most ichthyologists. In an age of increasing specialization his knowledge spanned many disciplines, which allowed him to reach insights that were beyond the ken of ordinary scientists.

In 1976, during my post-doctoral year at the British Museum (Natural History), the Piltdown Man hoax was often discussed. The perpetrator of the hoax has never been identified but some of the suspects include the respected French philosopher and palaeontologist Pierre Teilhard de Chardin, and even Sir Arthur Conan Doyle, the creator of the fictional detective Sherlock Holmes.

gives off radiations and small particles (of helium), thereby changing into a special kind of lead. The time that uranium takes to do this is known—it is many millions of years. By measuring the amount of lead in the uranium, the time that has passed can be estimated. When this takes place right inside a rock, the amount of helium (a gas) also gives a confirmatory figure. There are several other methods as well, one involving 'isotopes'.*

It is interesting to note that while, with all advances in technique, readjustments of estimated past time occur, they are on the whole of a comparatively minor order, so that it appears likely that we really do know a good deal about the relatively enormous stretches of time that have passed in our making.

Almost everyone today accepts that our sun is a star, that in the universe there are countless billions of other similar stars, and that our sun started, somewhere and somehow, as an enormous mass of very hot gas. This, whirling and moving at an enormous speed through space, gradually cooled. Portions flew off at intervals, and these are now the planets, of which our earth is one. These smaller masses cooled more quickly than the sun itself. Originally, of course, our earth was so hot it was almost all gas. As it cooled liquid first formed at the centre; then the surface became

* A method that is proving of great value in dating remains of once-living organisms has been evolved in recent times. It is based on the fact that carbon in the structure of living organisms, both plants and animals, has been found (1946) to contain a constant small amount of a radioactive isotope of atomic weight 14. Compared with uranium this has a short life, the period of half change being only 5,600 years. Because of its presence the carbon in organic remains such as the bones of a skeleton, or of a tree trunk preserved by some means such as being buried in a swamp, will steadily show less and less radioactivity as time goes on. The amount is so very small that its measurement demands great skill and many precautions. It has been possible to test the method by the use of remains of accurately known age, and in the hands of an expert it yields remarkable results. A striking application of this method has recently caused a considerable scientific sensation. Early this century the biological world was aroused by the discovery in deposits at Piltdown in England of the bones of a skull claimed by many experts to be of an early type of man since named the 'Piltdown Man', dating back close on a million years. While some doubts were expressed about its validity, most British experts accepted this view, and the bones remained a treasured possession in the British Museum. The carbon-isotope method has led to the discovery that this skull is made up of bones of different ages, all comparatively recent, none really old. The whole thing was a deliberate fraud, there never was any 'Piltdown Man'. A book has recently (1955) been published giving the whole story.

solid, still entirely surrounded by a gigantic dense atmosphere of whirling vapours and gases.

All the enormous amount of water now liquid on the earth was then gas. There came a time when the whole mass had cooled so far that the cold of outer space caused 'rain' to form in the dense clouds that covered the whole earth. At first this rain never touched the earth, it was too hot, but eventually it did reach the solid crust, only to sizzle off at once again as gas. For a long time, probably thousands of years, all over the whole earth it never stopped 'raining', literally pouring, a process which caused quite rapid cooling. One can well imagine that there must have been continual 'storms' of violence undreamt of today. In passing, we may note that at present the main part of our earth is still liquid and very hot under its solid crust. There is, of course, abundant liquid water and the atmosphere of gas. The earth is cooling all the time, and it is steadily losing water and air to outer space. If the earth survives long enough there will come a time in the far-distant future when any water or 'air' that may be left will all be solid. One way and another, life as we know it now, free life on the surface of the earth that needs water and air, can only be a passing phenomenon in the infinite time span of the universe.

The sciences of Geology and Palaeontology go together and scientists in those fields have divided the time of existence of the earth into different eras, systems, and periods, which have for convenience been given names.

The table overleaf (p. 14) giving a Geological Time-scale is a summary of what is more or less generally accepted.

On the earth there is a sharp distinction between dead, or 'inorganic', matter and living things which nobody has yet been able to bridge. The earliest forms of life on the earth were doubt-less preceded by the formation in some fashion of 'organic' matter; that is, non-living compounds containing carbon and other elements essential to living organisms of the type we know on the earth, that in some fashion came to be alive. Nobody has as yet succeeded in pushing any types of non-living 'organic' compounds over the borderline to 'life', but it is not impossible that suitable compounds are constantly being produced in nature, that the transition to living matter may still occur, so that even if

JLB's succinct description of the evolution of the planet would probably have been the first introduction that many lay people had to these ground-breaking scientific concepts (unless they chose to peruse the relevant sections of encyclopaedias of the time). JLB was writing at a time when South Africa was in the grip of a Nationalist government bent on imposing conservative, often anti-evolutionist religious views on the populace, and in the throes of implementing the repressive regime of apartheid.

His far-reaching opinions on the eventual extinction of life on Earth came as a shock to some readers, especially those who regarded life as sacrosanct and infinite according to the religious teachings to which they were exposed.

The time frames on the Geological Time-scale given by Smith have inevitably been revised in the light of more recent research. Perhaps the most startling change is that early humans (*Homo sapiens*) may have arisen as early as 90,000 to 115,000 years ago (rather than 25,000 years ago) based on discoveries in Border Cave in northern Zululand and elsewhere.

It is still widely believed that all life started in water. To a certain extent even animals that live on land are still 'aquatic' in that they were only able to make the trek from water to land because they took some of their alma mater with them in the form of their blood and cell fluids.

JLB's idea that radioactivity might have triggered the transformation of inorganic matter into organic life was well ahead of its time and may still prove to be true.

GEOLOGICAL TIME-SCALE

Eras	Systems	Periods	Years Ago	Types of Life
ORIGIN OF EARTH			3,000,000,000 at least	
PRE-CAMBRIAN	Eozoic Archaeozoic Proterozoic	. .	1,700,000,000	First lowly forms of life.
PALEOZOIC	Cambrian . . .		500,000,000	Invertebrates
	Ordovician . .		400,000,000	Invertebrates
	Silurian . .		350,000,000	Vertebrate fishes
	Devonian . .		320,000,000	Rhipidistia Coelacanths Various fishes Amphibians
	Carboniferous . .		280,000,000	Primitive plants on land Amphibians
	Permian . .		220,000,000	Amphibians
MESOZOIC	Triassic . .		190,000,000	Reptiles Mammal-like reptiles
	Jurassic . .		150,000,000	Birds
	Cretaceous . .		120,000,000	Flowering plants Mammals
CAINOZOIC	Tertiary	Eocene .	70,000,000	Mammals
		Oligocene .	50,000,000	Mammals
		Miocene .	25,000,000	Mammals
	Quaternary	Pleistocene .	1,000,000	Ape-man Stone-Age man
		Holocene . .	25,000	Modern man

all life on the earth were to be obliterated, there is at least a chance that it might start all over again.*

It is universally accepted that life started in the water, and the first living things are presumed to have been very lowly, something small and soft, like the simple, tiny protozoa that zoologists

* The constant presence and proportion of the radioactive C_{14} isotope in living matter inclines me to believe that the 'creation of life', or the 'animation' of matter, probably took place in suitable non-living matter under the influence of a special type and density of radioactivity. It may well eventually be possible to deduce what this was and to carry out the process in the laboratory, though the living matter so produced may not necessarily be the same as that which originally appeared on earth.

know so well, minute living blobs of jelly. 'Inspired guesses' based on faint signs in ancient rocks put the first appearance of living matter on the earth at about 1,700 million years ago. These first forms of life developed slowly and gave rise to other types, some more advanced, and by 450 million years ago there were numbers of 'invertebrates', backboneless creatures of many types, some quite large, in most of the waters all over the earth.

The first true vertebrates, or backboned animals, are estimated to have appeared by 400–350 million years ago. They must have developed from some ancestor without a true backbone, and they certainly were peculiar creatures, for they had no scales and no true jaws, just soft, sucking mouths. It is in one sense incorrect to speak of some at least as 'backboned', for they had no true bone but vertebral columns only of gristle. Some of them, however, had bodies covered with heavy bony armour, and these have left excellent fossil records.

There is evidence that at the close of the Silurian period and over the beginning of the Devonian some striking change was at work, for it was then that fishes something like the modern types we know first appeared. They had true bony jaws and overlapping scales, and a skeleton at least partly bony. Their fins were peculiar, rather like small paddles with a fringe of soft rays, so that they were named 'Crossopterygii' or 'fringe finned'.

These fishes represented a tremendous step forward in evolution in more ways than one. Not only did they at that very early stage show important features that have remained predominant in fish life to this day, but one group of them gave rise to forms that colonised the land and were indeed our own ancestors.

It would be as well to realise that up to the Devonian period the land was very different from that of today. There was abundant animal and plant life in the water, but apparently hardly any on land, which was bare, mainly rock. Indeed, only about this time did plants start to creep ashore, so that up to then there had been nothing to tempt creatures to leave the water. It is strange to think of static life like plants being able to move out and colonise a different medium, to come out of water and march across the land, but it was done. And as such things go, plants can move quite rapidly in that way. If you look at a pine forest you will see how trees can march across country, for there will be younger

Further research over the past 60 years has pushed the date for the first appearance of life on Earth back to over 2.7 billion years, and maybe as far back as 3.9 billion years! The date of appearance of the first multicellular life is now an extraordinary 900 million years ago (mya), that for backboned animals about 540 mya and fish (placoderms) about 505 mya.

Giant placoderm　　　www.everythingdinosaur.com

About 460 mya the fish split into the cartilaginous fishes (which became the modern sharks, rays and skates) and ray-finned fishes, which diversified into most of the modern fishes that we know today.

The Crossopterygii ('fringe-finned fish') are now considered to fall within the subclass Sarcopterygii ('lobe-finned fish'). Living sarcopterygians include the coelacanths and lungfishes, which gave rise to the tetrapods (four-legged animals).

JLB's description of the invasion of land by plants is another example of his broad approach to science. Few scientists in South Africa at the time, perhaps with the exception of Robert Broom and Keppel Barnard, could match his comprehensive understanding of many different scientific fields, including evolution.

Recent research confirms that most coelacanths known from the fossil record were relatively small, less than 50 cm long, compared to modern coelacanths, and lived in freshwater lakes and swamps, although some large fossil marine coelacanths, over 3 m long, have also been found (inevitably in Texas).

However, the fossil record is very biased and it is likely that many fossils left by larger coelacanths that lived in the sea have not been found, or were never formed, or were destroyed by continental drift and other global geological events.

JLB is right in saying that the migration from water to land was 'the greatest step in the history of life'. The crucial transition from water to land took place between 380 and 365 million years ago. At the beginning of this period all backboned animals were fishes living in water but, by the end of it, there were many transitional forms as well as the first true amphibians and reptiles.

and younger trees stretching far out, developed from seeds carried by the wind.

Insects apparently went out on land about the same time as the plants, but before backboned creatures were living out of the water almost a hundred million years more passed, and that is a very long time.

In those early days there was apparently abundant life in most of the waters, but the vertebrate fishes such as the Crossopterygii appear to have lived mainly in fresh-water swamps. Now, fish need oxygen just as we do, but they get theirs from the water, which dissolves a little from the air. If water loses its oxygen, fish cannot live in it, and we have all seen what happens when dams and vleis start to dry up. Rotting vegetation in water uses up all the dissolved oxygen, and you find fish dying and dead all round the margins of such bodies of water. For the same reason one putrefying fish will kill many others.

There have apparently always been floods and droughts for long or short periods. In a short sudden drought there would be heavy mortality of swamp fish. In a slow drought fish would not be killed off so quickly, and it is apparent that sometimes in those early days, fishes gasping in putrefying pools, managed to live by absorbing a certain amount of oxygen directly through surface blood-vessels, probably in the mouth and gill cavity. Over long ages certain types probably learnt to gulp air and to breathe at least partly that way, first by necessity, then by choice. This would have tremendous consequences. First of all, when a drought came and pools began to dry up, such fishes could live long enough in air to flop out and perhaps reach other and better water, and so survive. Over long ages fishes doing this probably came to spend more and more time out of the water, possibly in getting from one pool to the other they found succulent food on the way. Gradually some fishes gave rise to creatures more suited than themselves to life on land, creatures that could live on land and in water, the so-called amphibians, of which the modern frog is an example. Fins began to modify and change to limbs, and so was taken this greatest step in the history of life as it affects us, the first real step that led to man.

This is where the Coelacanth comes into the picture. There were two main lines in the Crossopterygii, named Coelacanths and Rhipidistia. No one could fail to see, even from only fossil remains, how closely they are related. Because somewhat earlier fossils of the Rhipidistia than of Coelacanths are known, most scientists today hold the view that the Rhipidistia came first and that Coelacanths developed from them. Because this view was expressed by one or two leading workers, even without any really positive evidence, it is commonly held even today. If it were true, however, since all those forms lived under the same conditions, it is almost incredible that there should be no sign of transitional types. But in fact, there are the Rhipidistia, and there are the Coelacanths, and nothing in between. It is at least as likely

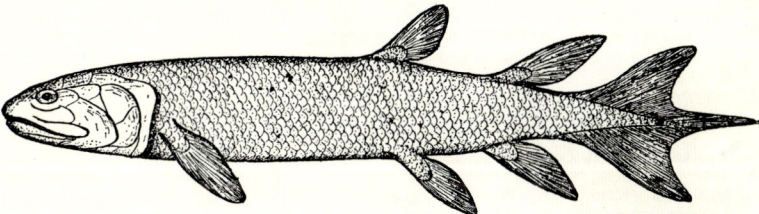

FIG. 1.—A Rhipidistian fish, *Eusthenopteron*, that is believed to be the ancestor of land animals. About 12 inches in length.

that both of these types came from some as yet unknown common ancestral form, and that form would be ancestral to man as well. I have a 'hunch' that one day the remains of something like this will be discovered in earlier strata.

Be that as it may, let us examine the history of these two important main lines of fishy evolution. Both lived in swamps. Their fins indicate that both were able to leave the water and flop about on land, either from necessity or desire. There is good evidence that the Rhipidistia gave rise to amphibia and hence to all other land vertebrates, including man, while the original stock, the Rhipidistian fishes themselves, all died out very long ago. The recent appearance of the Coelacanth has taught us a lesson, and we had better be careful and say rather that all the available evidence indicates that the Rhipidistia became extinct long ages ago. In other words, while they themselves were not strong enough to survive profound changes, by some accident they gave rise to creatures that found a new way of life on the land. It was in one sense their own weakness that led to the survival of and ultimate

O.F.—2

In fact, the most recent research on the genome of the coelacanth reveals that the lungfishes (Rhipidistia) are closer to the direct line of origin of four-legged animals on land, including humans. Although we have only two legs, like other bipedal animals (such as chimpanzees, kangaroos, birds, and many extinct dinosaurs), our two arms count as limbs and qualify us as 'tetrapods'.

Coelacanths were not, however, far from the ancestral animals that gave rise to the tetrapods and they do share some characteristics with four-legged animals, such as the structure of their inner ears and their lobed 'limbs'. More correctly, we should say that tetrapods share some characteristics with coelacanths, which evolved first.

The discovery of the first animal that bridged the gap between amphibian-like fishes and the fish-like amphibians is a classic example of scientific enquiry. A team lead by Neil Shubin and Edward Daeschler from the University of Pennsylvania thought carefully about the best place to look for exposed fossils, chose a rocky area of the right Late Devonian age in the Canadian Antarctic, and, in 2004, made a sensational discovery – a fossilized tiktaalik.

It was the perfect 'missing link', almost exactly splitting fishes and amphibians, with a crocodile-like head on a salamander-like body and the scales and tail of a fish, but, unlike fish, it did have a neck.

National Science Foundation

Tiktaalik rosae

To date fossils have been found belonging to over 90 species of extinct coelacanth. No fossils of the two living species are known (which is further proof of the unreliability of the fossil record). Fossils of at least three extinct species of coelacanth have been found in South Africa, *Serenichthys kowiensis* (360 million years before present), *Coelacanthus dendrites* (scales only; 175 mybp) and *Whiteia africanus* (160 mybp).

distinction for their descendants. The Coelacanths were quite different. They were tougher. It is not yet certain whether the first Coelacanths of 300 million years ago originated in fresh water or in the sea. It is often quite a problem to decide this particular point for any one fish. Fossils generally result when animals are buried in mud which later hardens to rock. This can happen, for example, at the mouth of a river which brings down quantities of mud and silt. In South Africa at least, when rivers come down in flood, fresh-water fishes are often carried out to sea and killed. If that happened in those far-off days, as it probably often did, near the mouth of such a river there could be laid down beds containing fish fossils, some at least fresh-water types, together with typical marine organisms such as shells. If no other fossils of these particular fishes happen to have been found elsewhere, they could easily be listed as marine. What has been established is that Coelacanths spread and lived over a great part of the earth, some definitely in swamps, some probably in rivers and estuaries and some in the sea. Some wonderful fossils have been found. For example, during quite recent excavations for a University Library at Princeton, U.S.A., some shales were encountered that were the remains of a swamp of about 190 million years ago. These contained an average of no less than twelve Coelacanth fossils per square foot. That swamp was certainly swarming with Coelacanths. I wish I could go back there and fish for Coelacanths in a Triassic swamp with a bent pin and worms as I have done for Kurpers. I wonder if there were suitable worms then. I am foolish enough to hope that somewhere there may still be such a swamp, and I wonder who will be the first to find it.

Looking back to the time before it was known that Coelacanths were still living, it is astounding to see that nobody seems to have realised the wonder of the Coelacanths, even as they were then known. There was so much that was remarkable about them even then. For one thing, the fossil series showed them to have had, from the very beginning so long ago, important characteristics, like jaws and overlapping scales, far in advance of their time, characters that are obviously good for survival, because even the most modern fishes have hardly improved on the Coelacanth pattern. As a distinct and characteristic line they survived longer than any other type of vertebrate, all unmistakable Coelacanths, living from beyond 300 million years ago to close on 50 million years ago, spanning the incredible time of 250 million years, and surviving terrific climatic changes and upheavals that wiped out countless other forms and types. What is equally as astonishing is that over those relatively vast ages of their existence they changed very little, less indeed than any other known vertebrate.

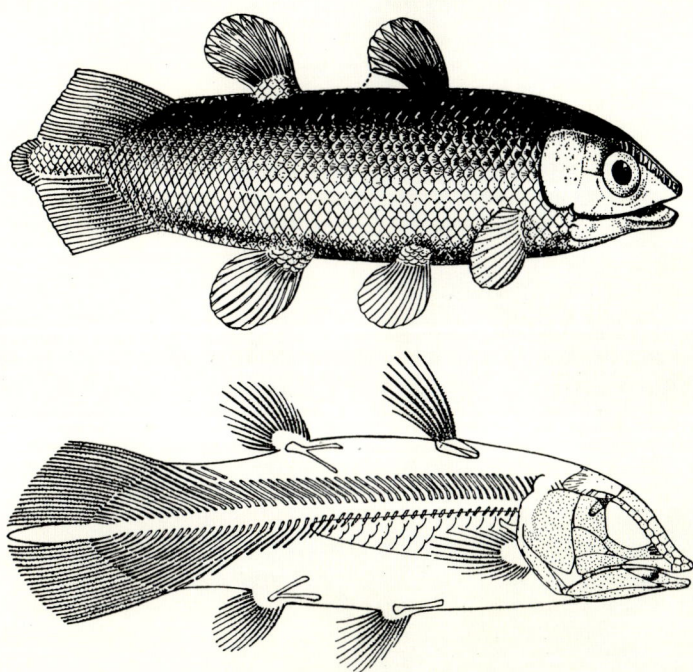

Fig. 2.—Reconstruction of an early Coelacanth from its fossil remains: below, much as seen from a fossil; above, as believed to have been in life. (After Smith Woodward.)

It is almost amusing that some scientists speak of Coelacanths as a 'degenerate side-line' because they have not given rise to other forms. This is a most peculiar view, since man then emphatically falls into the same category, for man will certainly not give rise to any other form save those he moulds himself. The power to master changing conditions can never be degeneracy. Some Coelacanths probably left the water, but their innate strength kept the line going when the supposedly more virile Rhipidistians

As JLB points out, coelacanths would have been regarded as amazing creatures even if their living representatives had never been found. Their extreme longevity and relatively unchanged anatomy over hundreds of millions of years, their evolution of advanced features such as bony jaws, overlapping, armoured scales, lobed fins, the fact that they give birth to fully-formed live young, and their pivotal role in tetrapod evolution, would have flagged them as significant beasts anyway (see Chapter 5 for further details).

One of the characteristic features of the coelacanth is the extra lobe in the middle of the tail fin, which is found in no other fishes. Marjorie Courtenay-Latimer identified this extra lobe as a diagnostic feature when she sent her first sketch of the coelacanth to JLB Smith.

had been wiped out (at least the evidence we have says so), and now of course it turns out that, after all, the old Coelacanth was even more powerful than we knew. The ancient line still goes on, after probably at least 30 million generations. Thirty million generations, just think of it!

Another characteristic of Coelacanths is that although they spread all over the world, they differentiated very little, that is, even those widely separated remained much the same. For example, fossils show that at one period there were Coelacanths living at the same time on places as far apart as Greenland and Madagascar that were so much alike that it is doubtful whether they were not actually the same species. Over all those vast ages only about twenty-five different genera are known, no less than ten appearing in the Triassic, which was the time that Coelacanths reached a climax in types and numbers. After that they seem to have declined, but it is as well to remember that we do not know much about those that lived in the sea, for fossils of purely marine creatures are not easily formed or accessible.

Over those vast ages the known Coelacanths were all much alike in body form. The variations were few. Judging by their teeth, they tried all sorts of diets, most must have been plain predators, catching other fishes, but some obviously lived mainly on shellfish, having powerful molars and dental plates for crushing rather than biting. Most of the extinct Coelacanths known from fossils were quite small, a matter of 5–20 inches in length. But as will readily be appreciated, the fossils are those of forms which lived where they were most likely to yield fossils, and that was chiefly in swamps. Water in swamps is not usually very deep, and life there is restricted, so that most fishes in swamps are on the small side. We almost certainly know more swamp Coelacanths than any others. It is therefore by no means certain that all Coelacanths of past times were small like that, and indeed in quite recent years the remains of one of near 5 feet long have been found in a rock stratum in Germany. It is plain that while we have a wonderful fossil record, it must be very far from complete.

So here we have the picture of the Coelacanths. This remarkable type appeared more than 300 million years ago, and has gone on, virtually unchanged as such things go, until the present time. In that long time countless other types of fishy creatures evolved, flourished, and vanished, many of them types that may have seemed more suited for survival than our old Coelacanth, but he has outlived them all. He goes plodding steadily on, his needs few and simple, and he will quite likely still be there when many of these 'active modern types', which are supposed to have driven him to the depths, will be gone and long forgotten. He reminds one of a solitary, tough old man, asking favours of none. Old man Coelacanth. Degenerate? Never!

One of the physiological characters that might have favoured coelacanth survival is their slow metabolic rate and energy-efficient way of living. They rest in caves during the day and drift slowly with water currents when hunting at night.

Painting of coelacanths hiding in caves in the Tanga Coelacanth Marine Park, Tanzania

They do, however, need to live in cool, well-oxygenated water as their blood haemoglobin count is the lowest of any vertebrate and they have a very small gill surface area for absorbing oxygen from the water.

South Africa's scientific glitterati during the period from the 1930s to the 1950s included Arnold Theiler (1867–1936, founder of the Onderstepoort Institute of Veterinary Research), HJ van der Bijl (1887–1948, first chairman of Eskom and Iscor), Dr JH van Eck

Basil Schonland

(1887–1948, involved in establishing Eskom, Iscor, Sasol and the Industrial Development Corporation) and Basil Schonland (1896–1972, founding director and funder of the Bernard Price Institute for Geophysical Research and the CSIR).

JLB's comments on 'Cinderella' museums in South Africa are still valid. A great deal of valuable cultural and natural history material was sent from South Africa to museums abroad for study, identification and curation. It was therefore a brave move by the young Curator of the East London Museum to keep the first coelacanth specimen in her museum. The 'glass slipper' fitted her perfectly!

Marjorie Courtenay-Latimer with the first coelacanth in East London

East London Museum

Chapter Three

CINDERELLA

IN South Africa today science is very much a part of everyday life, and South Africans are playing an increasingly important role in the world of science, a number already having achieved international repute in their respective fields. It is not generally realised that this has taken place only comparatively recently, most indeed within the last generation.

It is not long since life in South Africa, especially in the south-eastern areas, was a grim battle for existence, not only with nature but also against raiding native hordes. It was the 'Mau Mau' on a much larger scale. At that time, and even long after conditions had become more settled, virtually all scientific work in South Africa was done by visitors and later by trained persons imported from other countries. As many of them were scientifically isolated in South Africa, it is easy to understand that they maintained constant contact not only with their homeland but also with the overseas institutions from which they had come.

In those days, outside the largest centres, there were no real scientific institutions in South Africa, and the few small, local collections of historic relics and curios could scarcely be termed museums. Whenever anything unusual was discovered, and there were many such discoveries, it was generally necessary to send it overseas for expert opinion.

It is easily understood, therefore, that there came to be accepted a general belief that scientific institutions in this country, such as museums, and the work done in them, were inferior to what was old and established overseas. These younger 'Colonial' museums served the purpose of housing local material, but even in that case it was generally felt, and openly expressed, that any article of great value should not remain in any such small establishment, but be sent to some long-recognised overseas institution like the

British Museum. In some parts of South Africa this view still has its adherents, even today.

Sentiments of this kind were at least partly responsible for the late foundation and initially slow growth of the East London Museum. In December 1938 the East London Museum was little known, being one of the youngest in the country. It had indeed quite a struggle for existence, being supported by only a small grant from the Government, and distressingly small material support from its own community, which at that time did not regard it as of importance or value. The total annual income then was less than seven hundred pounds, which had to cover salaries and wages, material, stationery, everything. It is almost incredible that anything like a musuem could have existed under such circumstances, for it had started without endowment of any kind and was poorly equipped. Like most such institutions in South Africa it was supervised at first by a series of honorary part-time Curators, but eventually Miss M. Courtenay-Latimer was appointed as its first full-time Curator.

While there were some who found it strange that a relatively young woman should have been selected for this position, it is plain that those responsible had perception and sound judgment, for they could scarcely have made a better choice. Miss Latimer showed herself able, capable, and energetic, and was soon at grips with the many difficulties that beset her ideas and ambitions for the Museum. She had great difficulty, not only in managing all that she desired with such limited means, but also in convincing the Board and especially its Chairman of that time, that their Museum could be developed into one of the best known in South Africa, and even beyond, as indeed it is now.

From the start Miss Latimer wisely concentrated on building up exhibits representative of the life of the area served by the Museum, and this she carried out with characteristic energy and enthusiasm. As Miss Latimer realised that angling is the chief sport and hobby in that area, she got the commercial fishing firms to collaborate, and especially from Messrs. Irvin and Johnson's branch at East London received a constant stream of valuable marine specimens which were mounted and exhibited at the Museum. She wisely made personal contact with the officers and crews of the trawlers, and infected them with some of her own

When Marjorie was appointed as the first Curator of the East London Museum in 1931 at the tender age of 24 years she received a salary of only £2 per month. She had originally trained as a nurse and wore her nurse's outfit to the museum for the first few months to save money!

In addition to its famous display of the first coelacanth, the East London Museum has other remarkable items in its collection, including the only surviving intact dodo egg, valuable mammal-like reptile fossils (assembled into skeletons by Marjorie) and a live beehive that I remember gawking at as a kid in the 1950s.

East London Museum

Marjorie Courtenay-Latimer and the dodo egg

Captain 'Harry' Goosen later recalled, 'I was watching while the men were dumping the "rubbish" in case there were some specimens for the Aquarium and that is when I saw the strange fish's tail sticking out and realized it was different from anything I had ever seen before. Just then I told the crew "not to damage that fish".'

East London Museum

The steam trawler Nerine

In 1987, while I was assisting Hans Fricke with a documentary on the coelacanth in East London, we interviewed a man who claimed to have been Marjorie's taxi driver. However, he turned out to be the uncle of the long-deceased cabbie. Such is the evil lure of fame!

The capture of a coelacanth at a depth of 40 fathoms (about 73 m – very shallow for coelacanths) on the gradually-sloping, sandy continental shelf within 5 km of shore (they prefer steep, rocky, off-shore canyons and caves) was an extremely lucky fluke.

Mike Bruton

Model of the Nerine *and the side trawl that caught the first coelacanth*

enthusiasm, so that they watched for unusual specimens of all kinds from the trawl, many of which were kept and brought to port. It became the custom to pile up the 'rubbish' so that she could scratch through it, and indeed she found many treasures that way.

It was therefore with no sense of anything unusual that Miss Latimer received a telephone message from the manager of Irvin and Johnson at East London in the late morning of the 22nd December 1938, to say that a trawler had brought in a pile of fish for her to examine. She called a taxi and with Enoch the native assistant of the Museum went down to the wharf some miles off. When she got there the captain had already left the ship, but one of the deck-hands took her to the pile of fish they had put aside, mostly sharks. Those she already knew and had got previously, but then, almost hidden, she noticed a large heavily scaled blue fish, and as a peculiar fin and the colour attracted her attention, she had the fish pulled out. It was a peculiar creature, like nothing she had ever seen before, and she stared at it in puzzlement for some time and examined its mouth and fins. She asked the old trawlerman if he had ever seen one before, but he replied that in his thirty years at sea in that work he had certainly never seen any fish of that type, and he pointed out that the fins were like arms, it looked almost like a big lizard. Miss Latimer thought it looked something like a Lung-fish, but in any case decided that it was obviously something rare which it would certainly be advisable to keep. The trawlerman said it was a lovely blue when taken from the water, but was a vicious brute, snapping its jaws fiercely. They had all been struck by its unusual appearance, for none of them had seen anything like it before, so they had called Captain Goosen to look at it, and when he touched the body, it heaved itself up suddenly, snapping its jaws viciously and had nearly caught his hand in its formidable fang-lined mouth. The captain ordered the crew to put it on one side so that Miss Latimer could see it, for by then he had decided to go straight in to port.

The fish was 5 feet long and heavy. As a matter of interest Miss Latimer got them to weigh it; 127 lb. It was a scorching hot day and the fish had a smell—all fish have on hot days—but the Coelacanth has one all its own, as we came to know only too well.

According to Miss Latimer, she had very considerable difficulty in persuading the taximan to consent to having the fish in his taxi at all, even though she had brought along old bags to put on the floor on which to rest any fish. The taximan was so reluctant that he stood aloof and distant while she and a native struggled and wrestled to get the creature into the vehicle.

One may sympathise with that taximan, while smiling at the incongruity, refusing to transport what was the most valuable zoological specimen in the world, though none of them knew it at that time. There are many fishes in South African seas that to anyone not an expert would appear much more strange than a Coelacanth, and, as has been explained, museum directors in South Africa have so many fields to cover that they just cannot be experts in every branch of science.

Getting this heavy fish to the Museum was one thing, what to do with it there was another. Miss Latimer had nothing in which to keep it, but first had a hunt through her pitifully few reference works to see if she could get some idea of what it was. But she found nothing; indeed, from its fantastic nature it would have been almost a miracle if she had. So, after making a rough sketch and taking measurements, she borrowed a hand truck, and with the native boy took it to a taxidermist who did that work for the Museum. (The 'Museum' did not even have a handcart of its own.) She also asked an expert amateur photographer to take some photographs, which he did, but for some reason the whole film was a failure.

According to Miss Latimer, the bony plates of the head, the scales, and the fins made her feel it was a Ganoid fish, probably a Lung-fish of some type, but she had no means of verifying this and relied on hearing from me. She told the taxidermist to keep all the parts he cut away in case they were important, and this he did; but by the 27th December 1938 they were in such an appalling state that not having heard from me by then Miss Latimer agreed to his urgings that they should be disposed of. Miss Latimer states that when she rather timidly ventured to tell the Chairman of the Board of Trustees of the Museum, the late Dr. Bruce-Bays, what she thought about this fish, he scoffed at her views, and crushed her by saying in rather harsh terms that 'all her geese were swans', and that if she wanted to keep the fish

When Marjorie first saw the coelacanth she described it as, 'The most beautiful fish I have ever seen. It had four limb-like fins and a puppy-dog tail'. A deckhand referred to it as 'a great sea lizard' because of its bizarre fins.

Model of the coelacanth on display in the East London Museum

The first coelacanth

In the diary entry of her father, Eric Latimer, dated 7th January 1939, he states, 'Margie is furious with Dr Bruce-Bays. She says she cannot understand him, he is most annoyed about the fish – says it is a Rockcod and she is foolish making such a song about it, when Dr Smith sees it he will laugh at her and he couldn't be bothered with the thing. Margie is very upset and worried – she persists that she is sure that it is something wonderful.' Bruce-Bays was the chairman of the Board of the East London Museum at the time and was described by Marjorie as 'a very sarcastic old gentleman'.

How wrong Bruce-Bays was – weeks later, during a public viewing of the coelacanth, the tiny museum, according to Marjorie, received over 20,000 visitors in one day (surely an exaggeration?). To this day the East London Museum has retained its reputation as a fine regional museum.

Newspaper heading in 1939

The taxidermist who first mounted the specimen arranged the lobed fins so that they pointed downwards as if they were limbs, which gave rise to the nickname coined by journalists, 'old four legs'. Recent research has revealed that the modern coelacanth does not use its lobed fins as limbs at all but hovers above the sea bottom.

Marjorie will forever be commemorated in the species and family names of the African coelacanth, *Latimeria chalumnae* and Latimeriidae.

East London Museum

Marjorie Courtenay-Latimer in about 1952

The sketch that she sent to Smith, although rough, has become part of zoology folklore.

he would authorise it to be mounted. Surely that was enough, he asked?

There will be some who are not impressed by the mere record of these events. But they may be assured to start with that to handle and treat a large smelly fish on a hot day is no pleasant or light task. It is a formidable job, and much more so for any woman. By great good fortune Miss Latimer was no ordinary woman. I know from vast experience just how much she did at this time and under what great difficulties, with no encouragement from any source, rather the reverse, driven solely by that inner fire which makes her what she is. I shall always admire and respect her for it. She merits the admiration of every true scientist, and gets it. It was only Miss Latimer's instinct for what is valuable and her force of character and determination that saved that specimen; one less determined would have been overwhelmed.

The morning following the arrival of the fish, Miss Latimer wrote to me, enclosing the rough sketch and notes. Little did she know what she was touching off. Its effects were those of an atomic blast, of two atomic blasts, for twice the Coelacanth sent a wave that went all round the world, and the backwash of the second still goes on.

Chapter Four

STRANGER THAN FICTION

ON the side of the lagoon at Knysna, some miles from the sea, we have a house with a laboratory, where I do not only considerable general angling but carry out regular and periodic investigations on the extraordinarily rich and varied fish-life of this large estuary. It is an exceedingly interesting body of water, with many unique characters.

In December 1938 we had gone from Grahamstown to Knysna. I had been unwell, and was not fully recovered even in the New Year.

About midday on the 3rd January 1939 a friend brought us a large batch of mail matter from the town, very much of a Christmas–New Year accumulation, and this was sorted out between us, each settling down to his or her letters. Mine were the usual mixture of examination results and queries, and numbers about fishes. One was from the East London Museum, in Miss Latimer's well-known hand, the first page very much the usual form, as will be seen on p. 30, asking for assistance in classification. Then I turned the page and saw the sketch, at which I stared and stared, at first in puzzlement, for I did not know any fish of our own or indeed of any seas like that; it looked more like a lizard. And then a bomb seemed to burst in my brain, and beyond that sketch and the paper of the letter I was looking at a series of fishy creatures that flashed up as on a screen, fishes no longer here, fishes that had lived in dim past ages gone, and of which only often fragmentary remains in rocks are known. I told myself sternly not to be a fool, but there was something about that sketch that seized on my imagination and told me that this was something very far beyond the usual run of fishes in our seas. It was as if my common sense were waging a battle with my perception, and I kept on staring at that sketch, trying to read into it perhaps more than it held. In this surge of violent thoughts and reactions, the

The Smiths' famous 'blue house' still exists although it is now the Ardeco Guest House.

One of the Smiths' neighbours was Leonard Thesen who could not understand what all the coelacanth fuss was about. In order to trivialize Smith's discovery, he painted a picture of a coelacanth on a curtain that he hung in his beach cottage, claiming that it was based on a coelacanth that he had seen washed up on shore in the 1920s. It was later shown that the painting was a copy of an image from *Old Fourlegs*.

Smith's keen knowledge of fossil fishes, and the characteristic external features of coelacanths, made it possible for him to realize immediately that Marjorie's rough sketch depicted this fish.

JLB's description of his reaction to Marjorie's drawing, and of his wife's anguished glance, vividly portrays the ambivalence that he must have felt at the possibility that the specimen was a representative of a long-extinct group of fishes. Few scientists have been faced with such a major dilemma, which was exacerbated by the poor communications at that time between Knysna and East London. As a skilled though amateur ichthyologist, he needed to be sure of his identification, for both his and Marjorie's sakes.

JLB Smith, Marjorie Courtenay-Latimer and the first coelacanth were depicted on a South African postage stamp issued in 1989.

world about me had ceased to exist, until I heard my name called, it seemed from far away, then suddenly again, close by, and more urgently, and loudly. . . . There was my wife, staring at me from across the table with deep concern on her face, as was also her mother across at the corner; both were looking at me intently. I found to my surprise that I was standing. My wife tells that she was engrossed in a letter when suddenly she felt that something was wrong, and looking up saw me on my feet, staring at a letter in my hand. The light was behind me and she could see a fish-like drawing through the thin paper. 'What on earth is the matter?' she said, and I came back to the present. Looking again at the letter and sketch I said slowly, 'This is from Miss Latimer, and unless I am quite off the rails she has got something that is really startling. Don't think me mad, but I believe there is a good chance that it is a type of fish generally thought to have been extinct for many millions of years.' My wife says that she did wonder if I had got a touch of the sun, for she knew that I usually weighed every word, and this was quite the most extraordinary thing she had ever heard from my lips. I passed the letter over to her, and the two women read the first page and then examined the sketch, while, the rest of my mail pushed aside, I sat and stared into what this might mean if my first deductions were correct.

Yes, those fishes of bygone days had always intrigued me, and I went over in my mind what this could possibly resemble. As will be seen (Plate 1, facing p. 32), the sketch was in some ways impressionistic and not a very good representation of the animal. But that tail, and the clearly large scales, and those limb-like fins! One alone in a sketch might be passed; but all together! At the same time what I suspected was so utterly preposterous that my common sense kept up a steady fire of scorn for my idiocy in even thinking of it.

I was afraid of this thing, for I could see something of what it would mean if it were true, and I also realised only too well what it would mean if I said it was and it was not. On that sketch alone I could never decide anything; I must see the creature itself. That would almost certainly mean a journey to East London, most inconvenient at that particular time, for in the University of South Africa I was that year an examiner, and the centres had been notified to send the papers to Knysna. In that capacity

I was bound to complete my work within a specific period, and an absence at that time would almost certainly have prevented fulfilment of that obligation, for a reason I could scarcely present as justifiable. The other alternative was that the animal should come to me, but in my chaotic state of mind at that particular time, its great size, $4\frac{1}{2}$ feet long, had not registered, nor the date of its catching or indeed anything but its possible identity. Was this the fulfilment of the peculiar premonition I had always had?

One is hesitant about saying such things, but I have some peculiar sense outside the ordinary, sometimes spoken of as 'sixth sense', which warns me of impending events, usually danger or trouble, sometimes very long in advance. Again and again I have realised later that this subtle anticipation has caused me to act so as to avoid serious inconvenience and disaster, and those who live with me respect these 'hunches' of mine, even when they involve what seem utterly irrational prohibitions, and even though the others point out that when we have obeyed my apparently ridiculous directions, we often have no means of knowing that anything would have happened had we not done so. I don't always know either, but bitter experience inclines me at least to obey them. One of my most constant and peculiar obsessions had always been a conviction that I was destined to discover some quite outrageous creature, I had no idea what, but had come to suspect it might be a true sea-serpent or something like that. This was so firmly fixed in my mind that just as my peculiar set of circumstances and qualifications had set the stage ready for the appearance of the Coelacanth, so in one sense had this premonition prepared me to deal with such a fantastic possibility as had now arisen, and, indeed, even while my common sense rejected it, to seek for it in an obviously impressionistic sketch by someone not an ichthyologist.

My wife was speaking again—we had been married only nine months and, as I am older by a good many years, she was constantly finding unexpected pockets from the past, and here was one. 'I didn't know you had worked on fossil fishes,' she said; so I briefly told about my incursions into that field. 'What makes you think this may be one of them?' she asked. 'Well, mainly the tail. As far as I know there is no living fish with a tail like that. It is characteristic chiefly of the earlier members of a group known

JLB's 'premonitions' were well known to his friends and family. He seemed to have an uncanny ability to anticipate future events.

When Margaret married JLB Smith on 14th April 1938 and entered his 'fishy' world she experienced a baptism of fire. Eight months later the first coelacanth was discovered and, within the first year, they had to work furiously to complete the description of the beast. On 25th June 1939, four days after they had submitted the manuscript to a scientific journal, their son, William, was born.

Margaret and JLB Smith in 1949

NRF-SAIAB

Some scientists have suggested that Smith did not have sufficient knowledge of fossils to identify the coelacanth from Marjorie's rough sketch and had to consult palaeontologists at Rhodes University College *en route* to East London to confirm his hunch. I have found no evidence to support this claim and there is no reason to think that Smith's reconstructed conversation with his wife in Knysna (which Margaret often recalled) is inaccurate.

At the time, the university did have, for a 40-year span (1929–1969), a legendary Professor of Geology, whose name, naturally, was Edgar Mountain, but there is no record of Smith having consulted him. JLB might have spoken to Jack Rennie, who joined the Geology Department at the University in 1931 and founded the Geography Department in 1937. Rennie was a family friend and did have an interest in palaeontology; he appears in one of the photographs with JLB and the Dakota in Grahamstown.

Edgar Mountain

Rhodes University

as the Crossopterygii; and the scales, the fins and the bony plates on the head all point the same way.' They studied the sketch closely, noting these points. Then my wife turned to the first page again, and she exclaimed sharply, 'Do you see when this was written?' and passed the letter to me again. Good heavens, 23rd December and this was the 3rd January, eleven days gone. Here is the letter, the sketch is on Plate 1.

> EAST LONDON,
> South Africa.
> *23rd December 1938*

Dear Dr. Smith,

I had the most queer-looking specimen brought to notice yesterday. The Captain of the Trawler told me about it so I immediately set off to see the specimen, which I had removed to our Taxidermist as soon as I could. I however have drawn a very rough sketch, and am in hopes that you may be able to assist me in classing it.

It is coated in heavy scales, almost armour like, the fins resemble limbs, and are scaled right up to a fringe of filament. The spinous dorsal has tiny white spines down each filament. Note drawing inked in in red.

I would be so pleased if you could let me know what you think, though I know just how difficult it is from a description of this kind.

Wishing you all happiness for the season.

> Yours sincerely,
> M. Courtenay-Latimer

As I have already said, the East London Museum at that time had a very small income and hardly any equipment; it was a kind of Cinderella among the museums of South Africa, it even had only a young woman, also a Cinderella as it happens, as Curator. What had happened to this thing in the meantime? If this was something really wonderful, what had happened to it? It had been handed over to the taxidermist, and as they obviously had no idea of its being something sensational, it would be a miracle if they had bothered to preserve all the insides. It was summer, the flesh and intestines would certainly be putrid by now, but perhaps the gills and any skeleton could be found, as they had very likely buried all that. I must act, and quickly.

Knysna is 350 miles from East London. In 1938 the roads were shocking and the telephone service was not what it is today.

Long-distance trunk-calls were not to be undertaken lightly. They took an awful time, and you often could not hear clearly. Knysna has always been curiously isolated from the outside world. It will be noted later that letters to and from East London took never less than six days—350 miles. Even today they take as many as five.

There was a shop with a telephone near by. It was close on 1 p.m., and on my asking, the Knysna Post Office held out little hope of my being able to speak to East London that afternoon at all, it might just manage to get through before 5 p.m. when the Museum closed, but Miss Latimer had no telephone at her house. It is amusing to look back on the reaction of both Post Office and the staff of that shop. East London! That was a long way for a telephone call! After lunch I went into the village (Knysna), and as soon as the Post Office opened sent the following telegram:

'MOST IMPORTANT PRESERVE SKELETON AND GILLS FISH DESCRIBED,'

I also made provisional arrangements for a telephone call to the East London Museum for next morning, and back at the house wrote at once to Miss Latimer. I remember composing and destroying half a dozen scripts, each shorter and containing less than the one before, for I was afraid to say too much. Here is the final draft.

Written from KNYSNA.
3rd January 1939

Dear Miss Latimer,
Thanks for your letter of the 23rd last which has just reached me. Your news is most interesting indeed, and I am very sorry that I am not in Grahamstown or I should have come over to see your fish within a short time. I shall be away for some time, and I am hoping that you saved the gills and viscera of the specimen, since they are most important. If all that was buried, you may still be able to save the gills at least.

I cannot hazard even a guess at the fish at present, but at the very earliest opportunity I am coming to see it.

From your drawing and description the fish resembles forms which have been extinct for many a long year, but I am very anxious to see it before committing myself. It would be very remarkable should it prove to be some close connection with the prehistoric.

Marjorie has been unfairly criticized for discarding the soft organs of the first coelacanth, the one part of an animal that is not preserved in the fossil record. However, she had no choice as her museum at the time focused mainly on dry public displays and was very poorly equipped for preserving large, wet specimens.

THE COELACANTH
December 1938

East London Museum

Montage of the first East London Museum, Marjorie Courtenay-Latimer, the first coelacanth and the trawler, Nerine

It must have been a tremendous relief to Marjorie when she received the letter dated 3rd January 1939 from Smith suggesting that the specimen 'must be of great scientific value'. Her reputation with the Chairman of her Board was at stake.

JLB had a rational, scientific mind but was in foreign territory as he contemplated the impossible – that a group of fishes that was thought to have gone extinct millions of years ago still had a living representative in our seas. On the face of it, it was preposterous, like finding a living dinosaur walking down the street!

Forty-four days elapsed between JLB's reading Marjorie's first letter and finally being able to examine the specimen in East London and confirm its identity – this must have been agony for him!

Meanwhile guard it very carefully, and don't risk sending it away. I feel it must be of great scientific value.

With kindest regards and best wishes for the new year,

Yours sincerely,

J. L. B. Smith

For the rest of that day I had enough worry for a lifetime. What was that fish? Had they saved anything of its insides? The night brought little rest.

As soon as the exchange opened next morning I was at the shop telephone, and spent an anxious three hours waiting for the call. Eventually it came. Yes, my worst fears were realised, all the insides had been thrown away and had gone off with the municipal rubbish cart. Miss Latimer could feel the agony in my voice; but I had no blame for her, for there were so many queer fish in our waters that no one but an expert could know if this one was what I suspected (and my brain said, 'Is it?'). So I asked her to find out at once where the municipal rubbish carts dumped their loads, because it could probably be worked out where those remains lay, and I had already decided in my mind that I might be able to get a plane to take me from George to go and scratch for them. I managed to telephone again next day. When I did so I learnt that all rubbish collected by the municipal service of East London was dumped out at sea. So that was that, and I could do no more about those insides. They were gone beyond recall. This was my first taste of the many frustrations Coelacanths were to bring.

I recollect that I asked the Post Office how much that call would be, and when I told the shopman and paid him, they were plainly astounded that anyone should be prepared to pay out so much for a telephone call about the insides of a fish. You may be sure they had heard everything.

My worries carried me along, my mind was in a chaotic state. Was this a prehistoric relic? If it was, the loss of the insides was a first-class tragedy. First of all, of course, I had to make certain what this thing was. It must not be forgotten that I was no expert on fossil fishes, just that in odd times my deep interest had led me to study what was known about them. My mind was busy all that time trying to assign that sketch to some clear type. It appeared to be something like a shark in its make-up, but so were those early Crossopterygians. I had to take into account

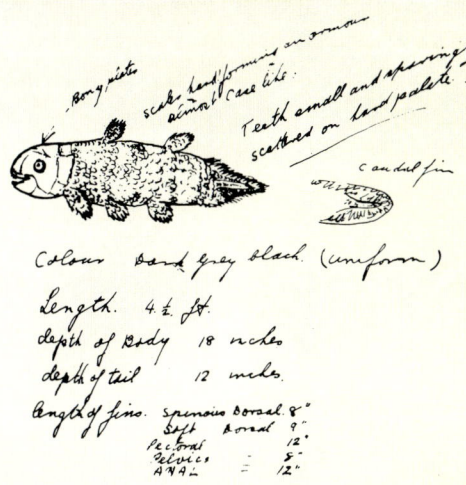

Miss M. Courtenay-Latimer   Miss Latimer's sketch and notes

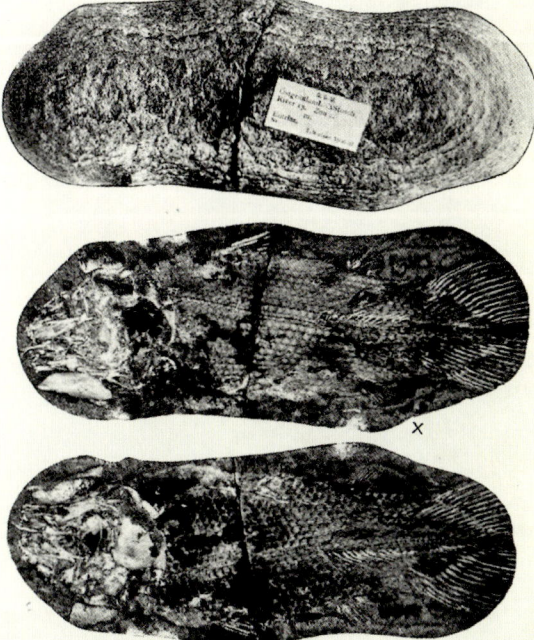

A Coelacanth fossil, which dates back 170 million years, found by Dr. E. Nielsen in Greenland. Above is the nodule enclosing the fossil, 7 inches long. When tapped by an expert at the point marked X the nodule split into the two halves shown here, disclosing the fossil. Note the amazing resemblance to *Malania* in Plate 2.

Plate 1

Marjorie was once engaged to be married to a young infantryman, but he was killed in a skirmish in the Eastern Cape and she never married.

Coelacanth fossils continue to be discovered around the world, the most recent in the vicinity of Grahamstown in South Africa.

The first coelacanth was caught off East London, South Africa, on 22nd December 1938, the second off Domoni on the island of Anjouan in the Comoros on 20th December 1952 and the third off Mutsamudu, also on Anjouan, on 24th September 1953. All three specimens belong to the species *Latimeria chalumnae*.

The first coelacanth on display in the East London Museum

The first three Coelacanths
Above, the East London Coelacanth, mounted, as first seen. *Middle*, the famous *Malania*, without first dorsal or extra tail. *Bottom*, the third Coelacanth, taken at Anjouan.

Plate 2

that it was only a sketch and clearly impressionistic, and so might be misleading. At the same time, it pointed directly to those Crossopterygian fishes of long ago. My peculiar photographic memory had recorded that the fossil Crossopterygii were described in Volume II of the *Catalogue of the Fossil Fishes* of the British Museum, published in 1891, but I had no such literature at Knysna, nor were any available there. Such works were at Grahamstown and Cape Town.

In the early days of my work on fishes, my collections were naturally incomplete, and as the largest were at the South African Museum at Cape Town, I had gone there several times to examine this material, and all too frequently this led me to question Barnard's* printed opinions. These tilts, whether refuted or established, he took with equal patience and good humour, and I had continual correspondence with him. I knew, therefore, at Knysna, that in time, as measured by the post, Cape Town was nearer than Grahamstown. On the 4th January I therefore telegraphed Barnard asking him to post me immediately that volume dealing with the Crossopterygii, which he did, as always, promptly, that being his way, and this arrived at Knysna on the 6th January 1939.

Miss Latimer's sketch and notes with that book left little doubt in my mind that if this fish of hers was not a Coelacanth, it was very like one. What a fantastic thing! Just imagine: a Coelacanth, still living, and all the greatest authorities of the world would be prepared to swear that all Coelacanth fishes had died out about 50 million years ago (it is estimated at 70 million today). Here was I in remote South Africa with the audacity to be convinced in my mind that this was a Coelacanthid fish. Even though I had done only spare-time work on fishes for less than ten years, I knew a good deal about them by then, and the careful and detailed papers I had published were known to scientists who worked on fishes all over the world; but I was still only on the way up.

Those were awful days, and the nights were even worse. I was tortured by doubts and fears. What was the use of that infernal premonition of mine if it was just going to lead me to make a scientific fool of myself? Fifty million years! It was preposterous that Coelacanths had been alive all that time, unknown to modern man. If that was a Coelacanth and it had been alive, then there

* Dr. K. H. Barnard (see p. 8).

We now know that the Cretaceous extinction event, when the coelacanths were thought to have died out, occurred about 65 million years ago. Contrary to popular opinion, the dinosaurs went extinct not because they were maladapted; rather, they suffered a severe case of bad luck, as their extinction was almost certainly caused by a giant asteroid that struck the Earth in the Gulf of Mexico. This catastrophic event, which would have changed the climate and interfered with photosynthesis by plants at the base of the food chain, was exacerbated by other major ecological events at the time, such as massive volcanic activity in India and changing sea levels due to continental drift.

Far from being evolutionary failures, the dinosaurs were a highly successful group of animals that survived for 155 tumultuous years, often as the dominant animals on land, in the seas and freshwaters and in the air. Furthermore, they live on today through the birds, crocodiles and tuatara, all of which are directly descended from them.

JLB's prediction that there would be an 'initial storm of scorn and disbelief' when he announced the discovery was an understatement. Barnard's response was a foretaste of things to come.

Newspaper headlines from 1939

JLB's observation that 'the trend in evolution was normally towards smaller size' has been confirmed by recent research. Prehistoric organisms that survived mass extinctions in the past were often smaller than the ones that preceded them. This is known as the 'Lilliput effect', named after the land of tiny people in *Gulliver's Travels*.

Big animals require more space and food and, as competition becomes fiercer and the number of animal species increases, it makes sense to be smaller. Furthermore, during the various Ice Ages, food became scarce, resulting in changes in the abundance and size, first of herbivores and then carnivores.

must be others living somewhere, perhaps off East London. But was it reasonable to think that such big fishes as this could exist near a place like East London and not have been found before? In any case, the fossil Coelacanths had all been pretty small fishes, 8 or 10 inches or 1 foot long—this fish was 5 feet long, enormously greater than any known before. The trend of evolution was normally towards smaller size, the giants had mostly gone. Yes, everything was against its really being a Coelacanth. From almost every aspect it seemed impossible, the answer must be 'No'; and yet every time I took out that sketch, it said 'Yes', emphatically 'Yes'.

I have been asked why I did not rush off at once to East London when such a wonderful thing lay there waiting. Apart from my examination commitments, I could never quite bring my mind to accept that it could possibly be true. It was too fantastic. Then, as the insides had been lost beyond recall and the rest was safely mounted, or being mounted, there was no longer any great urgency. Thirdly, I didn't want to go until I had girded up my mind to the stage where I felt I could face the situation if it did prove to be true. I was afraid to go then, really afraid, and I am not at all ashamed to say so. I wanted to put off going to look at it until I had built up a reserve of inner strength to stand the terrific strain it would mean if it were true. If I found that I could call it a Coelacanth, or something like one, I expected to have to endure an initial storm of scorn and disbelief from the whole world of science until all the facts could be given to prove it was so; and that would not be easy to face. So I remained at Knysna, worry and lack of sleep stripping still more flesh from my skinny frame, my mind never away from that fish and East London.

On or about the 7th January I wrote a cautious letter to K. H. Barnard to tell him something of my belief, but requesting him to treat the matter as strictly confidential and not to mention it to anybody else.* Barnard's reply to that first letter of mine was as

---

* The disturbed state of my mind at that time is revealed by the recent discovery that I apparently did not keep copies of my letters to Barnard from Knysna about the Coelacanth. I have written to Dr. Barnard asking for those letters in order to reproduce them here, but he has replied that they are not in the Museum files and that it must be assumed that they were destroyed. The dates and contents of my letters to Barnard quoted here are therefore compiled partly from his letters in reply and partly from memory.

usual prompt, but it was couched in such incredulous and face-
tious terms that it served only to increase my fears of the reactions
from a wider field. I had to go and look at the sketch, the notes,
and the volume again, as I did a hundred times a day. They
acted as a soothing drug on a maniacal mind. I was not easy to
live with those days.

On the 9th January 1939, having heard no more from Miss
Latimer, I wrote again as follows:

KNYSNA.

9th January 1939

Dear Miss Latimer,

Your fish is occasioning me much worry and sleepless nights.
It is most aggravating being so far away. I cannot help but mourn
that the soft parts of the fish were not preserved even had they been
almost putrid. I am sorry to say that I think their loss represents one
of the greatest tragedies of zoology, since I am more than ever convinced on reflection that your fish is a more primitive form than has
yet been discovered. It is almost certainly a Crossopterygian allied
with forms that flourished in the early Mesozoic or earlier, but which
have been extinct for many millions of years. Comparatively little is
known of the internal structure of such fishes, naturally nothing of the
soft parts, since fossil remains are all that help us to know what they
were like. Your fish has the general external features of a Coelacanthid, fishes common in early times in northern Europe and America.
Whether or not it is a new genus or family I can determine only on
examination, but I feel sure that it will make a great sensation in the
Zoological world. I have been anxiously awaiting a letter from you,
because I hope you understand that the thing must on no account be
stuffed until it has been examined. It is very important that the
structure of the skull shall be determined and the relations of the
bones of the jaws. You do not happen to have noticed whether the
air-bladder was partly ossified or not? I asked you to see if you
can possibly send the skin, etc., to me by passenger train so that I
may examine them. Even if you have to have a special box made, it
will be cheaper to send that way than for me to come to East London
at present. If the skin is properly packed it will come to no harm,
and I have a large preserving tank here that has held larger specimens
than that. You should really have some such thing at the Museum.
It can be made very cheaply by a plumber out of stout galvanised
iron.

If you judge it quite impossible to arrange to send the skin I shall

One of the benefits of finding living examples of an animal that was thought to be extinct is that scientists can test the predictions made by palaeontologists for extinct animals. The loss of the soft organs from the first coelacanth specimen meant that JLB would not be able to carry out this fascinating exercise.

Even before he had seen the specimen JLB was planning the dissection and figuring out which anatomical features would be important to examine.

JLB's suggestion that the skin and skull of the specimen should be sent by rail to him in Knysna was never carried out. In hind sight it was probably not a good idea, and Marjorie was wise to ignore it.

It is extraordinary that JLB was so convinced that the specimen was a coelacanth, based on Marjorie's simple drawing and description, that he was prepared to give it a name and assign it to a new family of fishes before even examining it!

Peter Jackson, a Senior Research Fellow in the Ichthyology Institute (1972–1984), described Smith as a 'lone wolf ... who cared little about the coolness with which his fellow scientists greeted his publicity seeking' in a biographical article published in 1997.

I do not agree that Smith was a 'publicity seeker'; to him celebrity was a very fickle mistress. He was, however, certainly streetwise enough to realize that the coelacanth could focus attention on his work and on ichthyology in South Africa, and therefore attract research funding. He exploited this potential to the utmost.

Many years later, at a conference of the South African Museums Association, I presented a paper impishly entitled 'The Coelacanth is a Mammal'. I did not propose, of course, that it was a lactating, fur-bearing animal, as opposed to a fish, but pointed out that it had tremendous potential to be 'milked' for marketing and public relations purposes, and that every museum should have its own 'coelacanth'.

have to come over. But I should like to avoid the trip for many reasons. (Examinations !)

I should like you to understand that any opinions about the fish I have expressed here are naturally provisional only, and can be confirmed only when I have seen the thing. But I think you will probably agree with me about the zoological affinities of the specimen. Well now I am anxiously awaiting your letter. I think it would be as well if you could telegraph me whether you can manage to rail the thing, only if you do don't forget to insure for, say, £100—I shall naturally be responsible for all expenses.

To honour you for having got this wonderful thing I have provisionally christened it (to myself at present) *Latimeria chalumnae*, and it may even be a new family.

Kindest regards,
Yours sincerely,
J. L. B. Smith

One curious feature of this whole affair was that at no time did I look upon it as anything but my own. There was no question in my mind that I had to take the full responsibility for the decision of the identity of this creature. Normally in a difficult situation of this type it is natural to consult others or to seek aid, but for some reason this never arose in my mind. It was perhaps due to that curious premonition that fate had prepared this occasion for me, and that, come what may, I must face it alone. It has since that time come to me from several outside sources that other zoologists were resentful that I took this on myself alone, and even more resentful that I carried out the subsequent investigation of the remains myself. Those criticisms left me and leave me quite unmoved. I was possessed or inspired, call it what you will.

Nevertheless, in those days at Knysna my soul was fearful and anxious, and away from that sketch and book I was sceptical of what my reason told me, but I knew I had to go on and take the decision alone. It was do or die. It was indeed characteristic of all my work on fishes that right from the very start I struggled alone, possibly because no help was available even had I wanted it, but certainly because I am what my wife calls a 'Lone Wolf' and work best on my own.

On the 10th January there came a further letter from Miss Latimer, dated the 4th.

EAST LONDON,
South Africa.
4th January 1939

Dear Dr. Smith,

I went straight off to see how the Fish Specimen was shaping, after you phoned.

It's terrible to think I only received your wire three days after it was sent,* on account of the holidays—but when the specimen came in and I found it to be something unique, I strove to do all I could to preserve it. As I found the work too much for me, I had it taken to Mr. Center and got him to do all the heavy work.

There was no skeleton. The backbone was a column of soft white gristle-like material, running from skull to tail—this was an inch across and filled with oil—which spouted out as cut through—the flesh was plastic, and could be worked like clay—the stomach was empty. The specimen weighed 127 lb. and was in good condition, only it was very hot and work had to proceed at once.

The gills had small rows of fine spines—but were unfortunately thrown away with the body.

Mr. Center has almost mounted the specimen now, and is not doing it badly at all—the oil is still pouring out from the skin, which seems to have oil cells beneath each scale.

The scales are armour like fitting into deep pockets. (By that I mean hard and heavy.)

The skull is in the skin and I have got Mr. Center to mount it with the mouth open. I have the tongue or hard mouth-plate here.

I have done every possible thing to preserve and not lose any points and feel worried to think in the end I allowed the body and gills to be discarded. They were kept for three days, and when I did not hear from you I gave the order for disposal.

Kind regards,
Sincerely yours,
M. Courtenay-Latimer

It is interesting to note that the final sentence in the letter indicates that when she did not hear from me within three days, Miss Latimer assumed I did not think the matter important. She knows me better now. We answer all queries and at once.

Miss Latimer received my letter of the 9th (see p. 35) on the 15th January 1939, and telephoned me the following day, giving

---

* In fairness to the post office it may be stated that Miss Latimer later realised that this was not correct, as will be seen from the date of her letter.

Marjorie's mention of a backbone 'of soft white gristle-like material', rather than bone, must have sent a thrill up JLB's own spine. This was the first indication that the coelacanth had features of both bony fishes and cartilaginous fishes; later research would reveal more shark-like characters. Was the coelacanth a shark in bony fish clothing?

Mike Bruton

The bony head of the coelacanth is an important diagnostic feature

The first specimen has continued to 'leak' oil for over 75 years. The oily nature of its flesh is one of the reasons why it is not at all palatable (I have tried fried coelacanth fillet; it tastes like cotton wool dipped in vinegar!). In the Comoros the flesh is not eaten but the oil is used as a mosquito-repellent or laxative. Fishermen in Madagascar and Tanzania occasionally eat the flesh and, in Madagascar, use it as bait.

It is amazing how many scientists who have worked on fishes, like Dr Gill, have 'fishy' names. Those whom I have encountered include Tom Pike, Dave Rowe-Rowe, Tom Hecht (German for 'pike'), John Bass, Lev Fishelson, Vic Springer and Geoffrey Fryer, but the fishy literature includes a host of Snoeks, Garricks, Souls and even some Fischers, Hookers, Castanets and Pechères!

Barnard did not assist JLB with the identification of the coelacanth, nor did he verify Smith's identification, so there was no need for JLB to acknowledge his help.

The first two coelacanth catches were made close to Christmas (22nd and 20th December), which made the logistics of their recovery very awkward.

various details in answer to my queries. The more I heard the more certainly did it all point to a Coelacanth, but even so, my mind would not really accept this; it was too fantastic, it just could not be. Nevertheless, when it was all assembled, the factual evidence appeared overwhelming. On the 17th January 1939 I wrote again to Barnard, telling him briefly but frankly this time that I believed the animal to be a Coelacanth. His reply of the 19th January indicated that he was now really startled and no longer facetiously incredulous. I wrote again about the 24th January, giving more information, and this time, convinced, he was apparently so overcome by the almost unbelievable nature of the whole affair that he disclosed it in confidence to the Director of the Museum, Dr. E. L. Gill, who had at one time worked on fossil fishes and therefore naturally had an exceptional interest in this.

This was indeed all that transpired between myself and the authorities of the South African Museum. Not long after the discovery had been featured in the press, a report was published stating that I had been in consultation with the authorities of the South African Museum all along over the matter of the Coelacanth. As I had nowhere acknowledged any such assistance, this naturally implied a reprehensible omission on my part. This implication shocked me, for I have always been most punctilious about giving the fullest acknowledgment of any type of assistance received. In this case there had indeed been an almost exceptional degree of isolation. I had deliberately chosen to carry the terrible responsibility myself, making it indeed very much my own funeral. I could see no clear way of rectifying this in a public fashion in any satisfactory manner, so just let it go, but it remains in print and may one day rear its head.

All along I had been frantic to see a photograph of the fish, but none came. For some curious reason something always went wrong with the attempts. Below are quoted several relevant letters:

KNYSNA.
24th January 1939

Dear Miss Latimer,

I have been waiting to hear from you again about that specimen. I should very much like to see a photograph as soon as you can send one. I doubt whether I can get over till near the end of the month, now that the fish is stuffed it does not matter very much.

I am still convinced that the fish is a Coelacanthid, but hope you will not give any information to the press till I have had an opportunity of examining the specimen in detail. Will you kindly make a special point of finding out from the skipper of the trawler if the fish showed any signs of life when it was caught. I have an idea that it might perhaps have lain somewhere in the ocean bed in some preserving ooze or mud these millions of years. Chemically it is possible. It would be very interesting to know if it was definitely alive or not. If it was there is always the chance of another, and you can offer the trawler people £20 for another perfect specimen for me. If by any millionth chance one should be obtained, please have a large tank made, buy as many gallons of formalin as are necessary, and inject strong formalin all over into the body, and of course telegraph me immediately. It is most aggravating to be so far away, and I am very anxious to come as soon as possible. If you can detach one carefully, kindly send me one of the scales, as they are important in diagnosis.

Kindest regards,
Yours sincerely,
J. L. B. Smith

This shows clearly that my mind was perpetually reacting against the fantastic idea that any Coelacanth should still be alive. Could one not have been preserved in some antiseptic bottom ooze? But there was no doubt, it had been alive, for it had snapped at the captain's hand, and had lived for several hours after being caught.

EAST LONDON.
25th January 1939

Dear Dr. Smith,

Bad luck seems to have dogged this fish—I went down to ask Mr. Kirsten whom I got to take photographs of the fish in flesh today and he tells me the entire film was spoilt.

I wish you could come over to East London. I seem to have no one to get interested and feel very despondent about the photographs.

Yours sincerely,
M. Latimer

At last some scales arrived. They hammered flat most of my doubts. Coelacanth, yes a Coelacanth for certain. Phew! What lay ahead?

Many coelacanths caught by traditional fishermen have remained alive for several hours after capture. The record was 'three days' for a specimen caught off Grande Comore but others have remained alive for up to 42 hours, and more usually 1–11 hours.

In fact, the first reward offered by Smith was £10. During the 1940s he increased it to £100, a very substantial amount at that time.

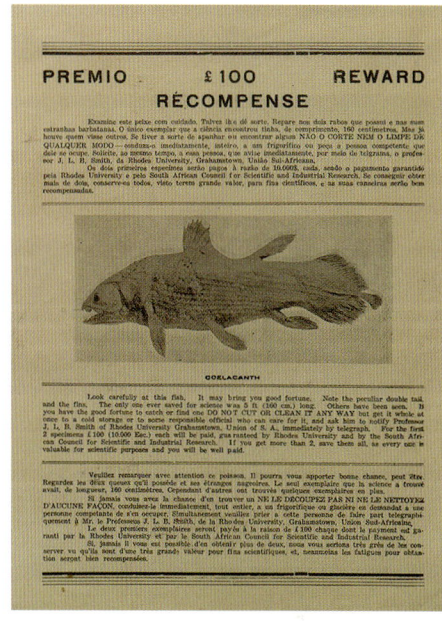

JLB Smith's £100 reward poster

JLB's comment that the coelacanth may have been an ancient specimen preserved 'in some antiseptic bottom ooze' indicates further that he still did not trust his own judgement.

Coelacanth scales are different from those of other fishes: they are elongate, heavily armoured at the posterior, protruding end, and have an inner layer of bone and an outer layer of hard keratin covered with sharp tubercles. They lack the middle layer of cosmine (a dentine-like material) found in the cosmoid scales of other crossopterygian fishes.

The flexible armour-coating provided by the overlapping scales protects coelacanths from predators (mainly large sharks), but they are still subject to predation as evidenced by the many specimens with missing scales or even fins. Studies on coelacanth life histories in the 1980s and 1990s revealed that predation is by far the most frequent cause of mortality, much more than the catches of traditional fishermen.

EAST LONDON.
1st February 1939

Dear Dr. Smith,

Thank you for your last letter. I have tried to get in touch with the Trawler but at present it is at sea. However, I have a promise that a message shall be delivered and I shall get all information again.

When I went down to fetch the specimen I was told it had been trawled 40 fathoms off Chalumna and it had been alive. I am enclosing three scales for you. You will notice each one fits into a socket twice its depth. They have not faded much.

Are you returning to Grahamstown—I shall most probably be able to take the specimen over to you then.

Yours sincerely,
M. Courtenay-Latimer

Written from KNYSNA.
7th February 1939

Dear Miss Latimer,

Many thanks for your letter and for the parcel of three scales. They leave little doubt about the nature of the fish, but even so my mind still refuses to grasp this tremendous impossibility. The discovery is going to be a real zoological sensation, and we shall have to see the trawler captain and crew in order to get their testimony, also the taxidermist. Your original letter to me will probably figure in my first report to the Royal Society. However, all this is confidential at the moment.

Thanks for your offer to bring the fish to Grahamstown, but the matter is so important that I must come over. My wife and I have decided to leave here a week earlier than we had intended, so as to be able to spend some days in East London. We hope to arrive about Wednesday the 15th next, and I shall probably telephone you immediately we arrive. It will save me time if you have or can have taken a full-plate size print of the photograph of the fish from which drawings can be made as basis. I only hope the taxidermist has not varnished the thing, as I must have details of the external structure of bones, etc.

No more now. It is curious that in spite of all this evidence, my intellect says that such things can't happen.

Kindest regards,
Yours sincerely,
J. L. B. Smith

I had perpetual nightmares all this while. Looking back, it is miraculous that my relatively frail health of that time did not crack under the terrible strain, but I was possessed and sustained by a curious belief that it was my lot to carry this through and that I should be able to do so. Most men find learning new things increasingly difficult after the age of thirty, and indeed I had experienced that myself in chemistry, trying to keep up with the progressive changes in theory. I started my study of fishes when already past thirty, and it was astonishing to discover that my brain soaked it up like a sponge, and even now it is still the same. I can only suppose it must be a kind of natural affinity. At any rate, before we left Knysna I had absorbed everything available that had been published about Coelacanths. I certainly knew a lot more than a few weeks before.

We left Knysna on the 8th February 1939, intending to go straight through to East London, but there was to be nothing easy about this, for we travelled in continuous heavy rain and were fortunate to reach Grahamstown, since by that time floods had rendered almost all roads impassable. Drifts and slippery mud made motoring in South Africa no light undertaking in those days. We had to wait a whole week before the roads to East London became usable, and after an awful journey reached there on the 16th February 1939.

We went straight to the Museum. Miss Latimer was out for the moment, the caretaker ushered us into the inner room and there was the—Coelacanth, yes, God! Although I had come prepared, that first sight hit me like a white-hot blast and made me feel shaky and queer, my body tingled. I stood as if stricken to stone. Yes, there was not a shadow of doubt, scale by scale, bone by bone, fin by fin, it was a true Coelacanth. It could have been one of those creatures of 200 million years ago come alive again. I forgot everything else and just looked and looked, and then almost fearfully went close up and touched and stroked, while my wife watched in silence. Miss Latimer came in and greeted us warmly. It was only then that speech came back, the exact words I have forgotten, but it was to tell them that it was true, it was really true, it was unquestionably a Coelacanth. Not even I could doubt any more.

And now, what lay ahead? I told Miss Latimer she could tell

JLB and Marjorie's accounts of his first sighting of the coelacanth specimen differ. She said that he saw it with her in her office where she had the mounted specimen on display whereas he said that Marjorie was out when he arrived. When she returned he said to her, 'Lass, your discovery will be on every scientist's lips throughout the world. It is a coelacanth!'

Mike Bruton

Coelacanth Gallery in the East London Museum

All JLB's intuitions and predictions had proved to be true – the specimen was undoubtedly a 'long extinct' coelacanth. Now the real work started!

It is significant that JLB aimed high from the outset with regard to his plans for the publication of the coelacanth description and identification. *Nature* was (and still is) the leading scientific journal in the world. JLB's other descriptions of the coelacanth were published in the *Transactions of the Royal Society of South Africa* and the *South African Journal of Science*, also prestigious journals.

When scientists in the British Museum (Natural History) in London first saw a specimen of the platypus from Australia in the late 1790s they couldn't believe their eyes and, at first, thought that it was an elaborate hoax perpetrated by a mischievous antipodean!

Illustration by John Gould

Duck-billed platypus

JLB's slender frame sometimes led people to underestimate his intellect and physical capabilities, and comments such as 'What! Is this skinny little fellow your expert?' gave him an inferiority complex that was at odds with his towering intellect.

her Board, but no one else, and they must for the moment please not make anything public. My plan was to make no kind of announcement until I could prepare a brief account and send it to some scientific journal—*Nature* of London was in my mind. I told Miss Latimer this; she agreed to inform the Board that evening, and we arranged to come to the Museum next day.

That night again I slept little, I was too excited. A real Coelacanth, and yet and yet, could such a thing be? Even though I had seen it and confirmed every single detail, one by one, I was like the old lady at the zoo, who, seeing a giraffe for the first time, said to her friend, 'I just don't believe it'. This was worse, far worse. My whole life seemed to hang on it. My wife has reminded me that I woke her that night at least half a dozen times, and each time I would say, 'Please forgive me, but is it really true about the Coelacanth? I haven't just dreamt it, have I?' And each time she solemnly assured me, sleepily, but with conviction, that it was true.

Despite all this preoccupation, early next morning I was out on the rocks hunting fishes in the pools, and when we went to the Museum I was still clad in field clothes: khaki shorts and shirt. Miss Latimer told me the Board were very excited and that the Chairman, Dr. Bruce-Bays, was coming in shortly to meet us. When he did arrive I was standing looking at the fish, listening to Miss Latimer, who was talking at the moment, and as her back was towards the door she was unaware that he had entered. He stopped dead and his gaze was all for me. I am slight and thin and had then hardly any grey hairs; in fact, despite all I have endured there are too few even now. His features did not change, but his eyes and that queer power of reading the thoughts in other men's minds told me exactly what was in his. What! Is this skinny little fellow your expert? In those clothes I must have appeared very young to that dignified and portly old man, far too young to be able to give so startling an opinion about this fish. He would have to weigh this matter very carefully indeed before permitting the Museum to be involved in any fiasco from youthful enthusiasm.

It is all very well to have a slender, youthful appearance and few grey hairs, but you pay for them. On Boards and Committees greybeards used to wonder who the devil this youngster was to open his mouth, and I had many battles and learnt a lot. You learn far more from those who resent or dislike you than you do from your friends.

After polite introductions and preliminary words, Bruce-Bays questioned me, quite sharply at first, but as I warmed to the subject, the tension eased and he was soon deeply interested in all I had to say. The many ramifications of the discovery soon convinced him that it was not just an old fish but something of very much greater importance. He forgot my apparent youth, my lack of flesh and my clothes, his doubts had clearly evaporated, and his parting words and handshake were warm, almost enthusiastic.

Miss Latimer had left no stone unturned to find out all she could about the circumstances covering the actual catching of the animal, and in response to my queries told me that the trawler had been working the usual grounds along the coast westwards of East London. On the 22nd December 1938, Captain Goosen decided to return, and on the way back thought it might be advisable to have a run on the bank off the Chalumna River mouth, an area that is normally poor but sometimes yields good catches. So here was another link in the story, one of the many fantastic chances. A trawler captain's impulse! If he had not acted on it? Captain Goosen sent me an account of that trawl. The net was shot about three miles off-shore, some twenty miles south-west of East London, the depth close on forty fathoms. The course was roughly elliptical, the axes about three and six miles respectively, the closest approach to the shore being about two miles. They ended the run about three miles off-shore where the depth was about forty fathoms, the average depth trawled. That particular area where they trawled was on the inshore part of a submarine shelf about ten miles wide, that slopes gradually to about sixty fathoms in depth at the edge, which is abrupt, and plunges to about two hundred fathoms. The bottom of this shoreward shelf is foul and trawling troublesome and difficult. In this case their catch proved to consist of about a ton and a half of edible fish, not the highest grade, about two tons of sharks, and—one Coelacanth!

I met Captain Goosen in 1987. He was a humble man who sought no fame for his role in the capture of the first coelacanth. His son, Pieter, later became a respected weather and shipping journalist in Cape Town. In late 1939 the *Nerine* was called into military service and converted into a minesweeper.

Captain Hendrik Goosen

The wharf where the *Nerine* berthed in December 1938 has since been dubbed 'Latimer's Landing' and has been developed as a leisure and water sport hub.

Latimer's Landing in East London

After the Second World War trawlers scoured the marine coast off East London looking for more coelacanths, but to no avail.

When we explored the marine environment off East London in Hans Fricke's *Jago* research submersible in 1991 we found it to be unsuitable habitat for coelacanths, even over the edge of the continental shelf.

Interestingly, the first coelacanth caught off Mozambique was also netted by a trawler over a sandy bottom in relatively shallow water (40–44 m).

News of the discovery of the first coelacanth off East London spread through the media like wildfire and it was soon being proclaimed as the 'Best fish story for 50,000,000 years'.

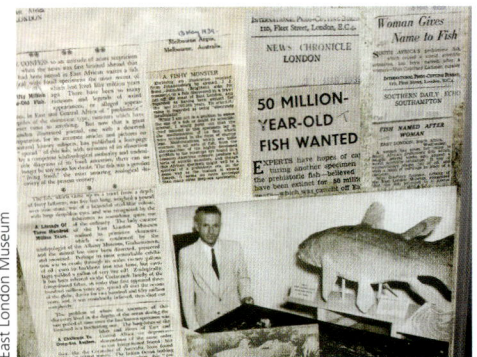

East London Museum

Further newspaper headlines on the coelacanth in 1939

In October 2000 a population of coelacanths was found living off the northern Zululand coast. When the *Mail & Guardian* announced the discovery, it was under the puckish heading 'Fossil fish bonk off SA coast'!

Mike Bruton

Mail & Guardian *headline in 2003 after further research revealed that the coelacanths discovered by mixed-gas divers in 2000 were not strays from the north but a stable breeding population of coelacanths off the coast of Zululand*

Chapter Five

JEKYLL AND HYDE

DESPITE the world-wide sensation I knew this discovery would produce, had the matter rested with me nothing would have been given to the press. I was hoping that I might be able to publish the first reports about it in a scientific journal. All the scientists I had known in my formative years had been scornful of the press, and they decried those who appeared to seek or who welcomed publicity. I had come to acquire the rather quaint idea that it was scientifically 'improper' to give information about scientific discoveries to the press, an attitude which in its bareness is a type of scientific snobbery, and generally found in the immature or in those who are unlikely ever to feature greatly in that way. Most young scientists encounter this problem and are worried by it, and its solution lies in the realisation that it is fundamentally the man in the street who pays for scientific research, and he is therefore entitled to know the results. The majority of mankind have not the opportunity of doing scientific work, but there is no question that almost everyone is deeply interested in it and eager to know about it. Another type of intellectual snobbery is the dictum that science has now passed beyond the understanding of the ordinary man. That, however, is very largely a matter of presentation. With the possible exception of higher mathematics, there is not a single branch of science whose broad outlines the ordinary man cannot appreciate if it is properly explained.

   The views I held about publicity at that time had to be pushed aside, for when the Board of Trustees of the East London Museum heard the full story from Miss Latimer, they were rightly eager to exploit the whole affair to the best advantage of the Museum. The publicity which this remarkable discovery would bring and the interest it would arouse would be of the greatest benefit to any institution, and I was in no position to refuse or to contest this view. At their request I agreed to receive a reporter and to give him information. He was obviously greatly impressed by the importance of the whole matter, and not only asked many questions but came back several times before he and I were satisfied that he had the whole story correctly presented.

He concluded by asking permission to have a photograph of the fish. This I emphatically refused, to his consternation, and he urged that the article would have much less value without a photograph, which I countered by pointing out that it would at least be world news from his pen. A good deal lay behind my refusal. Once you decide an organism is new to science—and that alone is a long story—it has to be named. A name alone is no use, you must give sufficient descriptive detail so that the species can always be recognised again, or else a good illustration, preferably both. If two people happen to describe the same organism as new, as often happens, the one whose name and account are published first has 'priority', i.e. the organism for ever thereafter bears the name he gave it followed by his name. Thus the Coelacanth is *Latimeria chalumnae* J. L. B. Smith. However, there is a type of scientific piracy, in that if you are foolish enough to publish a picture of some rarity that is unnamed, you take the risk of someone else getting in before you with a name. Thus an unnamed picture of the Coelacanth in a newspaper could have led to its being known for ever after something like *Neoundina moderna* J. O. L. Roger.

When my hand had been forced in the matter of immediate publicity, I had privately determined for the reason given above that as far as I was concerned, although the press could be given the fullest details, there was to be no photograph for publication. I wanted to do the thing properly, and intended that the first picture of the Coelacanth should appear with a brief description in some scientific journal—as it happened *Nature* of London. I had previously ascertained from Miss Latimer that she had definitely not permitted anyone to photograph the specimen, and as far as she could determine no one had had the opportunity of doing so without her knowledge. She pointed out it was unlikely that anyone would have been prepared to take any risks in securing photographs, since up to then no one had had any idea of the fantastic nature of the creature.

As Smith points out, when scientists discover a new species they need to ensure that the first illustrated description of it appears in a respected scientific journal, as this description will, for ever, remain the definitive report on the holotype, i.e. the specimen on which the new species is based. The new species name will also be based on that report.

If an *informal* report on a new species contains all the information (and illustrations) required for a new species description, this can cause havoc in the taxonomic literature.

The coelacanth saga is peppered with intrigue, deceptions and backstabbing, mainly perpetrated by people who wanted to get onto the coelacanth bandwagon. Many coelacanth researchers have been impacted in this way.

In an email to Mark Erdmann (discoverer of the Indonesian coelacanth) in 1998 I stated, 'I do agree, though, that the coelacanth tends to bring out the worst in people; I have experienced this many times, sometimes in fairly nasty ways'.

This reporter was certainly persistent, and after finding me unmoved carried his attack to Miss Latimer, whom he got to consider the matter, and she suggested we might make some compromise. It was only with very considerable reluctance that I agreed that he might take photographs, but on the express condition that they were to be published only in the *East London Daily Dispatch* and nowhere else. I insisted on this undertaking, and it was given by this man in the presence of Miss Latimer, my wife, and myself. On that, the specimen was carried outside and he was permitted to take several exposures. I asked him to show me all the negatives, but as it happened I never saw any, so did not know what good ones he had.

Two of the pictures, and they were excellent, duly appeared in the *Daily Dispatch* on the 20th February 1939, rightly labelled as the only ones in the world.

Some days after we had returned to Grahamstown I had a telephone call from a Durban newspaper, when it was mentioned that they had been offered a photograph of the Coelacanth by a person in East London. Considerably perturbed, I promptly wrote to inform Miss Latimer.

The next surprise was a telephone call from a friend in England to say that I was taking an awful risk in permitting any photographs of the animal to appear unnamed. I heard that they had been sent to various newspapers over there, most of whom had just ignored them, thinking it was a hoax. There came also a cable from London urging me to attach a name to the Coelacanth. As has been indicated I had long since intended to apply the name *Latimeria chalumnae*, which I now attached to the fish. Although greatly disturbed by all that had happened over the photographs, I was so desperately occupied at that time that I did not manage to find time to investigate how it had all come about.

Some years later I read in some paper that this particular journalist had been greatly admired for his clever 'scoop' in getting the pictures of the first Coelacanth so promptly, that he had made a good deal of money, and was indeed still drawing royalties from them. To improve things a bit, it went on to tell how he had had the foresight to photograph the Coelacanth when it was on the quay!

Owing to the peculiar circumstances of that time, there was no real urgency in the matter of the announcement of the discovery of the Coelacanth, and since odd points kept cropping up about which he wanted further information, as mentioned before this was delayed since not only did I give several interviews on the matter to this same reporter, but after he had completed his account I insisted on checking the final draft. The full statement therefore appeared in the press of South Africa only on the 20th February 1939. At East London there was in addition an announcement that the animal would be on view to the public on that and on the following day. We were early at the Museum, and all morning long lines of curious sightseers thronged the grounds and filed past this curious fish, so roped off that it could not be touched, and at my special request under constant guard.

I had told Miss Latimer and Bruce-Bays that scientists everywhere would be clamouring for details of its structure, and that despite the loss of the soft parts and skeleton, it was desirable that it should be examined as soon as possible. At my request they recommended to the Board that it should be sent to me for study at Grahamstown, and it was agreed to do this.

On the 20th February 1939 we returned to Grahamstown. It was a chaotic return. A brief account of the discovery appeared in the Grahamstown local press on that Monday the 20th February. It was accorded far less prominence than the report of a sports day of a local school. It was said later that when the Press Association message arrived, the editor had consulted a local zoologist and had been advised to be cautious. The story sounded really rather too sensational.

Several friends plied us with questions, but most people eyed us strangely. I was quite irrationally still fearful, because although my intellect was completely satisfied with the irrefutable evidence my eyes had seen, completely satisfied that the fish was indeed a true Coelacanth, it seemed too impossible, too fantastic, that this could have happened. A Coelacanth. Alive! Every night I had a nightmare, dreaming that I had found a Coelacanth, and it was confused and troublesome because I realised it was impossible. Then I would wake and ponder on this curious dream until suddenly I would realise that it wasn't a dream, but true. I had that happen to me hundreds of nights in the years that followed.

The report in a Grahamstown newspaper was probably published in the enduring *Grocott's Mail*, the oldest family newspaper in South Africa (for which I have written many columns on matters scientific).

Even though museums and science centres are increasingly using interactive electronic displays to entertain and educate the visiting public, there is no doubt that humans have a fascination for the 'real thing'. I can remember, as a young scientist in the late 1970s, staring in awe at, and gently touching, a piece of real moon rock in the Smithsonian Air & Space Museum in Washington, DC.

JLB was in an invidious position: he was employed as a chemistry lecturer but, as an amateur ichthyologist, he had taken on the responsibility of single-handedly dissecting and describing one of the most valuable zoological specimens ever found. *En route* he had to teach himself about the anatomy of a group of fishes that had never been dissected before.

It is healthy for scientists to be sceptical about one another's findings and conclusions. In fact, good science is all about doubt, uncertainty and testing ideas, not about certainty and dogmatism. The most significant advances in science have often been made when the 'current wisdom' is unceremoniously discarded and replaced with a new (but still imperfect) understanding. Science is never complete, it is always 'work in progress'.

Sometimes it got all mixed up, for I would dream I had dreamt it, and when I did wake up it took a long time to sort it all out. This sounds fantastic, and it was.

The East London Museum sent the fish with a police guard on the 22nd February 1939 by rail to Grahamstown, and it arrived on the 23rd. It was taken to my house and put in its special room. It had a curious, powerful, and penetrating odour, an odour that in the coming weeks was always to pervade our lives, awake or asleep. From the start the whole family was rigidly drilled and kept on the alert. The house was never left alone, night or day, and if a fire should occur, the fish must be the first thing to be got out, and at once. Every waking moment was full of worry for the safety of that specimen, and I dreamt of little else.

In sending the specimen to me the Board of the East London Museum had stipulated that it was not to be exhibited to the public in Grahamstown. This caused some ill-feeling, for several institutions wished to have the specimen for a period for their own special purposes. For about two weeks after our return, the back-wash from the impact on the world beyond had not yet reached Grahamstown, and there was little about the Coelacanth in the local press, only in papers outside. They got hold of some fantastic stories, among them one that this fish (which weighed 127 lb.) had dripped ten gallons of oil!

There were many curious incidents in these first few days. Several colleagues asked to see the fish, and came to my house. After I had shown it to them, one, an Englishman, said to me, 'But you are surely not expecting people to believe so astounding a thing on your word alone. You will surely be sending it to the British Museum for them to make sure.' He was astonished when I said that I doubted if anyone there knew so much more than myself to justify such a step, and that I was quite satisfied it was a Coelacanth. I added that within a week or two I expected to know a good deal more about the intimate details of a Coelacanth than any other person in the world.

A Government scientist I had known for many years called to see me at my office at College. He put his hands on my shoulders and said earnestly, 'Doc., what has made you do this thing? It is terrible to see you ruin all your scientific reputation in this way.' I asked him what thing. He replied, 'Calling this fish a Coelacanth.' I said it was a Coelacanth. He shook his head in sorrow. 'No, man,' he said; 'I have just been talking to X [a scientist], and he says you are crazy, that it is only a Rock Cod with a mutilated and regenerated tail.' I dealt gently with him and my lack of concern shook his doubts, but he was not convinced.

Cables, telegrams, and letters from near by had almost drowned us, and soon the overseas correspondence developed into a flood. All scientists were frantic for information. It was an incredible time.

There was among many others a trunk-call from the editor of a well-known daily paper about the fish! 'Dr. Smith, are you quite positive that what you say is true?' 'No!' 'No! Then how could you have said it?' 'I didn't. What I said and what I say is, that as far as my knowledge, experience, and observations go this is a true Coelacanth.' 'What is the difference?' My answer was, 'If you showed me a flower and said "Is that blue", even if it looked blue to me as a scientist my attitude would be "I should say it is blue" not "It is blue".' Somewhat bewildered words from his end concluded the interview.

Possibly because I had been so incredulous myself, it was staggering to receive no incredulity from overseas scientists. One prominent American scientist wrote to say that he had been called up late at night by the editor of an important paper who told him that they had got a report from South Africa that a live Coelacanth had been found. He supposed it was just hot air. This man asked who had said so. He replied a man named Smith. 'J. L. B. Smith?' 'Yes.' 'Well, then, I think you should be safe to go ahead and publish.'

I set to and from a general preliminary examination of the chief external features prepared an outline description of the creature. This with a photograph was sent to *Nature* in London, and appeared on the 18th March 1939. If anyone anywhere had any doubts, that article killed and buried them. There never were any more, not even here in my own country.

I had sent a scale of the Coelacanth to a scientist correspondent-friend in the U.S.A. In replying to express thanks he said that this had been received in a solemn scientific meeting. He showed it to a colleague, and the excitement that followed just about disrupted the whole affair.

O.F.—4

When American marine biologist Mark Erdmann and his wife, Arnaz, on honeymoon in Indonesia in September 1998, encountered a coelacanth in a market in Manado, they didn't immediately realize the significance of their discovery and allowed a friend to post information about it on the Internet.

Mark subsequently lamented that he had been caught up 'in a web of politics and turf battles which basically committed me to secrecy for the past year after the find'.

Mark Erdmann and Arnaz Mehta

Rik Nulens

Peter Boehme

The coelacanth has been called many names over the years, including 'missing link', 'living fossil' and 'Lazarus species' (see Chapter 5 for explanations).

Pencil drawing of the coelacanth by Peter Boehme

The publicity surrounding the whole discovery led many people to write, telephone, or call about the oddest things. One woman wrote to say that she had seen in the papers that I was interested in old things. She had a violin that had been in the family for over a hundred years, if she sent it would I tell her if it was valuable? Others had sailors' fish-monstrosity fakes, rare and valuable shells and other ancient curios, while one man offered me a share in a project to hunt for treasure in the middle of Durban, to be based on an ancient supposed pirate's map of buried loot. Many rare and presumably prehistoric creatures were reported at that time, mainly fishes. In our most difficult period I was wakened near midnight by an excited call from Knysna to tell of a wonderful creature one of the deep-sea fishermen had got— it had a face like a monkey, short legs, and an eye in the top of its head. Would I come at once and see it; yes, right away. I asked a few questions and suggested it was a 'Jakob', a curious shark-like creature, but not exactly rare. I did not go and later when the fish came, it was what I had suspected.

It was in this period that odd reports came to attach the term 'Missing Link' to the Coelacanth, a label that especially later was to prove exceedingly troublesome. There were letters from apparently ultra-religious people who roundly reproved me for ignoring the Bible in my preposterous statements about millions of years, and did I not know that the theory of evolution was evil and an anti-religious invention of the devil put into some men's minds to enable them to divert others from the path of true thought? These came from a wide area of the globe.

Meanwhile, for the eager world of science I was faced with the task of preparing a detailed description of what remained of the animal, and, of course, ample and accurate illustration was essential. I bore a heavy teaching and administrative burden in the Chemistry Department, which gave me hardly any free time during official hours. In view of the world-wide interest in my researches on this creature, I could well have done with some relief to expedite that work, but perhaps foolish pride kept me from asking, and as none was offered I somewhat grimly kept my teaching at normal intensity. Visiting scientists, and others by letter, expressed their astonishment at this situation. I gave no response, but it really was a trying ordeal. Each day I rose at 3 a.m. and worked at the animal until 6. Then I cleaned up and went for a four-mile walk over the hills. On my return I would write up my observations from the notes, and when I left for College at 8.30 a.m. my wife took these and typed them. On my return at lunch-time I would go over them again, and she would retype during my absence in the afternoon. On my return at 5–5.30 p.m. I would start in again and usually worked until about 10 p.m. I do not suppose I averaged four hours' sleep any night during the week, but slept late on Sundays. It was the same old mixture my life had always been, turbulence and trouble, only more intense. We had no social life, business and financial affairs took a back seat, and our food reached its destination over and between sheets of manuscript. We had no conversation, no thoughts, no ideas nor eyes, for anything except Coelacanth, all day and all night. We could never forget it, certainly not with that smell. It was an equally severe strain on my wife, especially as a child was due in about three months' time.

I had no compunction about doing the whole investigation my-self—I had earned it and I could do it, but I worked on that stuffed and mounted creature with very mixed feelings. It was a wonderful thrill to be the first to see the finer details of a skull of a living Coelacanth, and yet the loss of all the soft parts was a perpetual tragedy that clouded the investigation. I pushed this aside, for it was not irremediable and only made me determined to find more and whole specimens. There must be others some-where, and at the back of my mind 'a cloud no bigger than a man's hand' had formed, the forerunner of the project that came to over-shadow all else in my life—the hunt for the home of the Coelacanth.

The work progressed slowly. The structure of the skull was the most important part, and I decided to open the whole of the one side. It was done very, very slowly and carefully, every fragment of skin and bone kept in place. By that time I had received from all over the world all the latest publications on Coelacanths and related fishes, and it was wonderful to uncover a tiny bone and then after a hunt through all the pictures and drawings to find the same or its equivalent in some fossil of a fish that had lived several hundred million years before. In some cases the structures were exactly alike; it was indeed fantastic, this peculiar feature of Coelacanths, their unchanging nature.

JLB's walks in and around Grahamstown were famous and I had the privilege of accompanying him on some of them. He walked fast and spoke little. In the 1970s, when I visited William and Jenny Smith at their lovely home in the Featherbed Nature Reserve in Knysna, I went on a hike with William. It was quite arduous as he is over 2 m tall, took giant strides, and never stopped talking!

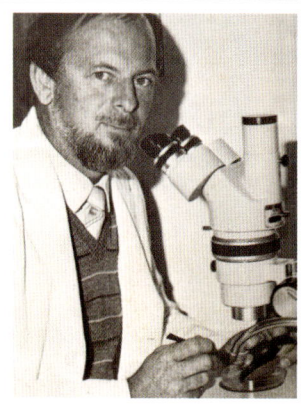

The author as a zoology student at Rhodes University (1968)

Rhodes University

It is extraordinary that JLB, trained in chemistry, was able to master the anatomy of a strange fish so quickly and competently. His anatomical description of the coelacanth was acclaimed worldwide and was superseded only by the extremely detailed descriptions and drawings published by a team of French scientists 20 years later.

The sensory canals are very important in the coelacanth as it hunts at night and has poor eyesight. They are used to detect minute changes in water pressure and movement and to orient the fish relative to nearby objects, including other fishes and potential prey. Later research would reveal that the coelacanth uses electro-reception to detect its prey.

The 'mysterious central cavity in the cartilage of the front of the head' is associated with the rostral organ, which has three openings into the outside environment and is part of the coelacanth's 'latero-sensory system' for detecting vibrations and currents.

The dissection showed the delicate nature of some of the bones just beneath the skin, which it is believed were developed to carry what are called 'sensory canals'. These are channels filled with a slimy substance, and served as sensory organs, probably able to detect small changes in pressure and to warn the fish it was approaching something solid, or that something was approaching it.

The structures on this head showed clearly that some of the bones were no more than modified scales, and that the teeth had developed from the tubercles on the scales.

The modern Lung-fishes have 'internal' nostrils, i.e. they open inside into the roof of the mouth, and some scientists asserted that the Rhipidistia had the same, and they even managed to convince themselves that the structures in the Coelacanth fossils proved that they had had them, too. All scientists who dealt with them certainly believed this. But I could find not a sign of them in this Coelacanth, and as its bone structure proved to be virtually identical with that of those older types, they probably had none either. It was not exactly a popular discovery in some quarters.

Highly technical detailed accounts of the relationships of the Coelacanth structures would be out of place here. Those who have any interest in these may read all such details in my monograph on *Latimeria*. Suffice it to say that I found several structures which had not been detected in fossils, including a mysterious central cavity in the cartilage of the front of the head, which led to openings that in a normal fish would be the nostrils, but which in the Coelacanth are not. We do not yet know what it is, how it originated, or what it is for. Nothing like it is known to occur in any other type of fish.

One special thrill in this slow and difficult work was to discover that the taxidermist in scraping the skin had missed removing a marvellous chain of fine sensory bones just behind the head. To hold these delicate and beautiful structures in my hand was a wonderful experience, just to realise that hundreds of millions of years ago these special bones had been in the heads of Coelacanths, and here they still were ! Because so many leading scientists were deeply interested in what would be found in this fish, at intervals I sent round a circular giving a brief résumé of my progress and discoveries. These were greatly appreciated, but in their acknowledgments the exponents of different schools of thought would each urge me to use the particular nomenclature they favoured for the numerous bones of the head. I allied myself with none, and gave the bones numbers that meant more to me than possible significant names.

Various curious things happened. I received a letter from the Curator of a museum in Australia, who was shown some scales of the Coelacanth by an Australian. We attemped to discover how he got them, but that mystery was never solved. Some others got into the possession of a scientist in Johannesburg, but we did not solve that either, for as far as could be determined from the time Miss Latimer took possession of the specimen, no unauthorised person was permitted to touch it or even to approach it near enough to grab a few 'souvenirs'. I had given a special warning about that souvenir danger. One 'explanation' of possession of the scales was that they had been collected on the wharf at East London after Miss Latimer had gone off with the fish. This is scarcely plausible, for nobody then had any idea of the importance of the fish, and anyone picking up scales on the fish wharf at East London would indeed be a phenomenon.

Meanwhile the full backwash of the effect of the discovery overseas had come back to the Union. One was an enormous picture of the Coelacanth, together with an article by Dr. E. I. White of the British Museum, in the *London Illustrated News*. I did not find it flattering to remote scientists like myself, and it expressed the view that the Coelacanth had come from the deeper parts of the sea. (See Chapter Six, p. 59.) Coupled with this at the same time was a letter to me protesting against the proposed use of the name *Latimeria* for this historic find, and referring to Miss Latimer in terms that were scathing, to say the least.

I must confess to an angry reaction to this letter and replied expressing the strongest disapproval of such sentiments and remarks, and also published the following in regard to this criticism in *Nature* (6 May 1939):

> Few persons outside South Africa have any knowledge of our conditions. In the coastal belt only the South African Museum at Cape Town has a staff of scientific workers among whom is an ichthyologist. The other six small museums serving the coastal area are in extremely poor circumstances, and generally have only a

It is common practice in scientific nomenclature to name a species after the person who first brought the specimen to the attention of scientists, so the use of Miss Latimer's name in the genus was fully justified.

In 1955 JLB named a wrasse from Aldabra island, *Pseudocheilinus margaretae*, after his wife, Margaret.

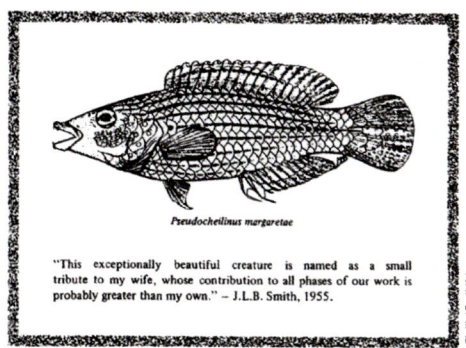

Pseudocheilinus margaretae

"This exceptionally beautiful creature is named as a small tribute to my wife, whose contribution to all phases of our work is probably greater than my own." – J.L.B. Smith, 1955.

NRF-SAIAB

Pseudocheilinus margaretae, named after Margaret Smith

In 1939 JLB published three papers on the coelacanth and, in 1940, a further two, including a 106-page monograph in the *Transactions of the Royal Society of South Africa*.

In 1968 an eminent American ichthyologist, Professor Carl Hubbs, wrote that JLB's description of the coelacanth was 'perhaps the most meticulously detailed account ever accorded a fish specimen, at least of a ... carcass'. This was significant praise indeed as, during the 1950s and 1960s, the French also produced incredibly detailed anatomical descriptions of the coelacanth.

JLB Smith (second from right) at a shark conference in the USA with Carl Hubbs (far left), Gilbert Whitley and LP Schultz in 1957

The clamour to display the coelacanth at the East London Museum is understandable as this single specimen would define the future of the museum and ensure that it would always have a high profile in the pantheon of regional museums worldwide.

Director or Curator, who cannot possibly be an expert in all branches of natural history. There are not uncommon fishes in the sea which to any of the latter would appear as strange as, if not stranger than, a Coelacanth. It was the energy and determination of Miss Latimer which saved so much, and scientific workers have good cause to be grateful. The genus *Latimeria* stands as my tribute.

Meanwhile I was continuing my terrible struggle to do my ordinary full-time University work and the detailed examination of the Coelacanth. On the 19th April I wrote in answer to Miss Latimer:

The fish is a terrible job. I could work solidly on it for 6 months. There will be over 50 plates alone and I cannot finish before June at the earliest. . . . I am hoping that I can send you the fish by the end of May perhaps.

On the 24th April came a telegram from the East London Museum:

'BOARD WISH FISH RETURN IMMEDIATELY LETTER FOLLOWS.'

That was a shock. The work was far from complete and I was almost frantic. Miss Latimer's letter that followed explained that since the tremendous effect the discovery had produced overseas had been coming back to the Union, it had caused the public in her area to clamour to see the fish again. There had been a succession of people who had travelled long distances to East London, and at the Museum they had been overwhelmed by complaints at its absence, many from influential people. It did not pacify them to be told that scientific work was being done on the fish. They wanted to see it. What a primitive instinct it is to stare at the unusual, but it does satisfy something.

I telephoned Miss Latimer, and eventually we compromised. I should return it on the 2nd May 1939. Where I had worked intensely before, the time remaining was a frantic nightmare. In the end I managed most of what I hoped, but not all. At the end of this terrible strain it was mainly with a sense of relief that the animal was handed over to its police escort. It arrived safely, and apparently the Museum was thronged for days by eager sightseers, who pressed in closely all the time. This widespread local resurgence of interest had several effects. The discovery of the Coelacanth had been prominently featured in South Africa, but it caused a far greater sensation overseas. As this continued to come back full blast to the Union, many of those connected with the Museum in East London came to feel that, in addition to its apparently exceptional scientific interest, the specimen might be worth a good deal as well. Since the Museum was in such poor circumstances,* rather than just keep such a valuable specimen would it not be better to sell it and use the money to develop the Museum? A few protagonists of the old 'Send it to the British Museum' policy also raised their voices, and all this induced Bruce-Bays to take action in the matter. Early in June he handed Miss Latimer the draft of a letter which he asked her to type (she was Curator, Secretary, Treasurer, and everything else). She read it with amazement and dismay, for it was to offer the Coelacanth to the British Museum of Natural History. She read it again and again, but then and there decided that she would never type it, that if it was sent she would resign, and promptly told several members of the Board about it and her resolve. A few days later Bruce-Bays came in and asked her if she had done that letter. She said she had not typed it, would never type it, and in what she considers one of the longest and warmest speeches she has ever made, told him in no uncertain terms what she thought of the whole thing. Miss Latimer expected opposition and even violence, but instead this mature, influential man was so overwhelmed by her personality and arguments that at the conclusion of this speech he quite meekly said it would be in order to leave the letter and the whole matter. It was so surprising a victory that Miss Latimer said it just took her breath away.

This did not end the matter, it came up again, and early in July 1939 I was asked to come to East London to advise the Board of Trustees of the Museum about it. This I did, pointing out to them how much the actual specimen would mean to their Museum, far more than any sum of money even greater than its value. It would constantly attract world-wide attention. That has been amply borne out by the course of subsequent events, and the Coelacanth was the first real step to fame for that Museum.

* Despite this, the Board of Trustees of the Museum had voted a suitable gratuity to Captain Goosen for what he had done.

It was unquestionably a wise decision not to sell the coelacanth specimen. The discovery of 'old four legs' put the East London Museum, and South Africa, on the world map and provided considerable impetus for the development of ichthyology, and the biological sciences in general, in South Africa.

Furthermore, the shackles of colonialism were further loosened with the decision to keep the specimen in South Africa rather than send it to a well-established European museum. Subsequently, all the fishes that JLB and his collaborators collected in Africa stayed in South Africa.

The intervention of the Second World War changed the pace of JLB's research and gave him breathing space to consider his future. His mind was strongly focused on finding another coelacanth with all the soft anatomy intact, while at the same time surveying the little known fishes of East Africa, where he predicted the coelacanth lived.

In the aftermath of the war, JLB's research benefited from the realization in South Africa (and other countries) that there was a strong need to invest in science and technology. Scientific research became a national priority and science a matter of national prestige. Soon after the War ended, South Africa established the CSIR; the National Science Foundation was founded in the USA and the Department of Scientific and Industrial Research in the UK.

Between 1938 and 1952 JLB made large collections of fishes along the South and East African coasts and published the first edition of *The Sea Fishes of Southern Africa* (1949). In 1946 he established, with funds from the CSIR and Rhodes University, the research Department of Ichthyology, where he worked until his death in January 1968.

Peter Barnett

Margaret and JLB Smith on a fish-collecting expedition in Mozambique in 1951

Mike Bruton

Display of JLB Smith's expedition paraphernalia in the SAIAB

After the fish had left Grahamstown we still had little leisure, for there was an enormous amount of work to be done in completing the manuscript of the monograph, virtually a book, with numerous plates and detailed accurate drawings, whose preparation took much time. The whole was finally dispatched late in June 1939, and it was only then that my wife was able to give any time to the purely mundane occupation of providing clothes for the infant that appeared five days later. He had a narrow shave from not only coming naked into the world, but of remaining so longer than usual.

Thus the course of the first Coelacanth—turbulence, trouble, and strain, the inevitable accompaniments of accomplishment. Any great event of this kind tends to become submerged by the little things, and it was only when I was able to detach myself and see the whole in its true perspective that the wonder of it all stood out like a shining beacon. It was a tremendous privilege to have been the first man to work on a Coelacanth. But I had to turn my back on that and look ahead, for now before me was the problem of finding more specimens, of finding where these incredible animals lived. The remote past had risen dramatically from the sea, at our very door; but did it really live there? From the first I doubted this, but had to make sure. Photographs and offers of a reward were sent to all fishing craft on South African shores, and daily I hoped to hear more. I spoke to the College authorities about the possibility of organising an expedition, but found no response at that time. In any case, the gathering clouds of war and the climax in September 1939, meant the end of all those dreams. Who cared about Coelacanths when bombs were going to fall? It was far more important to kill Germans than to find Coelacanths.

BOOK II
TROUGH AND CREST

Chapter Six

NO DEEP-SEA REFUGEE

AFTER the initial shock of the discovery had passed, one of the first problems to present itself was that of where these Coelacanths lived. They had, of course, been in existence incredibly long before any type of ape or man appeared, and all through the many thousands of years it had taken modern man to evolve and develop they must have been living as well. Yet right up to 1938 no scientist had ever seen or even suspected the existence of a living Coelacanth. As has been explained, it had been comfortably settled that all Coelacanths must have died out at least 50 million years ago, and they occupied merely a remote niche in the consciousness of most scientists, except for a very few most highly specialised workers. Now, in the shock of the discovery, scientists all over the world were busy considering the problem of how such a large and curious-looking creature had managed to escape notice all this time.

One obvious way out was the theory advanced by Dr. E. I. White of the British Museum, who not long after the discovery was announced, published an article, mentioned earlier, in an illustrated periodical, in which he stated:

> Our living Coelacanth, although trawled in only 40 fathoms, almost certainly was a wanderer from deeper parts of the sea to which its kind have retreated in the face of fierce competition with the more active modern types of fishes. This opens up the interesting possibility that other remarkable relic-forms may also inhabit the more inaccessible depths of the oceans.

I could never understand how this view could find acceptance. To me one glance at the Coelacanth disposed of any idea that it lived in the 'inaccessible depths of the ocean'; yet a number of scientists all over the world apparently accepted this with a sigh of uncritical relief. It explained the whole thing! It is astonishing

Dr EI White, the pre-eminent fish palaeontologist in Britain at the time, became a thorn in JLB's side as he ridiculed the amateur ichthyologist's efforts. But White was wrong in his assessment that the coelacanth was a deep-water fish, which led to futile expeditions by Jacques-Yves Cousteau in 1954 and 1963, in collaboration with Jacques Millot, to search for the fish in deep water off the Comoros. The depth preferences of the coelacanth in the Comoros are about 100–800 m yet the average depth of the oceans is 3,688 m; coelacanths are therefore relatively shallow-water fishes.

JLB was infuriated when White pre-empted the publication of his detailed description of the coelacanth in *Nature* by publishing a popular article entitled 'One of the most amazing events in the realm of natural history in the twentieth century' in the *London Illustrated News* on 11th March 1939.

JLB's predictions as to the habitat preferences, feeding behaviour and leisurely existence of the coelacanth were spot-on. They represent an extraordinary example of how a scientist (and angler) with a keen eye for detail can decipher the living habits of an animal from its anatomy. This was done despite the fact that the first coelacanth had been caught in a very atypical habitat and the second in an unknown habitat off Anjouan in the Comoros.

JLB's knowledge of living fishes allowed him to make a far more accurate prediction on the likely habitat of coelacanths than White, who had focused on the study of fossils.

to see how far this (at least to me) extraordinary theory penetrated. It even made scientists ignore facts. For example, it had been reported in the press all over the world that the first Coelacanth was taken by a trawler near East London at a depth of 40 fathoms (80 yards). This was clearly stated both in my monograph and in many other publications. In addition, it is well known that in South African waters trawlers have long operated at depths up to and even exceeding 300 fathoms. Yet a leading overseas scientific treatise published not long after the discovery, in discussing the Coelacanth, stated that the fish had been taken by 'A South African trawler dredging deeper than usual'. Then again, only a few years ago, a costly deep-sea expedition worked over the great ocean deeps of the world hoping to catch a Coelacanth there.

This idea that the Coelacanth might live in the depths has always seemed inexplicable to me, for when I looked at that fish, even the first time, it said as plainly as if it could speak: 'Look at my hard, armoured scales. They overlap so that there is a threefold thick layer of them over my whole body. Look at my bony head and stout spiny fins. I am so well protected that no rock can hurt me. Of course I live in rocky areas, among reefs, below the action of the waves and surf, and, believe me, I am a tough guy and not afraid of anything in the sea. No soft deep-sea ooze for me. My blue colour alone surely tells you that I cannot live in the depths. You don't find blue fishes there. I cannot swim at speed for more than a short distance; I don't need to, because by levering myself from convenient concealment among rocks or from a crevice, I can pounce so swiftly on any creature passing by that it hasn't a hope. When I spot any quarry that stays quiet, I don't need to give myself away by swimming. I can stalk it by crawling quietly along gullies and channels, pressing close against the rocks for added concealment. Look at these teeth and enormous jaw muscles. Once they clamp tight on anything, believe me it can't get away. Even big fishes have no hope. I just hang on until they die, and then feed at leisure, as my kind has done for millions of years.' All this and more the shape and form of the Coelacanth presented to my eye with its experience of living fishes.

Even though we have a far from complete knowledge of the life in the great depths of the ocean, a good many fishes from there have been caught, and they have given scientists a pretty fair idea of the life in the utter darkness of those cold depths. They do not give me the impression that life is especially easy down there, and almost all the fishes are black, while crustacea and others are red. Blue is no colour of the moving life of the abyss.

In any case, however, the Coelacanth just did not fit. Its scales alone ruled it out, for truly deep-sea fishes have no need of scales, certainly not scales like those of the Coelacanth. It is by no means certain and not even likely that the fishes that live in the depths, or their ancestors, went there to escape competition with other fishes. My work has repeatedly shown the enormous stretches that even small feeble fishes have colonised, and all the evidence indicates that fishes tend to move and seek new places to live, just like any other creatures. All the types we know from the deeps are derived from ancestors who lived in waters of ordinary depth, and though most deep-sea fishes are of course greatly modified to suit the special conditions, all are clearly related to surface forms, none of which are any markedly better equipped to withstand 'competition' than the ancestors of the deep-sea forms. In the depths bodies are soft, bones are light, eyes are enormous or have become obsolete, and huge jaws are filled with long fangs, often barbed. There is no valid evidence to support the idea that any of them retired to the depths to escape competition.

When the Coelacanth was caught, that haul of the trawl brought up several tons of sharks. As is well known, the bag (Cod-end) full of fish is hauled aboard by a winch hung up over the deck, and the lower end opened by jerking a rope, when the fishes cascade into a heap on the deck. All but the most hardy are squeezed to death in the net, certainly all at the bottom, and the fall and the pressure of the heap above finish off the rest. Not many fishes, perhaps occasionally an odd shark, are ever alive. It is further characteristic that when deep-sea fish come up to the surface, even without being squeezed in a net, most die long before they even reach the top of the water. It so happened that the Coelacanth was at the bottom of the pile with which he was caught, and it was some time before that great weight above, mainly sharks, was hauled away. At the end of all this the Coelacanth was still so much alive that it snapped viciously at the hand of the captain, who had been called to examine this strange

After the tragic crash of the Boeing 747-244B airplane, *Helderberg*, into the sea off Mauritius in 1987, the Ichthyology Institute staff in Grahamstown were able to view footage of deep-water fishes taken at a depth of 4,900 m by the remote-operated vehicle that was used to search for the airplane's flight data recorders. The footage revealed fishes with the typical array of highly modified characters that JLB ascribed to deep-water fishes: soft-bodied, scaleless, colourless, and with large eyes or no eyes.

JLB's prediction that the coelacanth is a tough, rocky reef-dwelling ambush predator was exactly right. It is extraordinary how wrong White's prediction was, but, then again, he had the full gravitas of the British Empire behind him!

During the 1991 *Jago* expedition, when we searched off the East London coast for coelacanths, we found that the wreckfish, *Polyprion americanus*, appeared to occupy the coelacanth's niche as an ambush predator over rocky reefs. Wreckfishes reach 2 m and 300 kg and live at depths from 40–300 m.

NOAA FishWatch

Wreckfish, Polyprion americanus

In fact, coelacanths are docile and non-aggressive. Besides, flailing human divers with bubbles erupting from their backs would not have been on their 'radar' as potential prey, and humans would probably also be regarded as too small to be potential predators.

However, a coelacanth off the Zululand coast that initially approached a diver did show a rapid escape response by suddenly turning through 180° and swimming off.

creature that had been caught. The crew all remarked on the fact that the Coelacanth still showed signs of life for several hours afterwards. No deep-sea creature could have endured all that and lived. So much, then, for any notion that this was a degenerate or feeble fish.

With all this clear evidence it was utterly impossible for me to accept the view that this fish lived in the depths. It seemed unlikely that it, or its ancestors like it, would ever have had to 'retreat' from 'competition' with other fishes. I know of no past or modern fishes that this Coelacanth, as a reef-haunting type, need fear, very much the reverse. To most reef fishes the Coelacanth would unquestionably be a terror, something like the larger and rightly dreaded Rock Cods. In any conflict between even the most vicious free-swimming types and the Coelacanth in his own environment of the reefs, I would back the old Coelacanth every time, and as a human diver among reefs I would unquestionably not like to meet a Coelacanth down below. Looking back, I find it as incredible as ever that the majority of scientists interested in the matter apparently accepted the idea that Coelacanths lived in the depths.

My complete disbelief in this 'inaccessible depths' idea did not, of course, solve the problem. The first question was whether this particular fish really belonged to the area where it was caught. Was it perhaps just very rare or had others been seen and not reported? Many people are diffident about taking or even reporting to museums what looks queer to them, for fear that it may be common and that they may be exposed as ignorant, and many rarities are lost that way. It takes some event to shake this, and every time a 'find' is reported in the press it brings a diminishing trail of other reports in its wake. When the first Coelacanth was exhibited at East London, several people said they had seen others. One man reported finding a fish just like it cast up on the shore north of East London many years before. He had been unable to do anything about it, as it was large and partly decomposed. A trawlerman said that in Natal waters, many years before, the net had brought up six large fishes that he felt sure had been Coelacanths, but the skipper had ordered them to be thrown overboard, as he doubted if anyone would eat such strange creatures. There were other stories, all rather vague, and in the absence of some characteristic parts, such as scales, or of photographs, it is impossible to estimate the accuracy of any such reports.

Despite careful inquiry all along the coast about East London, no evidence of any certainty was obtained. No line-fisherman in those parts could recollect ever having caught or seen any fish that could have been a Coelacanth. Nor had any trawler apparently ever certainly caught one before, anywhere about East London or indeed anywhere in South African waters, certainly not in any human memory. And many trawlers constantly sweep the ocean floor over great areas all along our shores, day and night, at all sorts of depths. While a trawler catches pretty well everything of any size in the parts it covers, line-fishing does not do the same. Repeatedly in much-fished areas I have caught by poison and other means fish, large fish, that never bite on hooks. Was there not still a chance that this Coelacanth might fall in that category?

All along the South African coast there are strong currents in the sea. The main stream is the Mozambique current that flows south and westwards, swinging close in to shore or farther out according to the wind. Though it does not change in direction, by its variation of position in relation to the shore it creates at times other almost equally powerful reverse currents. In such conditions line-fishing is extremely difficult and commercially impossible in any but fairly shallow water, so that if the Coelacanth happened to live among reefs at the 100-fathoms mark or deeper, it might possibly have escaped notice. On the other hand, there would almost certainly be regular strays to shallower water, or sick or dead fishes washing up, that would most likely have been noticed had they been about, as they would almost certainly be large. Most fortunately for me, shortly after the discovery, the South African Government Fisheries vessel came to East London, and working for some time over the whole area where the Coelacanth had been found, tried by every possible means to catch another or to find further traces, but failed to do so. The whole weight of the evidence therefore seemed to be unquestionably against the possibility that the Coelacanth could be living in the sea anywhere near East London, even at fair depths among reefs.

It is curious that most of the very primitive types of fish that still live on today are found in fresh water. This possibility had to

During my career as a fish biologist, several unsupported claims were made to me of coelacanths being washed up on South African beaches, at Gonubie near East London, at Robberg near Plettenberg Bay, near Knysna and elsewhere.

The furtive reference to 'other means' here is the use of explosives to catch fishes (now banned), which JLB employed extensively during his fish-collecting expeditions.

It is almost impossible for trawlers to fish over the rocky reefs and canyons that coelacanths prefer – hand-lining is much more efficient in these habitats. In 1939 a British scientist, JR Norman, predicted that 'long-baited lines perhaps hold out the best chance of catching fresh specimens'. More recently, the use of deep-set gillnets (for catching sharks) has proved to be a devastatingly efficient way of catching coelacanths, especially off Tanzania and Madagascar.

JLB's prediction that there are no living freshwater coelacanths in South Africa was, of course, correct. Our freshwater ichthyofauna does, however, include a tiny, ancient fish of Gondwanic origin, the Cape galaxias, *Galaxias zebratus*.

NRF-SAIAB

Cape galaxias, Galaxias zebratus

Freshwater and estuarine coelacanths are well known from the fossil record and it is likely that all three extinct coelacanth species known from South Africa were from these habitats.

be considered for the Coelacanth, too. As far as South Africa itself was concerned, certainly anywhere near East London, that could be ruled out, as everyone who knows this area will agree. Most fresh-water rivers in South Africa have no constant flow. Their courses are often so steep that in floods they run strongly, but for most of the year they are reduced to a series of disconnected and usually not very extensive pools. In floods a representative part of their fishes, sometimes great numbers, are carried into the sea and die, and in turn wash up on the shore. You can get a good idea of the fishes in the rivers that way.

In dry periods the pools hold a fish fauna extensive neither in numbers nor in kind, and they are constantly subjected to intensive fishing, not always legal or conservatory in nature. Taking all the evidence into account, while a few minute types may still remain unknown, that a large fish like the Coelacanth could be living unsuspected in our South African fresh-water rivers seemed just about impossible.

As is described elsewhere, my life in the months following the discovery was troubled and difficult, and there was little time for reflection. The problem of the origin or habitat of the Coelacanth was rather like a hovering storm, ever present, nagging at my mind. It seemed obvious that the chief hope of finding others lay in an expedition with a vessel well equipped to explore the life about the reefs, where ordinary line-fishing would not serve. Having no funds of my own and no sources appearing likely in South Africa, I made tentative approaches to several large overseas institutions; but the results were indefinite. There were rumours that this or that institution or body was preparing an expedition to come to South Africa, but nothing further. Meanwhile a picture of the Coelacanth and the offer of a reward for any further specimens had been sent to all trawlers and fishing vessels. On all these the Coelacanth came to be known as 'Old Fourlegs', and indeed bears that name to this day.

Although we remained hoping almost daily for further news, the Coelacanth storm slowly subsided, and eventually the gathering clouds of international tension and war finally disposed of my hopes for any expedition of our own or from elsewhere for that time.

All through the war years we constantly sought news and

evidence of Coelacanths all along our coasts. My wife and I walked many hundreds, probably thousands, of miles in all, showing the picture and telling the story to people of all classes, callings, race, and colour. But we got nothing of any value. Before the war ended it came to look as if the Coelacanth could not possibly live normally anywhere near where this one had been caught, and that it must have been a stray. There must surely be others. The problem of finding out where they lived became even greater.

If my deduction that the Coelacanth lived about reefs was correct, it was clearly such a predator that it ought to take a baited hook, or would at any rate be likely to be seen by fishermen sometimes somewhere. If that was the case, why had its existence not been reported before? That could be the result of many different causes. Coelacanths might, for example, live only about reefs where nobody fished. That might be because nobody lived on the shores where such reefs were, or the reefs might be far out at sea, as margins to banks where there was not enough dry land for anyone to live. It might even be that those reefs were constantly lashed by rough seas or powerful currents, or both, so that nobody could ever fish there at all. This would, of course, mean that if the Coelacanths lived in such a place they would never have been caught by any human agency, and would not easily be tracked by any means except that of going and finding the exact spot. It would certainly be a formidable task to cover all such places. On the other hand, it was equally possible that they had been and were being caught regularly in some area, but by primitive peoples to whom they would just be fish and who would not realise their significance. And in what part of the shorelines of the world were any or all of such conditions more likely to be found than in East Africa? Nowhere in all the temperate and tropical oceans was there at that time so great an area whose marine fauna had been so little investigated and which was so little known as East Africa. The whole area is full of reefs, rocky and coral reefs, some enormous, many hardly known. Add to this that the set of the current from north of Madagascar is always southwards. I could see no reason why the Coelacanth should not live normally in some remote and probably uncivilised part of that vast area. As I surveyed all the facts and evidence, it seemed very

O.F.—5

JLB's use of posters and rewards to acquire rare specimens for study was unusual at the time but eventually reaped rich benefits. This method also raised the profile of ichthyology among non-scientists and led to a fruitful collaboration between anglers and scientists that continues today.

Once again, Smith's keen deductive powers led him to the right conclusion – that the main populations of coelacanths lived over rocky reefs off remote islands on the East African coast, which were fished only by traditional fishermen.

In *Old Fourlegs* JLB Smith makes some remarks that would be considered to be racist today. We make no excuse for these remarks, nor do we endorse them, but our strategy has been to republish the text of *Old Fourlegs* in its original form as an historical document that recounts the story of the dramatic discovery of the first and second coelacanths.

JLB's prediction that coelacanths would be found off Madagascar was correct but the first specimen to come to scientists' attention from there was caught only in September 1995, nearly 43 years after the first Comoros coelacanth, although they are likely to have been caught from time to time by traditional fishermen in both localities over the centuries. To date, 13 coelacanth specimens are known from Madagascar, the third most after the Comoros (215) and Tanzania (63 documented, but probably more have been found). Madagascar has also yielded a rich bounty of fossil coelacanths.

likely. This one, caught near East London, could easily have come rambling down the coast in the warm Mozambique current, as quite a number of tropical fishes constantly do.

The peoples of East Africa have from the earliest known times been ardent fishermen; but save for an Arab, Forskal, who lived in the Red Sea area in the eighteenth century, none have been ichthyologists or had any pretensions to scientific knowledge. The vast majority, especially those of Bantu origin, are even today of a low order of intelligence, and restrained from a more brutish existence only by the threat of force. As I wrote in 1946 in a report to the South African Council for Scientific and Industrial Research: 'There may well be places in East Africa where Coelacanths are commonly caught and used as food, and nobody would be any the wiser.' What applied to East Africa applied with equal force to the 3,000-mile-long coast-line of Madagascar. Numerous Coelacanth fossils had been found on Madagascar. There must be stretches of coast there that no enlightened scientific eye has ever seen, and the tantalising vision of savages feasting unsuspected on succulent Coelacanth steaks on a Madagascan shore did not seem too fantastic.

And so my eyes were turned to East Africa, but not with any joy. To search every reef in that vast area would take many years of effort. It would need time and money, plenty of money. I was no longer young, and as for money, I was a scientist, not a wool farmer, not even a millionaire.

Chapter Seven

OBSESSION

My illustrated monograph on the first Coelacanth appeared in February 1940.* As I thumbed through the pages of the first advance copy, my feelings were mixed. Pride in its achievement strove against the grim recollection of all it had cost. The book certainly gave plenty of information. One scientific friend, not an ichthyologist, remarked of it: 'Great Scott, if you could write so much about only parts of a fish, what would you have done with a whole one?' Truly, all that work still lay ahead.

War, war, war! Scientific work, other than for war, declined steadily. Who cared about fishes except as food for the forces? My double life went steadily on, we had to train scientists so they could make explosives to blast other men, but the proportion of women in the University classes steadily rose. All this time my Coelacanth monograph lay on my table, and my brain was constantly obsessed by the problem. In 1944 the men began to return and life became more difficult than ever, with shortages of staff, extra lessons, and vacation classes for returned servicemen.

It became increasingly difficult to be enthusiastic about hammering science into the heads of men who from all they had endured in conflict could not but regard the academic life with some scorn. After all, if you had been accustomed to soaring through the skies in your own plane, to killing wherever you could, and to daily narrow escapes from death, things like valency and equivalent weight just couldn't mean a thing.

Even in those dark days of war the fascination of fishes went on biting more and more deeply into my soul, and by 1945 I came to realise it could not be long before I cut loose from chemistry somehow. I had not sufficient means to live without assistance and could see no clear way. I had heard that the Prime Minister had intervened to make life easier for a few prominent scientists,

* In the *Transactions of the Royal Society of South Africa, 1939*.

The heading of this chapter, 'Obsession', is appropriate as the coelacanth had, by now, become an all-consuming passion for JLB Smith.

It is a pity that he did not elaborate more about the role he played in studying and developing explosives during and immediately after the First World War. The expertise he developed in this field came in useful when he developed home-made 'bombs' to catch fishes during his East African expeditions. The powerful pressure wave produced by an underwater explosion stuns or kills fishes without damaging them unduly.

Here we read, for the first time, that JLB was seriously considering abandoning his career in chemistry in favour of a new vocation in ichthyology.

JLB's mention of formalin is apposite. Formalin, a 40% solution of formaldehyde in water, is commonly used as a preservative of fishes and other museum specimens, as it kills bacteria. However, it is also highly toxic to humans, as well as being carcinogenic. I remember meeting the famous South African herpetologist, Dr VFM Fitzsimons, towards the end of his life. He had been ravaged by the effects of formalin and was virtually blind.

We can expect that, with his knowledge of chemistry, JLB would have taken adequate measures to reduce the risk of formalin poisoning, but this would have been difficult under the rough field and laboratory conditions in which he worked. Could exposure to formalin have aggravated his health problems? Some of his ex-chemistry colleagues think so.

but Grahamstown is very far away from the centre of things. Then came a letter from a complete stranger, the late Bransby A. Key, of Johannesburg, inviting me to write a popular book about fishes, saying that a thousand pounds had been made available for this purpose.

By a curious coincidence this letter was dated 26th September (1945), which is the joint birthday of my wife and myself (our young son grew up with the quaint idea that all married couples had birthdays on the same day). Partly as a result of my own early struggles in the study of fishes it had come to be my ambition to produce a book of this kind, and several years before, without any hope of funds to publish, I had set out to produce one. It soon became clear that its cost would far exceed any funds I might hope to raise, so that the whole thing, text and illustrations, was packed away.

After receiving this letter, I got out and examined that earlier manuscript. It was interesting to observe how much I had progressed in the meantime, for I could see clearly that what I had composed then was not good enough. My ideas had enlarged and crystallised since that time, and what I had in mind was much more ambitious and comprehensive.

In reply to Key it was possible to give him an almost complete outline of the book I envisaged, and to say that a thousand pounds was not enough. He replied at once that a satisfactory plan and a competent author were far more difficult to find than the money; that could be raised, and so a Board of Trustees was got together. Before the end of the year (1945), the project was in full swing.

If chemistry and fishes had been equally balanced before, fate was now loading the ichthyological pan. Here were not only my beloved fishes, but a work I had longed to do, something big, with a definite aim and end.

About this time we heard the first rumours of the foundation of the Council for Scientific and Industrial Research, which would co-ordinate all scientific research and administer funds in its aid. Would it be able to help me in the move I was now more than dreaming about? Would it help me to fishes, to change from sulphuretted hydrogen to formalin? Meanwhile, work on the book went on apace. Our house became more of a laboratory and a studio than a home. I sought out a number of young artists and trained them to draw fishes as we wanted them, which is, accurately and as they really are. Many failed and left, but those who came through the trial period did good work.

The year 1946 was one of the most difficult of my life. We had huge classes to handle, some had to be duplicated. My wife, also a chemist, just could not escape the appeals for help, and had to teach as well. It was almost as bad as the time of the Coelacanth. We became just machines. We had people all along the South African coast, trawlers and fishermen, sending a constant stream of fish for illustrations. In June-July 1946, for purposes of the book, we took five artists and a photographer and spent a month in and about Lourenço Marques. There was a Dr. Jekyll and Mr. Hyde flavour even about this. We lived in an extremely ancient derelict house, the furniture mainly boxes, but in a select locality, not far from the Governor-General's palace. Lourenço Marques was startled by the succession of notables who visited the ancient structure, and the photographer's antics below the coconut palm in front provided free entertainment for all the urchins of the neighbourhood. I spent most of my time collecting specimens, on the bay, about the islands, and along the coast, while my wife culled the market, made friends with the Portuguese deep-sea fishermen, ran the house with servants directed by gestures, tried in vain to keep her young son clean, controlled the artists, and showed our work to impeccably dressed and often uniformed visitors. It was fish, night and day, and we could not speak a word of Portuguese. This was so maddening that we decided we must learn that language, which we eventually did, but only after a grim battle lasting five years.

When we returned to the Union in July 1946, I learnt that there would be at least a possibility of a Research Fellowship from the Council for Scientific and Industrial Research, which by then had been established. I consulted the various authorities concerned, and eventually, in September 1946, gave notice of resignation from the Chemistry Department as from the end of that year. It was a most difficult thing to do, as I have always been greatly attached to chemistry and enjoyed teaching, and it meant severance of close contact with students, whom it has constantly been a very real pleasure to handle, teach, and advise.

The Research Fellowship from the Council for Scientific and

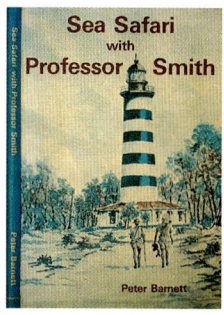

Peter Barnett, the photographer that JLB took on the 1951 expedition to Mozambique, published a vivid account of this expedition, *Sea Safari with Professor Smith*, in 1953.

JLB Smith was one of the first scientists to receive a research grant from the newly formed CSIR in 1946. His bursary of £300 per year was sufficient to cover his basic research costs and allowed him to become a full-time ichthyologist. It may have been one of the best investments the CSIR ever made! Today the South African research evaluation and award system, managed by the National Research Foundation in Pretoria, is one of the best in the world.

The other research unit that the newly formed CSIR funded was the Telecommunications Research Laboratory (TRL) at the University of the Witwatersrand, which was equally successful. One of its staff, Trevor Wadley, became South Africa's greatest inventor. He invented the Barlow-Wadley Broadband Radio, Ionosonde, Wadley All-Wave Communications Receiver and Transistorized Receiver and, most importantly, the Radio Tellurometer, the most accurate distance-measuring device in the world for over 30 years.

Trevor Wadley demonstrating his Radio Tellurometer In England in 1962

The new research Department of Ichthyology was housed in a quaint corrugated iron building dating back to the South African (Anglo-Boer) War (1899–1902). The new 'JLB Smith Institute of Ichthyology' building in Somerset Street, with its magnificent facilities, was constructed after he died and was officially opened by Basil Hersov, President of the South Africa Foundation and Head of the Anglo-Vaal group of companies, on 26th September 1977, the 80th anniversary of the birth of JLB Smith and Margaret's 61st birthday.

Industrial Research did materialise, and in 1947 my new life started. At first I had only one room, then more space was made available, and eventually the University agreed to have a separate Department of Ichthyology. This is now housed in one of the original military buildings, whose present contents would certainly startle those who lived there first. This is a curious and probably unique department, whose chief support comes from the South African Council for Scientific and Industrial Research.

At the time now described (1946), we had little in our heads but the growing 'book', but although this was more than enough to occupy our full time, the Coelacanth was never out of our minds. With the passing of the active phase of the war, from all parts of the world odd letters had begun to come in—Coelacanth! Eventually, in October 1946, I wrote to the President of the C.S.I.R., and said that as interest in the Coelacanth was reviving, it was inevitable that a search for more would be started, and it was naturally expected that South Africa should take the initiative.

He replied by return, expressing approval and suggesting immediate steps to further the project. I submitted a more detailed memorandum, and eventually a small committee was nominated by the C.S.I.R. to go into the matter.

In March 1947 the following notice was issued to the press:

LATIMERIA CHALUMNAE

The discovery of a living Coelacanthid Fish in South African waters, off East London, at the end of 1938, is an event still in the forefront of the minds of biologists. The published account of the mounted animal is as exhaustive as the material permitted, but all zoologists desire information about the soft parts of the creature which in this case were lost before they could be examined. The outbreak of war put a stop to preparations for expeditions to seek further specimens of this remarkable fish.

Now that the war is over, general interest in this project has been shown in various countries and in South Africa in particular. The South African Council for Scientific and Industrial Research has appointed a committee to consider how best to organise a marine expedition on a considerable scale. This expedition would aim not only at securing more Coelacanths, but would also explore and accumulate data in various fields of science in the relatively poorly known region of the Mozambique Channel.

The Committee is under the Chairmanship of Dr. S. H. Haughton, Director of the South African Geological Survey, and a member of the Council for Scientific and Industrial Research. The Honorary Secretary of the Committee is Professor J. L. B. Smith, of Rhodes University College, Grahamstown, and the committee requests that all Societies, Institutions, and private persons interested in the project should communicate with him.

Some bright soul sent a statement to the press in England about the proposed expedition, and said that volunteers were wanted, quoting my name in reference. Letters almost drowned me. I gained the impression that the British Isles were just bursting with young people wanting more adventures, and had to have a contradiction published. There ensued considerable correspondence with interested bodies and institutions all over the world, and two meetings of the Committee were held. The projected expedition was named the 'African Coelacanth Marine Expedition', or 'A.C.M.E.' for short.

I wanted to hunt and find Coelacanths, and knew exactly how I proposed to set about it; but some of the others had ideas of their own, and soon it became clear that a large-scale oceanographical investigation was to be hung on to the Coelacanth. I did not greatly mind as long as this led the way to the Coelacanth, and the area to which I had pointed, Madagascar and the Mozambique channel, certainly needed investigation. Just about nothing had been done there. When the details of the vessels and equipment necessary for all this came up, finance pushed up its ugly head. I intended to use explosives, and this raised further difficulties, for fishery interests might be antagonised. It was a long surging battle, in which I saw danger to my desire to find Coelacanths. The vessels were the chief problem, and there was at least a possibility of one being loaned by a group like the British 'Discovery' Committee. But when this suggestion was put to the Prime Minister, it was rejected by him. Despite my astonishment, it became clear that nobody else expected a Prime Minister to give reasons.

Before the end of 1947 it had become quite plain that the large-scale project that the Committee had visualised earlier was financially impossible, and I submitted my original plan, one much less ambitious and far less costly but at least as effective from the

The proposed 'African Coelacanth Marine Expedition' (ACME) had an echo many years later when the 'African Coelacanth Ecosystem Programme' (ACEP), with equally broad goals, was launched by the Department of Science & Technology in 2001, with the South African Institute for Aquatic Biodiversity (SAIAB) as the lead agency.

Unlike the later ACEP, which was highly successful and led to a major interdisciplinary oceanographic research programme in the Western Indian Ocean, the ACME project fizzled out due to political meddling and indecision.

The leaflet project is another example of JLB's 'out of the box' yet practical thinking. It must have felt strange for the new leaders of the CSIR, bent on launching South African science onto the international stage, to approve this zany project.

With characteristic zeal, JLB arranged for the reward leaflets to be distributed throughout East Africa and in the Western Indian Ocean islands.

The £100 reward was paid by Eric Hunt to fisherman Ahmed Hussein Bourou, who caught the second coelacanth south of Domoni off Anjouan Island in December 1952.

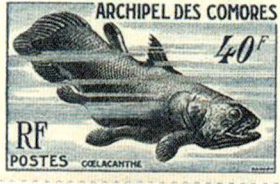

The first coelacanth stamp issued by the Comoros in 1954, signed by JLB Smith

In 1953, in support of the initiative to secure more coelacanth specimens, the French set up fish-embalming stations in the Comoros.

purely Coelacanth point of view. One essential part of my plan was a descriptive leaflet (Plate 3), showing a picture of the Coelacanth, giving a brief description, and offering a reward, in English, Portuguese, and French; and I proposed that this should be distributed everywhere along the coasts of East Africa, Madagascar, and all islands in those waters.

After a period of indefinite suspense, we drifted to the end of the A.C.M.E. project, it just fizzled out. Early in 1948 it was dead, and I have never really discovered whether it was international tension, finance, or the effects of the views of overseas scientists in higher quarters that finished it off. At any rate I was left up in the air with a sense of frustration. As a scientist I can never view with any pleasure the apparent ease with which some politicians appear to contemplate war, and the spending of countless millions on destruction and death, while they will in peace-time hedge and jib at a few thousand pounds for a scientific project. However, even if the Government would not help I was determined to go on, by myself if necessary. There was one way in which I could reach out and cover vast and remote areas without going myself and without great expenditure, and that was by means of the leaflet. So I told the C.S.I.R. I wished to proceed with that idea, which they approved, and both they and Rhodes University College agreed to guarantee £100 each as a reward for the first two Coelacanths obtained.

These leaflets were printed in Lourenço Marques, and distributed by every possible means. The Portuguese authorities sent numbers to every part of their shores, with instructions to officials not only to distribute them among all classes, but to explain them where necessary. This was done with characteristic promptitude and thoroughness. Our port authorities in the Union and those of Lourenço Marques agreed to hand leaflets to the captains of ships going north, and to ask them to leave some at every port where they touched. Batches were sent to every major port along the East African coast, with requests that they should be distributed among the fisher-folk. With a Portuguese official, my friend Carlos Torres, who speaks English, French, and Portuguese with equal facility, I visited the Consul for France in Lourenço Marques, and gave him an account of the whole matter, explained the leaflets and the object I hoped they would attain. I told him that in my opinion the Coelacanth was most likely to live somewhere about Madagascar or in its area, and asked if he would be kind enough to send a batch of the leaflets to the authorities there, as well as to write in explanation and to request them to distribute the leaflets as widely as possible in their territories. He became most interested and promised every possible assistance. A day or two later he informed me that the leaflets had been sent by air to Madagascar, together with a letter giving a full explanation of the whole project.

I felt that even if I could not go and look everywhere myself, money talks, and the leaflet would have thousands of eyes constantly looking on my behalf. Again and again we got evidence that the leaflets had gone far and wide, though we heard nothing of those sent to Madagascar, except later that officials had seen them. There is no evidence that the leaflets were distributed widely there, possibly because it was felt that it was too crazy an idea that Coelacanths should live round those shores. After all, had not competent scientists in Europe satisfactorily settled that Coelacanths had fled to the depths of the ocean? It seems fairly certain that none were sent to the Comores or, if they got there, they remained unused.

Even though I had high hopes for the leaflet, I was preparing for many years of laborious searching myself. Especially in the course of my later work on South African fishes, it had become clear that for a full understanding of them and of their peculiar faunal components, it would be necessary to study the fishes of East Africa as well. The more I investigated, the clearer did it become how very little work of any real value had been done in that vast area. I felt there could scarcely be any more fortunate combination of effort than to go hunting Coelacanths as well as other fishes in all that huge virtually unknown region, full of wonderful reefs and channels, just the parts where I judged that Coelacanths should live, and, big as they were, still remain unknown. The South African Council for Scientific and Industrial Research lent a ready ear to my proposals, and provided, and has since gone on providing, funds for this exploratory work. In addition, the Portuguese authorities were most co-operative and furnished invaluable aid on a considerable scale.

In 1947 and 1948 we carried out expeditions over the whole of the southern regions of Mozambique. The fishes of the southern-

Traditional fishermen in the Comoros and Madagascar would almost certainly have been catching five or more coelacanths a year (based on subsequent catch statistics) during JLB's 'leaflet campaign' in the 1940s and 1950s. It is less likely that coelacanths were being caught off the African mainland at the time as there were few hand-line fisheries over deep reefs there.

Mike Bruton

Traditional Comorian fisherman in a double-outrigger galawa

The first coelacanths subsequently captured off Mozambique (1991) and Kenya (2001) were caught by trawlers and those off Tanzania (2003–present) have mainly been caught using deep-set *jarife* gillnets or ring nets, although some have been hooked on hand lines by traditional fishermen.

The expeditions that JLB Smith launched in 1947 and 1948 would not have been contemplated by an ordinary man. They covered vast areas of relatively unknown and hostile terrain where transport, logistics and communications were a nightmare, yet he and Margaret carried out first-class research and brought back to Grahamstown vast collections of carefully labelled fishes. Margaret also did colour paintings of the live colours of many of the fishes that they caught. Along the way they even managed to edit the manuscript of the *Sea Fishes of Southern Africa* book!

Peter Barnett

JLB and Margaret Smith on a fish-collecting expedition in Mozambique in 1951

Because of the southward-flowing, warm Mozambique Current, many tropical fishes from East Africa also live along the southern African coast. In order to understand the taxonomic relationships of southern African fishes, JLB needed to know more about the diversity and classification of East African fishes.

most part of that area were included in my South African fishes volume, printing of which had started in March 1948, and we completed our work early in 1949, the book itself eventually appearing in July that year.

By that time the public of South Africa had become interested in our work, and our basic support from the C.S.I.R. was supplemented by gifts of supplies and money from private persons and firms.

From then on we carried out a series of expeditions, going steadily farther afield each year. Always we carried and distributed the leaflets, spoke of the Coelacanth, and asked questions. In 1948 I met a native in the Bazaruto area of Mozambique who picked on the fish at once. Yes, he had once caught exactly such a queer fish, he got it one evening in the deep channel south of Bazaruto Island, but had never seen or heard of another before or since. In the water it was like a big Garrupa (Rock Cod), but when he got it out the big scales and the peculiar fins stamped it on his memory as unique. He spoke of its oiliness, the soft flesh, and the absence of bone, things about which he could never have known except from an actual specimen. He could not say if the tail was the same, but it was near enough. This was the only reasonably hopeful sign we got in all that long search up the east coast of Africa. It remained the only one.

Before the book on South African fishes appeared, the publishers told me they expected it to sell well, and that it would probably be necessary to have a second edition within a year. The volume was issued in July 1949, six weeks before we were due to go on an expedition to East Africa. The whole edition sold out in three weeks! It was a situation! We were in the throes of our final arrangements for departure when we received a frantic call from the publishers to prepare for another edition. I do not yet know how it was done, but it was. The proofs followed us in batches over a long stretch of Mozambique, and were corrected under what one may term somewhat unusual conditions and in places where such work had certainly never been done before. A book of that type takes almost a year to print, and that second edition came out the following year.

Meanwhile we went steadily on, gradually extending our knowledge of East Africa, its reefs, conditions, and fishes, and continually exposing undreamt-of scientific wealth. There was so much that I almost got to wishing there was no Coelacanth urge to divert me from what had now become a most fascinating pursuit, unravelling the marvellous fish-life of East Africa.

In 1950 the 'Discovery' organisation wrote to say that the research vessel *William Scoresby* would be going on a voyage, and would call at and look for Coelacanths in South Africa. She was the vessel which from letters from overseas I had hoped we might get from this organisation for our A.C.M.E. expedition, but Smuts had refused to permit us even to ask.

The *William Scoresby* had bad luck. She arrived in April 1950. The engineer had to be shipped back from Cape Town, the vessel had to be dry docked. Her chief scientific officer, Robert Clark, came by land from Cape Town to Grahamstown to tell me their plans. They hoped to find Coelacanths at East London, and had come well equipped with lines, nets, and special traps, and requested my co-operation and advice.

I warned him of the difficulties he would encounter, currents and foul bottom, and advised consultation with trawlermen. I cancelled a trip to Lourenço Marques so as to be able to visit East London and go out with them for part of this time, and arrived there after they had worked for a few days. Very few things upset Robert Clark, but even he was a bit depressed. They had no Coelacanths. There was no point in my going out to sea with them, for they had no traps, no lines, or whole nets either. The patchy bottom and swift apparently opposing currents at variable depths had beaten them. The traps were lost, the lines ripped away, and the nets all torn. Instead of taking me out to sea, they took me down to the saloon and opened a bottle of champagne, of which my share was by choice only the smell.

Later in 1950 we worked in the area about the island of Mozambique, most of the time at and about Pinda, truly the most wonderful haunt of varied fish-life. It is a jungle-covered peninsula, wild and remote, with a lighthouse at its northern tip. The reef is enormous, at least five miles by eight in extent, and there is every variety of bottom, from sand to coral, sheltering fishes of all kinds. It was a hard life, supplies were difficult, water is very scarce, it is hot and we were plagued by packs of man-eating lions that terrorised the whole area. Almost every night they tore open

The sorry tale of the *William Scoresby*'s search for coelacanths is a further indication of how difficult it was to find and catch these fishes. South African Prime Minister Jan Smuts refused JLB Smith's request for funds to participate in the *William Scoresby* cruises.

No coelacanth has ever been caught on purpose by Western scientists, despite multi-million dollar efforts to do so.

Lions continue to predate on people in southern Africa. In October 1984, during one of our expeditions to the Okavango Delta in Botswana, a German woman tourist was eaten by a lion in the Moremi Game Reserve near our camp.

JLB's campaign to incentivize captures of the 'Hundred Pound Fish' paid dividends during his search for the coelacanth but it might have hampered later conservation efforts to de-commercialize the fish so as to reduce fishing pressure. During our coelacanth conservation campaign Hans Fricke and I persuaded CITES (Convention on Trade in Endangered Species) to list the coelacanth on Schedule I, which means that it may not be traded for commercial gain.

The tidal range in some parts of Mozambique is so wide that fishermen attach their gillnets to vertical poles stuck in the sand and retrieve the captured fish at low tide using homemade ladders!

the natives' flimsy huts and savagely choked their last frenzied screams. It was horrible to hear the triumphant roar that accompanied a kill; we even had one of the brutes come and cough at us early one morning from the top of a thicket-clad cliff as we worked on the reef below. In the morning we would find their pug marks near our bedroom window. It was not pleasant.

The Portuguese had done their work well. Even in remote lighthouses such as this the leaflet was posted for all to see. Again and again some headman would show it, stuck on the pole of his hut. As a treasured possession, it would sometimes be produced from the inside of a fisherman's garment. There could have been few even in those remote parts that had not heard about the Coelacanth; they all knew that ten thousand Escudos were offered for this fish. To the natives all along the coast the Coelacanth is now known as 'Dez Contos Peixe', i.e. 'Hundred Pound Fish'. In our travels we came to realise that most of them doubted whether there really was anyone so crazy as to pay that vast sum for just one fish, and we tried to convince them. Added to that, experience taught me that while the average native could recognise a picture of a fish he already knew well, it was the exception for one to be able to recognise an unknown fish from a picture.

In 1951 we worked over one of the wildest and least-known parts of East Africa, the northern territory of Mozambique, between Port Amelia and the Rovuma. This was one of the most arduous of all our undertakings. Along the coast are a number of islands, densely bushed but waterless and uninhabited. There are no communications and no supplies; from the point of view of our work the normal conditions were difficult, as there is constant high wind with occasional storms, currents are fierce, and there is a rise and fall of 14 feet at spring tides. We lived and voyaged all along those terrible shores in a small vessel provided by the Portuguese; indeed, we could never have done this work but for their aid. Taking advantage of the few occasions when the wind abated a trifle, we would rush from one island shelter to another, often only precarious. Fishes, well, they were there in millions, wonderful fishes so unsophisticated that they almost climbed aboard themselves, and we got marvellous collections; but never saw a trace of anything like a Coelacanth, nor had any of the few humans we encountered, white or black.

In all this time I was still the only scientist in the world to believe that the Coelacanth came from somewhere about tropical East Africa. My views were looked on as obsession rather than as logical deduction. I was plainly crazy even though this was leading to the discovery of marvellous scientific wealth in modern fishes. Constantly very conscious of all this incredulity, at the time now described I became increasingly puzzled and worried, because just north of Mozambique the great westerly current of the Indian Ocean divides, part going north and part south, the latter our powerful Mozambique current. If the true home of the Coelacanth lay anywhere south of that division of the current, it was easy to understand how one at least had wandered down to our waters, as many other tropical fishes constantly do; but despite all our searching we had so far failed to find that home.

The fact that the Coelacanth did get to East London made it far less likely that its home lay anywhere north of the level of Cape Delgado, opposite which the current divides. As we had so far failed to find that home in the area of the southern branch along East Africa itself, while it increased my uneasiness it also focused my attention on Madagascar, which was also washed by that same branch.

All along that part of the East African coast, and exactly opposite, lay the thousand-mile long stretch of Madagascar, not so far away. Even if my mind had not constantly been drawn across that channel by the Coelacanth, there were always vivid reminders of our nearness to Madagascar. In that wild part we found ruins that puzzled us at first, extensive ruins of forts that faced the best landing-points on the shore. Eventually we came to learn that they were the sole remains of an extensive colonisation of this northern area of Mozambique by the Portuguese of earlier times. Those hardy pioneers had scarcely become established when fleets of sailing canoes drifting silently ashore in the dead of night discharged hordes of raiding natives, who killed, ravaged, and pillaged, then sailed away as silently as they had come. They were the Saccalaves, hardy seafaring natives from Madagascar. So near was Madagascar, where the seas and reefs could not be very different from those of Mozambique, and I was comforted by the feeling that even though Madagascar lay beyond my present reach, my leaflet had gone there, and I hoped, that as in Portuguese

JLB wasn't far wrong in his assessment that most coelacanths were likely to be found south of Cabo Delgado in the far north of Mozambique. All the coelacanths from the waters of the Comoros, Madagascar, Mozambique and South Africa are from more southerly latitudes, with only those from Tanzania and Kenya being further north.

The ancient island of Madagascar was an obvious home for coelacanths. It has been a stable yet isolated island for over 60 million years and has a very distinctive terrestrial fauna, including endemic lemurs, chameleons and, until a millennium ago, giant elephant birds.

Marc Staub Wikipedia CC BY-SA 2.0

The endemic panther chameleon from Madagascar

Because of the continuous nature of the Indian Ocean environment, the marine fish fauna off Madagascar has lower levels of endemicity than its land animals, and many of the fish species there are shared with other Western Indian Ocean islands and the African mainland.

The verbal record of a coelacanth catch off Bazaruto Island is interesting. During visits to Vilanculos and the San Sebastian Peninsula near Bazaruto in the early 2000s I showed pictures of the coelacanth to traditional fishermen and questioned them about it, but none recognized the fish.

We now know that the coelacanth has fairly narrow habitat preferences and environmental tolerances and would not be able to travel 'half the globe', although they do make very efficient use of ocean currents. Their need for highly oxygenated water, and rocky caves or canyons in which to shelter during the day, would prevent them from crossing large, deep ocean trenches.

How coelacanths reached Indonesia from Africa (or Africa from Indonesia, if they originated in Asia) has still to be explained.

territory, many natives would have seen it and be aware of the rich reward this fantastic fish would fetch.

Under all the increasing weight of scientific treasure we got in these seas, even in my obsessed state, the shadow of the Coelacanth slowly receded, though we never ceased to talk and to show its picture on the leaflet. My wife was even more persistent than myself, she always had that quest in the forefront of her mind, and never let anyone forget it. We ended that series of expeditions with little hope that Coelacanths lived normally anywhere in Mozambique waters, for we had covered virtually every possible and likely spot. Even if that native had got a true Coelacanth at Bazaruto it could well have been a stray, as I believed the first at East London must have been.

Northern Mozambique was far enough from East London, where the first Coelacanth was caught, but now it began to look as if it must have come from somewhere even more distant. That was far enough for a 'degenerate' fish to travel in all conscience, without my now supposing it to have travelled still farther. From a place as remote as northern Mozambique a fish of the Coelacanth type meandering along the coast, even with the aid of the current, would probably take several years to reach as far as East London, every minute of the way beset with dangers. While there were those who regarded the old Coelacanth as degenerate or 'wooden', I did not, and I had no doubt that if he wanted he could travel half the globe.

We went on; we never ceased to hope, relying on the leaflet to do our work in Madagascar, and unaware that it had not reached the Comores. Whenever I planned any expedition and studied charts, always my eyes and mind would stray to the Comores, those mysterious blobs in the blankness of the seas, like drops left behind from a dripping Madagascar torn from the body of Africa.

In that wild part of northern Mozambique I have described, those Comores were a constant obsession. Again and again I stood looking across that blue water. They lay south of that critical current divide, in the southward arm. Yes, they obsessed me, and they were so tantalisingly near, much nearer than Madagascar; Grand Comoro was barely 200 miles away, scarcely more than a day in our small vessel. We had no compass, for coastline navigation in our boat did not demand one. I knew how the currents ran and their speed, and could have used my watch, the sun, and the stars to find the way to the Comores. Those islands are all high and can be seen a long way off, but my wife showed an unusual lack of enthusiasm whenever I raised the project. Nevertheless, I was sorely tempted just to go and look, but there were too many obstacles. It had been difficult to reach this virgin part at all, and it was proving so rich it would have been almost a crime to have gone somewhere else on a mere chance when every moment where we were was yielding rich results. We had very little water and, in any case, I could not take their vessel to foreign territory without the prior consent of the Portuguese authorities, who were nearly as far off from us as the moon. In that wind-lashed sea, among those remote islands, we were quite cut off from civilisation. Its isolation was emphasised by the way in which the only occasional native fishermen we saw, whether ashore or afloat, fled at our approach. This amused our crew, who always laughed with delight at their flight, and they explained that those men were almost certainly fugitive canoe-tax defaulters.

The seas around the Comoros are treacherous, with vicious currents and strong winds, and it would have been a hazardous trip there and back in a small boat. Thank goodness for Margaret's 'lack of enthusiasm'.

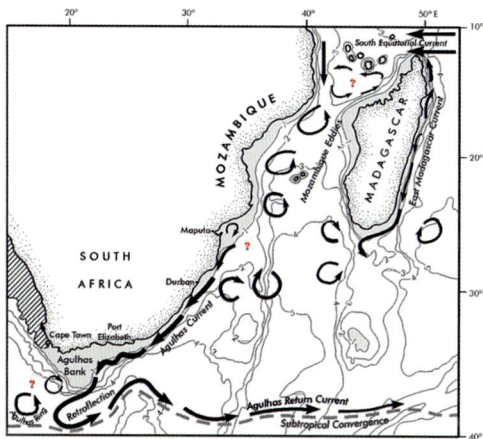

Currents and gyres in the Western Indian Ocean

When JLB flew to the Comoros in 1952 in a South African military airplane to poach the second coelacanth from under the noses of the French, he showed little concern for geo-political tensions!

Hans Fricke

The author and Jürgen Schauer extracting samples for DNA analysis from the first coelacanth in 1987

I visited Zanzibar in December 2008 and 2010 and retraced the steps that JLB and Margaret followed during their visit in 1952. Dhows still ply the waters between Zanzibar and the Comoros together with large motorized vessels. Stone Town still has its distinctive 'old world' flavour with outdoor markets selling delicious cockle, mussel and octopus kebabs. Slow-moving hand-drawn carts now compete with maniacal scooter riders for space in the narrow streets.

Today Stone Town is probably best known as the birthplace of Africa's greatest rock star, Freddie Mercury of 'Queen', who was born there as Farrokh Bulsara on 5th September 1946. Farrokh would have been a six year-old kid playing on the beach when the Smiths visited Zanzibar and met Hunt.

Carl Lender

Freddie Mercury

Chapter Eight

DUNNOTTAR DILEMMA

ERIC HUNT first came into this story in 1952 in Zanzibar, where we were working at that time as part of an extensive expedition covering Zanzibar, Pemba, part of Tanganyika, and Kenya. We had been greatly assisted in all phases of our work by the authorities of each country, and at the close of our time in Zanzibar, at the request of the authorities, we held an exhibition of our discoveries for the public, and this was crowded out all day. Hunt came late in the afternoon with a friend who knew my wife, and so was introduced. He wanted certain information about fish, as he did a good deal of commercial fishing.

We had a pile of the Coelacanth leaflets there for people to take, and Hunt spotted these and was soon immersed in one. My wife noticed his absorption and asked if he was interested. His reply left no doubt that he was, and he asked if he might have some of the leaflets to take to the Comores. Comores! My wife jumped at this, as may well be imagined; we might indeed have been working at those very islands at that moment had not the Kenya authorities been so anxious for us to come there. What did Hunt know about the Comores? Well, he had a schooner and lived by trading between Africa and the Comores, and knew them well. Her quick reaction led him to ask my wife at once if she thought there was any possibility that the Coelacanth might be at the Comores?

More than a possibility she told him, and that I had long believed that Coelacanths would likely be found somewhere about Madagascar—for one thing, fossils were well known there, and as for the Comores, well, they had long been something of an obsession with me. She told him how I wanted to go there, and how I had tried to find out about their natural history, but there just didn't seem to be any. She told him, too, something of our experiences the year before when we had been working at the Querimba

Islands in the northern part of Mozambique, and how I had very nearly set out to reach the Comores in the tiny vessel we had then. As far as the leaflets were concerned, they should already have gone to the Comores, because a big batch had been sent to the French authorities in Madagascar some years before, and they had been asked officially to distribute them. However, she was pleased to give Hunt a batch of the leaflets, and he remarked that if it should happen that they caused a fuzzy-headed Comoran to get the £100 reward for catching a Coelacanth, the Governor would indeed be 'Tickled to death'. He himself would certainly be thrilled to have a part in anything like that. After studying the leaflet, Hunt asked many questions, and in addition my wife gave him a good deal of extra information that might help him. She was impressed by his quick grasp of essentials, and finally showed him the account and pictures of the Coelacanth in a copy of my book on South African fishes, which we had there and which he studied closely for some time. Finally, he said that he was confident he would recognise any Coelacanth he might come across, and my wife told me she thought he would, too. She considered Hunt to be 'all there'. In addition to this, he had undertaken to try to get specimens of a certain peculiar small fish he had seen in the Comores, but which we did not recognise from his description. It was arranged with people in Zanzibar that he was to be supplied with formalin for that purpose. My wife emphasised the importance of this, but as things turned out it was never sent to Hunt, a not unusual type of failing in those climes.

This was September 1952 and shortly afterwards Hunt took the leaflets to the Comores, showed them to the authorities and spoke about the Coelacanth. One visiting official had apparently seen and heard about the leaflets before, but appeared to have the impression that they represented an insane idea and a useless search, for fish of that type lived in the deep sea, and it was certain that they had never been seen anywhere round there. Hunt, however, had little difficulty in interesting the Governor, who had them sent round to all the islands of the Comores and distributed by native runners, who also as far as possible explained their import to the natives. When you know natives of the type found there, you wonder how anything like that could be explained to them, except that money talks everywhere, especially big money

o.f.—6

In fact, the coelacanth was well known to Comorians and it is highly likely they would have recognized it from the reward poster. JLB's comment that 'you wonder how anything like that could be explained to them' is surprising as he had intense respect for the knowledge of traditional fishermen. What is not surprising is that Hunt struck gold within a few weeks.

Caught coelacanth on a beach on Grande Comore

At the time all transport between the four Comorian islands would have been by boat so the process of distributing the leaflets would have taken several weeks.

Sharks are still harvested extensively off the Western Indian Ocean islands and the East African mainland, and many species are now threatened with extinction. They are vulnerable to overfishing as they are slow growing, long lived and produce relatively few young per breeding. The threat to sharks has increased since the introduction of deadly deep-set gillnets.

White Shark Projects

Great white shark, Carcharodon carcharias

The 'man-size Parrot-fish' to which JLB refers is almost certainly the Napoleon or humphead wrasse, *Cheilinus undulatus*, a large, distinctive fish with a greenish-brown body that reaches 2 m. In French it is known simply as 'Napoleon'. I encountered these fish in the Red Sea, where they readily take food from divers.

Wolfgang Grulke

The author with a Napoleon wrasse off Ras Mohammed in the Red Sea in 2001

of that order, it would speak in a loud voice even in their simple lives. Hunt came and went. He sometimes carried local produce, and dealt also in dried fish and sharks. There is an enormous trade in salted sharks in the tropical western Indian Ocean, for they are greatly relished by all races and command an astonishing price. This was one of our problems at Zanzibar, for when we found a rare shark on the market, its price was so ruinous we could not afford to buy it as a specimen. I soon hit on the idea of hiring sharks and other big fish for the special purpose of photographing them, of making notes and taking measurements, and then bargaining with the owner for parts like the teeth and sometimes the head and skin. There was not much they did not eat. The sharks are salted but not sun dried; this is a special process not used in South Africa. The smell from the big concrete underground shark salting-pits of Zanzibar is pretty grim. When the wind shifts to the north you have it all day and all night. The food of those near by tastes of it.

At the close of our work in Zanzibar we went to Pemba; then to Kenya, where we spent several months working over a wide area of the coast. This expedition was most absorbing, but exacting and exhausting. The whole area proved so rich that we almost killed ourselves in that hot and humid climate trying to squeeze the utmost from the time. We had got together an enormous collection, certainly more than ten thousand selected specimens, with numbers of great rarities and many fascinating forms new to science. The last few weeks that took us into December 1952 had been especially trying, with little wind, the nights close and still, and since malaria was rampant, we lay naked and sweating under nets, vainly trying to sleep. All you got was a kind of vague and patchy, formless, clammy doze.

Throughout this expedition we had continually carried on the Coelacanth hunt, talking and giving out leaflets. It had been my good fortune to solve the mystery of the identity of a strange fish, a man-size Parrot-fish which was only occasionally seen. It was a curious creature with a big hump on its head, but at that time its scientific identity was unknown. I hunted for this creature continually and offered a reward for one, but in vain, until the day before we were due to leave Shimoni finally for Mombasa, when by the greatest good fortune I spotted over two hundred of these peculiar large fishes in a deep channel off Pungutiachi Island (South Kenya), and by means of considerable exertions managed to capture no less than eight of them, the largest weighing over 130 lb. We preserved a complete head, and the skull of another, and took these with us. Near mid-December we embarked in the Union-Castle* liner *Dunnottar Castle* at Mombasa. All our collections, including the Parrot-fish head preserved in the cold store, went with us.

The day before we left Mombasa, a reporter from the *Mombasa Times* came aboard with one of our Coelacanth leaflets, and was full of questions. We gave him the main story, which he clearly found fascinating. One of his last queries was whether we believed that the leaflet would really find the fish we sought, and we said we hoped it would. The issue next day prominently featured the Coelacanth story, quite a scoop for that paper with what lay only a little time ahead.

On the way south we called as usual at Zanzibar, and there my wife went ashore to visit the market, and to renew acquaintance with various people of the most diverse social strata she has a way of gathering to her net, all of them very useful in our work. As she neared the wharf, there was Hunt on his schooner, and she waved to him. He came round to meet her launch, and told her then that he had only recently returned from the Comores. She went aboard his schooner, he was awaiting a friend to go out on a trial run, the engines had just been refitted, wouldn't Mrs. Smith like to come as well? With little time and much to do, she was unable to accept this invitation, but inevitably she asked Hunt if he had found any clues to Coelacanths. No, he had none, but was as keen as ever and had a whole lot more questions to ask about the creature, keen, searching questions which showed that he had done a good deal of thinking about the matter in the meantime, and which made my wife even more confident of his ability to recognise a Coelacanth for certain if he ever saw it. The leaflets had been distributed in all the islands, and the Governor was both interested and co-operative, because, of course, a thing like that

* The Union-Castle Mail Steamship Co. Ltd. may well be termed a National Institution as far as South Africa is concerned. From this Company and from all its officials, ashore and afloat, we have received constant consideration as well as assistance on a considerable scale, representing in all a substantial contribution to our work and to science.

Today it seems peculiar to return from a scientific expedition on a Union-Castle liner but that was JLB's only means of long-distance marine travel at the time. Travel on Union-Castle liners was so routine in the 1950s and 1960s that, in 1964 during my last year at school, I travelled from East London to Port Elizabeth aboard the Pretoria Castle to take part in a tennis tournament!

Union-Castle liner poster

Margaret's legendary ability to connect with people was something that I witnessed at first hand during my travels with her. After one airplane flight from Cape Town to Johannesburg she so endeared herself to the other passengers that they cheered her off the plane!

It was remarkable good fortune that the Smiths met a schooner skipper as capable as Eric Hunt, as he played a pivotal role in securing the second coelacanth specimen.

In April 1998 I was invited to be a guest lecturer on marine biology on a cruise of the Cunard liner *Queen Elizabeth 2* from Cape Town to Southampton. During the voyage, my wife Carolynn and I had the rare pleasure of sitting at the table of Captain Ronald Warwick. The trip was an enjoyable but taxing experience as I had to give five one-hour talks, and answer endless questions, about marine life.

The Cunard liner Queen Elizabeth 2

Tim Dyer

would certainly put his territory 'on the map'. Had he got the formalin? Not yet, but it was promised and he expected it any time. She wished him luck, and started to leave when he said, 'If I do get a Coelacanth and haven't any formalin, what do I do?' She said he should not even think of such an awful possibility, but he said (doubtless knowing East Africa better than she did), 'Yes, but just suppose there isn't any, tell me what I could do? There is no refrigeration at the Comores.' So she replied, 'Well, heaven forbid it should happen, but the only way would be to use salt. Like those smelly sharks.' And he replied, 'O.K., thanks. Anyway, when I get a Coelacanth, I'll send you a cable.' And they both laughed with amusement. So did some imp of fate also laugh, for only ten days later all those things happened: a Coelacanth, no formalin, salt, and the cable.

My wife's encounter with Hunt brought him and the Comores to the forefront of our minds, and they were like a hovering cloud in our consciousness, receding only slowly as we left them far behind. We often spoke of both Hunt and the Comores, but during the short voyages from port to port, we were hard at work all the time describing and figuring our rarities. Not only this, but that whole voyage south was as usual strenuous, for every port brought interviews and visitors, both officials, friends, and press. Captain Patrick Smythe gave us every assistance and consideration, excusing even our reluctance to sit at his table on the ground that we were too exhausted to be polite to strangers, and it was on that account arranged for us to have a table to ourselves.

Before dawn on the 24th December 1952, my wife and I were up on the bridge to gaze on the lights of Durban that lay shimmering in the haze that hid the land. It was a real thrill to see our own country again. Most of the pilots up the east coast are old friends, but it was especially nice now to meet one of our own. We were leaning over the rail beside the searchlight on the bridge as we slid through the harbour mouth, scanning the anglers who as usual lined the piers. I drew a deep breath of contentment and said, 'It's wonderful to be back again. It will be a long time before anything gets me back to the tropics again.' That same imp of fate must have laughed again, and even more loudly, for within six hours I was frantic to be able to do just that, and quickly.

We docked at 7 a.m., and no sooner was the gangway down than we were beset by friends and the press, some of the latter new to us.

In mid-morning we were in the lounge with Stanley Dagger, our good 'Elastoplast' friend, who keeps us supplied with the most essential field dressings. One of the junior officers came up and said, 'Telegram for you, sir', and I took it absentmindedly—it had rained telegrams that morning. This bore the red 'Urgent' tab, one of a number. I had noted a young man who came with the officer and he now introduced himself as a reporter new to me. It had rained reporters, too, but I asked him to sit down for a moment while we continued our conversation. In a lull I slit open this telegram and read it casually. At first the words had no meaning, then I found myself on my feet staring at it, for two words stood out: 'Coelacanth' and 'Hunt'. 'What's the matter?' asked my wife in alarm. 'Hunt's got a Coelacanth,' I said. She jumped up, took the telegram and read it. It had been redirected from Grahamstown to Durban. It read:

'REPEAT CABLE JUST RECEIVED HAVE FIVE FOOT SPECIMEN COELACANTH INJECTED FORMALIN HERE KILLED 20TH ADVISE REPLY HUNT DZAOUDZI.'

Dzaoudzi, where on earth was Dzaoudzi? It sounded like Somaliland or some such place, we had never heard of any place of that name before, not in the Comores, anyway. One of the younger officers came up at that moment with some message from the Chief Officer, and I asked him if he knew where Dzaoudzi was. No, never heard of it, but he could soon find out for me, and he went off, while my mind groped in chaos asking if this could be true. 'Five-foot Coelacanth, Hunt.' Yes, Hunt would know a Coelacanth. A five-foot Coelacanth. Could it be true? The young reporter had his ears cocked, and he asked some questions that remained unanswered. The young officer came shooting back. 'Dzaoudzi is on a small island called Pamanzi in the Comores, sir.' So by Heaven, Hunt had got one in the Comores. It was those Comores, after all. Good for Hunt.

Comores, hot as hell, was how he had described them, and had said there was no refrigeration plant there. Formalin! How much? If it was the amount that had been planned for Hunt, it wasn't nearly enough for a five-foot Coelacanth. What a predicament.

Most mortals, after an exhausting expedition to the tropics, would have settled for a delayed trip back to the Comoros to retrieve the second coelacanth. However, Smith's sense of ownership of the specimen after his extensive leaflet campaign, and his passion verging on *terrebilità* to be the one to describe it, dictated that he must go back to retrieve it immediately.

This coelacanth (a male) was caught by Ahmed Hussein Bourou south of Domoni on the east coast of the island Anjouan at about midnight on 20th December 1952, 800 m from shore at a depth of 160 m. It weighed 37.5 kg and measured 135 cm in length.

Dzaoudzi is located on the small island of Petite-Terre (or Pamanzi) off the main island of Grande-Terre (or Maore) in Mayotte. The Dzaoudzi Pamandzi International Airport (as it is now known), where JLB landed, is located on the southern end of Petite-Terre island in Pamandzi.

Map of Mayotte

When I visited Mayotte in 2008 our airplane ruptured a brake fluid pipe on landing and skidded across the runway. I spent a miserable 7 hours huddled in a stiflingly hot airport waiting for a replacement plane as I was *en route* to Grande Comore to join Hans Fricke for a *Jago* dive. Even in modern times coelacanth research can be tedious and fraught with difficulties.

What was the date? Killed on the 20th! Already four days since it was caught. Dzaoudzi was almost certainly not the kind of place where you could just go and buy formalin.

I was too distracted to speak very much, for with all these perplexities my mind was slipping about like a rubber-clad foot on a muddy street, and would not get settled firmly in one place.

It is rather remarkable that neither in my mind nor in my wife's was there in those early tense moments of realisation any

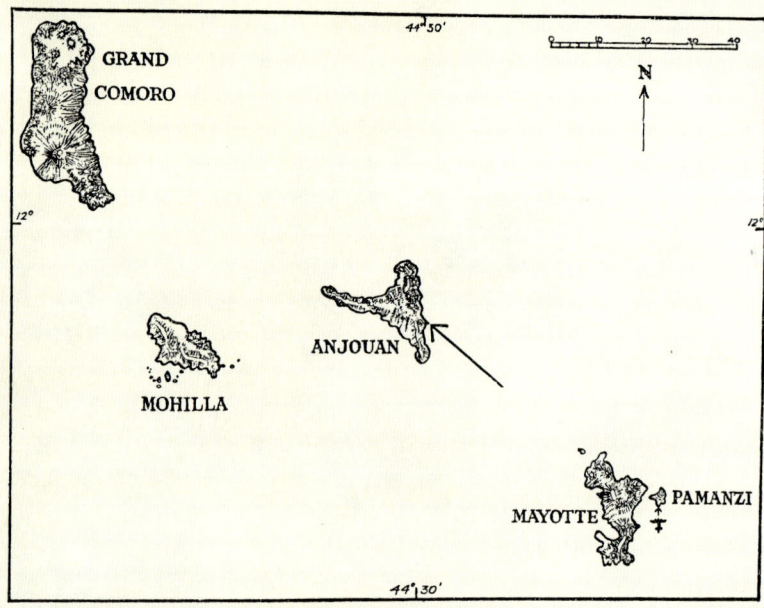

FIG. 3.—The islands of the Comoro group. The arrow shows where *Malania* was caught.

vestige of pleasure or rejoicing. We had endured so much over the first Coelacanth that we just had a feeling that Coelacanths meant trouble, at least to us. And now this had come in a fashion and at a time that looked as if it was going to be no better, probably worse. There were so many difficulties sticking out clearly ahead, we could hardly see anything else.

Dagger soon after took his leave, to be replaced almost at once by Guy Drummond Sutton, Frank Evans, and then Dr. George Campbell, all Durban men and old friends. I took the reporter aside and told him that he had come upon one of the most astounding events and that it would certainly be world news. If he used it properly he could have the scoop of a lifetime. I gave him quite an amount of information, warning him to use nothing but that, and none of what he had heard earlier except what I repeated now. As I went to join our party he went away, but he was so much younger than I that his thoughts were easy to read and they did not flatter me. The scepticism I had felt in him was possibly responsible for the manner in which that interview was reported and headed of all things 'Sea Missing Link'. As it happened, I did not see this until much later, but early next morning that same young man came aboard seeking further news, and was met by my wife, who had just seen the report. She asked him at once why on earth that phrase 'Missing Link' had been used, and assured him that I would be most displeased, as a Coelacanth was emphatically not a missing link, and I had certainly not used such terms. He said that it was a catchy phrase and sounded good as a title. It was not until Natalie Roberts wrote it up next day that the whole affair was reported in what I regarded as its true perspective, but those unfortunate words 'Missing Link' went out in advance all over the world, giving the whole affair a false aspect that it did not need. It was interesting enough as it was.

After this reporter had left I went back to our party and told them what had happened, and this of course attracted their immediate and concentrated attention. They wanted to know what I was going to do. I said that at the moment I just did not know, the situation would have to be weighed carefully. Was I going to try to fetch it? I couldn't say at that moment, it was going to be very difficult to do that, anyway. I knew that part of East Africa and the route, and flying there is no easy matter. I doubted if any private plane could do it even if we had the funds to hire it, which we hadn't. It would almost certainly have to be done by a Government plane if at all. At that moment I was not prepared to say anything definite, not even if I was going to try to go and fetch it. It needed thinking out carefully.

'What a pity Smuts isn't alive,' said one of the men. 'He would have helped you.' 'Smuts!' burst simultaneously and explosively from my wife's lips and my own. 'Smuts!' Our reactions were so

JLB grappled with journalists throughout his career. He expected them to uphold the same standards of accuracy and sincerity that he had set for himself but was often disappointed. He also abhorred sensationalism in science reporting.

According to Peter Jackson, who was an ichthyologist and an accomplished pilot (and had flown from England to South Africa in a small biplane), there were no civilian aircraft available in South Africa at the time to take Smith to the Comoros to fetch the fish. Those aircraft that could have done the trip before the War had all been commandeered and the aircraft industry after the War had yet to make, import or charter suitable aircraft for long-distance flights.

In 1945 Smith asked the then South African Prime Minister, Jan Smuts, to provide a military airplane to collect fishes killed by a submarine disturbance off Walvis Bay in South West Africa (now Namibia). Smuts refused even to meet Smith.

Sadly, there is still hardly any scientific presence in the Comoros. The Centre National de Documentation et de Recherche Scientifique (CNRS) in Moroni, Grande Comore, has tried gallantly to promote research in the islands but its museum, library and research facilities are sorely depleted.

Sign for the Musée des Comoros in Moroni

CENTRE NATIONAL
DE LA RECHERCHE
SCIENTIFIQUE

violent that they stared at us in amazement. 'No, sir,' I said. 'I once wanted to ask Smuts to help me in almost exactly such a case, and he wouldn't even see me. Smuts! No.' They collapsed like pricked balloons. Smuts was off.

My wife suggested that if it proved impossible for me to go in time to save the fish, it might be done by the French authorities in Madagascar. She reminded me that I had met the chief of their scientific organisation at a conference in Johannesburg not long before; what was his name? 'Millot,' I said, 'Dr. J. Millot'; and that I had had him in mind, but as far as I knew he was likely to be in Paris and not in Madagascar.* I knew that in our files we certainly had nothing but an address in Paris. Was there not any French scientist in or about Madagascar who could help, she asked? My knowledge of scientists was naturally chiefly in the ichthyological field, and although there had been a few publications about sea fishes from Madagascar, those were mostly odd papers produced by scientists in France. There had been no man of any prominence in my field resident at Madagascar for at least fifty years. There was no avenue of approach that way— as far as I knew, Madagascar was a complete blank for any purpose now.† In the case of the Comores hardly any marine biology had been done there. From my unique knowledge and experience I could predict probably 90 per cent. of the fishes that must live about the Comores, but an ichthyology of that part just did not exist. It was most unlikely that there would be any competent marine biologist there. If there had been, Hunt would surely have mentioned it.

It was clear that I had to seek for no ordinary help. This was a time for desperate measures, something at the highest level. The Prime Minister was the obvious mark, but the very idea of again trying to ask a Prime Minister for a plane to hunt dead fish, even a dead Coelacanth, made me shy away like a once-wounded animal from a gun. My natural reluctance springing from the Smuts episode was increased by the fact that Dr. Malan was almost certainly seeking brief rest from the many heavy respon-

* This proved to be the case.
† We learnt later that a French Fisheries Officer was actually in the area at that time, but the *Zoological Record* listing his publications did not appear until 1955. In any case, he was not at headquarters in this critical time (see p. 159).

sibilities he carried, and he was far from young. To all suggestions about Malan I said firmly, 'Not until we have exhausted all other● possibilities.' My wife was the most persistent on that theme, and when I repeated, 'Only as a last resort', she prophesied that I would go to him in the end, which I did. Trust a woman to have the last word.

Frank Evans said there was a Sunderland flying-boat at Durban which would be just the thing if I could get it. He knew the local chief and would go off and have a word with him, and my wife went with him to send Hunt a cable which I had drafted, as follows:

'IF POSSIBLE GET TO NEAREST REFRIGERATION IN ANY CASE INJECT AS MUCH FORMALIN POSSIBLE CABLE CONFIRMATION THAT SPECIMEN SAFE. SMITH.'

There is little pleasant in the recollections of those particular hours of the 24th December 1952. Everything with Coelacanths so far had been troublesome, and here was quite the worst situation I had ever encountered. There were so many difficulties that there seemed no way out. This precious fish was so far away, in one of the worst places in the world for safe preservation with probably only a mere speck of formalin. I had probably been to more remote areas of the coastal regions along East Africa than any other man, and if only this fish had turned up in some part I knew, how much easier it would have been, even in foreign territory like Portuguese East Africa. But here it was in quite unknown foreign territory, so little known that it might almost be another planet, and I had no knowledge of conditions and no personal contacts in that part. The tense uncertainties of the situation in some aspects resembled those of the first Coelacanth, and yet how different it was this time. The doubts about its being a Coelacanth the first time were due to a battle between facts and my common sense, the fish itself was accessible enough. Now the doubt about identity came from the difficulty of getting to the fish, and although he was exceptional, the whole thing rested only on the word of a layman I hardly knew and on that opinion I must stake a great deal. I was not even at home, with the re-sources of my Department and organisation, but on a vessel in

Catalina military flying boats operated from Durban in 1942, and 262 Squadron of the Royal Air Force established a flying boat base at Bayhead in 1943. This squadron flew reconnaissance and anti-submarine patrols out to sea from Durban Bay and from lakes St Lucia and Umsingazi in Zululand.

Sunderland flying boat – a British patrol bomber developed for the Royal Air Force by Short Brothers, and named for the port of Sunderland in northeast England

262 Squadron was eventually assimilated into the South African Air Force as 35 Squadron with the Zulu motto *Shaya Amanzi* ('Strike at the Water'). The squadron received Sunderland V flying boats in 1945 and continued to fly out of Durban until 1957, when all maritime reconnaissance duties were taken over by land-based Avro Shackletons.

JLB's fiercely independent streak and his unwillingness to transfer the burden of responsibility to others comes through strongly here.

The notion that the coelacanth was an issue of 'national prestige' seems quaint but the reality is that this point of view is still held today, quite rightly in my opinion. In 1989 South Africa issued four postage stamps commemorating the discovery of the coelacanth and in 1998 the South African Mint issued a gold coin depicting 'old fourlegs'.

International Coelacanth Trust commemorative cover

Hans Fricke in the Comoros in November 2008

Durban due to leave soon, all our enormous volume of baggage and valuable collections to be cared for, and Christmas holidays just on us. My mind rebelled. I was already unutterably weary, and the weakness that plagues us all urged me to throw the whole thing away and leave it alone. It welled up like a flood, and heaven knows there was enough excuse. This really threatened to overwhelm me, largely because I realised it would be necessary, absolutely necessary, to ask things of others, something that always makes you vulnerable, and I was deadly tired. What a situation! But I have faced many, and knew that I could not give up, I must go on. I had to remind myself that there was a lot more in this than my own personal feelings, it had long since passed beyond that. It was a matter of national prestige, for the whole Coelacanth affair was South African, everything in it was tied up with South Africa. It was our responsibility and our misfortune that my wife and I alone at that time realised the full implication, and that it was my obligation to go on with it. I had to go on even if it killed me in trying to do it, and when I said so to my wife, as my partner in this venture she calmly agreed with that view. I had to go on.

Our old friend Dr. George Campbell stayed to lunch, and it was a comfort to have him near. Captain Smythe had heard that something had happened and came over to our table for a time, when I gave him a brief outline of the situation and my predicament. He at once offered all possible help, and meant it. When he had gone I turned again to the meal, seeing and hearing nothing, my mind far away. The others realised my abstraction and went on quietly discussing the matter, leaving me to my thoughts, and I ate, absently, at least I believe I ate, for I sat on, isolated, my mind going round and round. Malan, the Prime Minister; H'm! Prime Minister; Smuts; H'm! I became quite oblivious to my surroundings, my mind went drifting off and back to that other time, and I went through it all again, as vivid as when it actually happened, yes. . . .

Chapter Nine

HIS OWN SHEEP

What a curious sequence of events it had been! Some years before while preparing certain parts of the manuscript of my large volume on our fishes, information was found to be lacking about certain types of fish and related matters of the waters of the South-west Cape, so I arranged to go out from Cape Town with one of the trawlers of Messrs. Irvin and Johnson, a commercial fishing company which had greatly assisted my work in many ways. Their vessels had added greatly to the knowledge of our fishes, constantly bringing rarities ashore; indeed, it must not be forgotten that it was a trawler of this firm which caught and saved the first Coelacanth.

This Cape trawler on which I was to go was the *Godetia*, one of the two new vessels that were more or less experimental, since they were very considerably larger than those previously used in South Africa. They certainly were very much more luxurious than anything of that type that I had ever seen before. I shared the skipper's cabin and had a real bed, away from noise and smell. After the other trawlers I had known, this one was almost incredible. I could not help thinking of the lumpy mattresses in narrow box-like bunks close under the iron deck. Just above one's head were steering chains that rattled and crashed on the iron plates every few seconds; indeed, unless the sea was smooth, without any intermission day and night. Those were the trawlers on which I had lived and suffered. To a landsman they spelled discomfort, smell, and nausea.

In this grand new vessel there was a real saloon and real lavatories, not just the heaving rail and the sea. It was an interesting experience, and different in many ways from the life on a trawler on our south coast, where the work goes on night and day, the crew often without proper sleep for days on end if catches are good, when they do not mind. In this cold Atlantic sea the trawling is

Irvin & Johnson (I&J), which has been a trusted name in seafood in South Africa since 1910, now owns and operates a fleet of modern stern trawlers as well as land-based seafood processing facilities and an abalone farm 200 km southeast of Cape Town.

I&J has for many years provided strong support for marine fish research, conservation and education. In addition to their ongoing assistance to ichthyology in Grahamstown, I&J sponsors the spectacular 'Ocean Exhibit' at the Two Oceans Aquarium in Cape Town.

Sea sickness, like other forms of motion sickness, is primarily a disturbance of the inner ear that affects the sense of balance and spatial orientation.

Motion is sensed by the brain through three pathways: the inner ear, the eyes and proprioceptors (in the deeper tissues of the body). When the body is moved intentionally, the input from all three pathways is efficiently co-ordinated by the brain. When there is unintentional movement of the body, as occurs on a rocking ship, the brain does not co-ordinate the information as efficiently and there is confusion among the inputs from the three pathways.

mostly in rather deep water, several hundred fathoms. At night here the fish leave the bottom and rise in the water, some right to the surface, so that a trawl net which scrapes the bottom all the time catches hardly anything then. As a result, trawling is profitable only by day, and as there is no anchoring in such deep water, the vessel just drifts, and all night long she rolls and rolls in the long Atlantic swells. A landsman finds this agonising and his only hope of sleep is to wedge himself in.

On our way out, some miles from shore, the skipper told me we should soon be met by 'Blondie', a large seal well known to the trawlers, easily recognised by a lighter patch on the side of the head. Seals always turn up when the trawl net is being pulled in, but apparently one or two of the more intelligent come and meet the vessels on the way out, stay with them all the time they work, and accompany them part of the way back. 'Blondie' had been known for many years. Sure enough, there was a hail from the bows soon after, and I saw this quite large seal puffing a welcome as it easily kept up, appearing alongside at intervals.

Most of the fish caught in a trawl are swept into a long bag in the middle of the length of the net, the apex of which is known as the 'Cod-end'. When the net is hauled up from the bottom, the water pressure becomes less, the gas in the air-bladders of the fishes expands, and usually before the rest of the net is up, the 'Cod-end' comes shooting to the surface and floats there, buoyed by thousands of these distended bladders. Often they are so much enlarged as to stick out of the mouths of the fishes, and look like great swollen red tongues.

As the mass floats on the surface, protruding tails fruitlessly beat the water, and all kinds of the smaller fish are forced through the meshes of the net. These provide a wonderful feast for the snorting seals, which splash around, grabbing with gusto and gobbling the dainties, seizing them crosswise, shaking them, throwing them into the air, to make them easier to swallow. They have clear-cut likes and dislikes, small Stockfish apparently being their favourite, while the more abundant 'Rat Tails' (Coryphaenoidid fishes) are scorned. In this they are like humans, who will not eat those perfectly wholesome but less-attractive-looking and quite unfortunately named 'Rat Tails'. It is not only the seals that eat their fill, but sharks soon gather round. At first they cruise cautiously at a distance, snapping up the fishes that have drifted away, but they soon come closer, and once their appetite is roused, go raging round, and unless checked will even ravenously attack the net, tearing great holes and causing loss of fish as well. Sometimes they become so insensate that they throw themselves out of the water on top of the floating mass, tearing at the enmeshed fish below. Trawlers carry rifles for these brutes, and during each haul the skipper generally accounts for a number, some of them 10 to 12 feet or more in length. One shot in the head and you see the ugly beast sinking slowly into the green depths, rolling over as it goes.

We got all the specimens I wanted on that trip, and some un-expected rarities as well. We were due back on a Monday. On the Thursday night the wireless news mentioned fish mortality at Walvis Bay, maddeningly only a few words, but they were enough for me; indeed they were as effective as an electric shock.

On the South-west African coast, mostly in late summer, fishes are sometimes killed in millions, sometimes in such vast numbers as to be a serious menace to health when those thrown ashore begin to decay, and their disposal is a great problem. This whole-sale slaughter is believed to be due partly to volcanic activity and mostly to quantities of sulphuretted hydrogen, released into the sea from bottom deposits. This stinking gas is poisonous to all animal life, but especially to fishes, for it removes dissolved oxygen from the water. 'One man's poison is another man's meat' is a transposition that applies here, for while this is hard on the fish and a great trouble to all who live on that coast, an event of that nature is an ichthyologist's dream, for every kind of fish in the sea is killed, and even though like most things in nature it is extravagant, it gives the scientist an opportunity he could other-wise hardly accomplish by any means himself. While interested people can pick out a few obvious oddities from among the piled masses of dead fish, it is essential that an expert should work through such an accumulation himself, for while the queerest-looking creatures may be merely common and scientifically well-known forms from deep water, an apparently ordinary-looking fish may well be a scientific treasure.

For many years I had been waiting for an opportunity like this, and here it was and I was much nearer than usual. Despite the

Male Cape fur seals reach 2.3 m in length and weigh up to 300 kg. During dives in Hans Fricke's research submersible *Jago* we saw seals at the extraordinary depth of 170 m off the Cape Peninsula coast.

Erhardt Thiel / Images of Africa

Cape fur seal, Arctocephalus pusillus

Fish and crayfish kills may also be caused by outbreaks of 'red tide', a bloom of algae that colours the water red or brown. Some red tides are associated with the production of natural toxins and the depletion of dissolved oxygen in the water, which may result in mass mortalities of fishes and shellfishes and even of marine birds and mammals.

Unusually for a politician, Field-Marshal Jan Smuts had a keen interest in science. He was an experienced mountaineer and went on several plant-collecting expeditions in the 1920s and 1930s in South and East Africa with staff from the Royal Botanical Gardens at Kew. He was quoted as saying, 'The intimate rapport with nature is one of the most precious things in life'.

His affinity with Nature was not, however, enough to persuade him to help a desperate scientist. Perhaps his preoccupation with mountains and terrestrial plants precluded him from having any empathy for aquatic life, as if rivers and oceans were on another planet. I have often found this to be the case among amateur naturalists during my career.

Wikipedia CC BY-SA 3.0 nl

Jan Smuts

hot sun, the sea at Walvis is cold and the fish do not decay seriously for several days. I slept less than usual that night and next morning sat at the wireless. More news came, it was one of the greatest killings for many years. This was Friday, the trawler was due back only on Monday, the catch was moderate, there was still plenty of room in the fish-holds. I naturally wanted to return at once, and so the battle started, a running naval battle, with all the odds normally on the skipper, but I kept on. Eventually he agreed to return early, but only by one day, and we were set to arrive at Cape Town by the mid-morning of Sunday. I was in a fever of anxiety and slept not at all, my mind was buzzing like a machine. Walvis Bay is less than a thousand miles from Cape Town, but it might in some ways be on the moon—at least, it was like that in those days. To go from Cape Town to Walvis by sea or road was ruled out on account of time, air was the only way. Nobody on the trawler knew anything of private planes, they doubted if there were any.

The moment we docked—Sunday—I went to the telephone. It took some time to find out, but it soon became pretty certain that no charter plane was available that could reach Walvis Bay. The Air Force was the only way. After incredible difficulty (Sunday!) I made contact with a responsible official in that force, but he was less than encouraging, saying that it would be impossible even to think of such a thing without the permission of the Commander-in-Chief who was in Pretoria, and even then the Government might not permit it. It was very clear that as time was the vital factor, to try to do anything via Pretoria was quite useless; it would have been even on a weekday, and this was midday, Sunday.

I was determined not to be beaten. Smuts! He was the key. He had the reputation of being interested in science. I had not met him, but was due to do so the very next day. The Trustees of my book of fishes (then in preparation) had decided to ask him to write a foreword. Our Albany member, T. B. Bowker, had arranged that I should meet Smuts, and an appointment had been made for the next afternoon, Monday. That would be too late. How could I see him that day?

I sat in that office at the docks going over in my mind whom I could get to give me the approach. It *would* be Sunday of course

—any other day would have been better. I decided to ask assistance from a man whom I had known for many years and who knew Smuts. I was soon putting the problem to him, to ask if he would telephone the Prime Minister to find out if he would see me. While he was enthusiastic about the whole idea his reply disturbed me, for he said that as he was only an Ordinary Member he dare not telephone the Prime Minister directly. I was astounded and said so. 'You don't know the Oubaas as I do,' he said. 'If I did that it would be as much as my seat is worth. He is a holy terror to the rank and file. But [he added] you are different. The Oubaas is very interested in science and I am sure if you get to him he will help you.' I asked whom I should telephone, but he advised me not to telephone at all as they might just put me off, being a Sunday. He said it would be best for me to go straight to Groote Schuur and ask to see the Prime Minister. I was more than dubious about the wisdom of this, as I knew what my own reactions would be in such a case, but despite my strongly expressed doubts he said that he really felt that it might succeed where a more formal approach might fail. It was only with considerable reluctance that I eventually agreed to do what he advised. It was then well past noon, and after clearing up affairs with the trawler and seeing to my fishes, I was taken to Groote Schuur by some friends. The Prime Minister was out. He had some overseas banker with him and they had gone to Muizenberg, but were expected back late in the afternoon. I put my case to two officials who received me. It is one of my assets and one of my troubles that I can read other men's minds. The younger of the two said little, but as many others had done, and many were still to do, he plainly thought me mad to go to such lengths for long-dead fish. The other was more cautious, though I could feel that he also thought it insane; still, you never knew which way the Oubaas was going to jump, he might rise to this and actually give this apparent lunatic the plane he wanted, so it would be better not to turn him away irrevocably now, but to wait and see. He told me that the Prime Minister did not like being troubled on Sundays, as I could well appreciate, but I emphasised that it was only the factor of time that drove me to this end, and that I had been advised against my judgment by one who should know to come in that way. I asked if I should go away and telephone later, but after consideration he

JLB had an uncanny ability to read people's minds. He was prone to long silences and then sudden outbursts that made it clear that he had listened to, and analysed, everything that had been said by others and had already reached a conclusion about a particular matter long before the conversation had ended.

During his early career Smuts was heavily involved in science. He became an authority on South African grasses and, in 1926, wrote up the ideas that he developed at Cambridge University in a book, *Holism and Evolution*. He was President of the South African Association for the Advancement of Science in 1925 and, in his presidential address ('Science from the South African point of view') included a discussion on the origin of the Cape flora.

Two new plant species are named after him: *Digitaria smutsii* and *Pteronia smutsii*.

In 1930 he was elected a Fellow of the Royal Society and, in 1931, served as President of the British Association during its centenary year, his presidential address dealing with 'The scientific world picture of today'.

said it might be as well to stay until Smuts returned. That was twice I had been advised to try the direct personal approach by those who should know, so I told my friends that I would wait and should telephone later, and they went off.

They put me in a room, but I could not sit still and found my way outside, pacing up and down beneath the magnificent trees that toned the sunlight almost to gloom. Every minute was an age. When would this man return?

It was an awful time. I can stand a good deal, but find suspense wearing, for while my imagination helps my work it clouds my life. I could see those precious piles of fishes starting to rot, being eaten by birds, washed away by the waves and buried by sanitary gangs. Some time later a kindly official came out to offer me tea, so I went inside, still steeped in anxiety, and ate and drank without any memory of what they served. This man questioned me again, astutely and with interest, and I got to speaking of my work. Quite suddenly I realised that he had become interested, even my ally, and would do all he could. Plainly concerned, he set out to alleviate my pressing tension and showed me round the lovely mansion, but the rarities and treasures meant nothing to me, I could see only those rarities and treasures rotting on that sun-drenched dazzling sand at Walvis Bay, and in the end he left me to start again my restless pacing beneath the trees. Would this man never come?

I had gone some way from the house, the sun was already so low that it was turning dark beneath the trees, when I heard the throb of a car which passed round the corner of the house out of sight. I had, from my reluctance to worry any man at this time, suggested that I should not be in the house when Smuts came, but that if he was willing to receive me I should come in. Slowly now I made my way towards the house, and went on pacing up and down the gravel path along the side. A good fifteen minutes passed. Suddenly my sixth sense told me I was being scrutinised, and from upstairs. Unobtrusively I scanned the windows, but they were screened and told me nothing. I waited tensed and anxious, and when the official came to seek me, he had no need to speak, for I could read his message in his walk and bearing. To his great regret he had to tell me that the Field-Marshal would not see me. He had told him the outlines of my story and had even ventured to press my case, but in vain. I thanked him, asked him to telephone, and my friends came soon after.

Tomorrow I should see Smuts, but it would be too late, my treasures would by then be rotten. Even so, I had no hard feelings for his refusal, only disappointment in his lack of vision. I reflected philosophically that I was probably once again paying the price for my youthful appearance, for my figure and looks belied my years. And yet Smuts must have known at least something of my work and reputation. With his scientific leanings he would certainly have been aware of the Coelacanth, more than aware indeed, for he had been kept fully informed about the A.C.M.E. project, when he had not wanted us to make any move to secure a vessel from overseas. It was a curious situation, for the C.S.I.R. fell directly under Smuts.

I went to the House next day and lunched with our member for Albany, Tom B. Bowker, who took me along and introduced me to the Prime Minister's Secretary. Yes, the appointment for 2.15 p.m. was arranged. The Prime Minister was entertaining some visiting American journalists to lunch and had not yet returned to his office, but it was a good time, for the House was quiet, and all should be well. Suddenly there was a stir, and the Secretary went off. He returned in a few moments and stood in the doorway, a curious expression on his face, looking at me before he spoke. His exact words I do not remember, but they were to the effect that he wished I had told him about my visit to Groote Schuur before letting him go in to the Prime Minister, and why on earth had I not tried to speak to him before going out there?

I gave him a brief outline of events and of the advice I had received and acted on. He was non-committal and diplomatic and asked me to wait for a short time, but he said that the situation was very complicated, and meanwhile the Prime Minister was receiving his American guests one by one. He vanished before I could get anything more, and I sat, amazed and bewildered, watching a series of thrilled Americans going and coming from heaven. Eventually I seized the Secretary and demanded to know what the position was, and after some parrying finally learnt that Smuts was annoyed that I had ventured to go to Groote Schuur, and that in consequence he now refused to see me. His Secretary plainly had some concern on my behalf, because he

O.F.—7

In common with many highly-strung and zealous people, JLB had no talent for waiting. This episode with Smuts must have stretched his tolerance to the limit.

JLB's detailed description of the frustrations that he experienced in dealing with Smuts, and with bureaucracy in general, is indicative of an acute inner 'power struggle' that he experienced between himself, the lone scientist, and the world beyond science. He sometimes portrays himself as a 'lone ranger' single-handedly fending off the scientific apathy of the rest of the human race.

In truth, there were many other scientists and science administrators in South Africa at the time fighting the same battle.

Robert Broom was one of South Africa's foremost palaeo-anthropologists. He discovered Mrs Ples and the remains of many other early humans at Sterkfontein, Kromdraai and Swartkrans caves, including *Paranthropus robustus*.

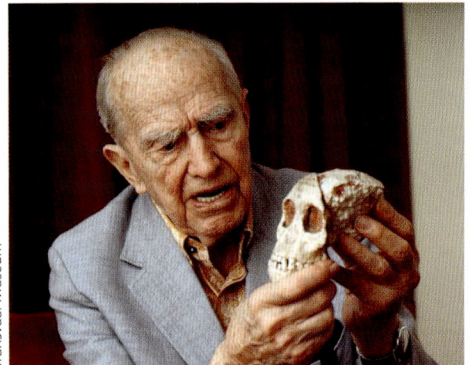

Transvaal Museum

Robert Broom with the Tuang child

Interestingly, Broom had some quaint ideas about evolution. In his book, *The Coming of Man: Was it Accident or Design?* (1933) he claimed that 'spiritual agencies' had guided evolution, as animals and plants were too complex to have arisen by chance. According to him, 'Much of evolution looks as if it had been planned to result in man, and in other animals and plants to make the world a suitable place for him to dwell in'.

Contrary to what JLB implies here, Broom was helped in his career by Smuts who, at the insistence of Raymond Dart, put pressure on the government to create a professional position for Broom at the Transvaal Museum in 1934 so that he could continue his research full time.

said that if only he could have a few words with the Prime Minister he might manage to right matters. He called a typist and asked me to write and express regret for having attempted to speak to Smuts at Groote Schuur, as that might help.

Eventually, getting on for 4 p.m. my patience was exhausted and I demanded to know the position. The Secretary said he was sorry, he could do nothing more, Smuts refused to see me at all. So I left, abruptly, this time not without deep annoyance in my heart, for I could not but feel that it was not my youthful appearance that was to blame, but my nationality, that if I had been a scientist from some other land I might well by now have been snatching treasures from the sands of Swakopmund. Thanks to him, they were gone. To be esteemed by foreigners was more important than the special plea of a scientist of his own country. I was not the only South African scientist who had been treated in this way, there was Broom. . . . It all brought to my mind again the fable of the shepherd and the sheep,* as true today as it was then so long ago. I was one of his own sheep.

Some time later I had a meeting with the Trustees of my book fund to finalise a number of issues. We had many reasons to be pleased, for the public had been responsive, the quality of the specimen work of the *Cape Times* was all that could be desired, and the estimates proved to be within our means. Towards the close of the meeting the Chairman raised the issue of the Foreword, saying that he supposed he could take it that we were all agreed that 'The General' should be approached to do us the great honour of writing the Foreword. 'If by that you mean Smuts,' I said, 'I regret to say that I will not have his name associated with my book in any way.'

This was clearly a bombshell, but I was adamant without many words. The only reason I gave was that in my view it would be more proper for the foreword to be prepared by a scientist, rather than by a politician who had no fundamental interest in the subject or the author. When this matter had come up for

* Some sheep caught by a storm sought shelter with a shepherd, who, scenting profit for himself, put them with his own flock, whom he deprived of the best food and the warmest beds to give to the strangers. The storm over, the strange sheep prepared to leave, when the shepherd entreated them to stay and asked, 'Did you not like the way you were treated?' 'Yes,' they said, 'but we saw how you treated your own sheep.'

discussion at a much earlier meeting, Smuts had been mentioned. I would rather have had some person prominent in science, I had in mind the President of the South African Council for Scientific and Industrial Research, at that time B. F. J. Schonland,* but at that stage this issue meant little to me personally and I had made no comment. Now, however, I proposed the President of the C.S.I.R. and this was accepted.

After that meeting, the Secretary of the Board of Trustees, the late Bransby Key, asked me to give him some moments, and as soon as we were alone raised the matter of Smuts, would I mind telling him what it was all about? After brief consideration I gave him the outlines of the story. To my astonishment his reaction was violent, and he called Smuts by names that had a Chicagoan stockyard flavour. He then told me how he had been treated by Smuts in much the same way, but with even less excuse. Another of his own sheep!

This decision had some repercussions. A subscriber, who had been led to expect that the book would bear the signature of Smuts, when he heard that it would not, came to tell me that the absence of that signature was bound to have an effect on sales. If it did, it was not what he expected. I was told that shortly after publication, in the larger centres those seeking to buy copies were at times compelled to form queues.

When the news came through that in the General Election Smuts's party had suffered defeat, it was a shock to many people. Even more startling to the world at large was the news that in his own long-standing constituency, virtually a stronghold, Smuts had himself been defeated, and by an opponent who outside political circles was till then not widely known. After the initial shock I was not really surprised. I knew why. His own sheep had turned against him . . .

* One of the most outstanding scientists produced by South Africa, born in Grahamstown, a leading physicist of our time, the first President of the South African C.S.I.R., virtually its creator, whose ability and unflagging energy put this organisation in the front rank within a short time of its inception.

Jan Smuts was undoubtedly a towering figure in international politics, the only man to sign the peace treaties that ended both the First and Second World Wars and to sign the charters of both the League of Nations and the United Nations.

Anglo-Boer War Museum

Jan Smuts in 1950

However, he had a reputation for putting the needs of foreigners ahead of those of his own people, of strutting the world stage without being popular in his homeland. In particular, his support for South Africa's participation in the Second World War had severe political repercussions. He lost the national election to the National Party in 1948 (as he had done in 1924) and died of a heart attack 28 months later.

George Gordon ('GG') Campbell (1893–1977) was a well-known surgeon and promoter of science in Durban. He was the brother of the poet Roy Campbell and cousin of the historian Killie Campbell. During a discussion on an expedition to northern Zululand in 1948 he kick-started the idea of establishing the aquarium in Durban, which JLB Smith strongly supported.

George Campbell

Dr Petrus Johann du Toit (1888–1967) was a leading South African veterinary scientist and Director of Veterinary Services at Onderstepoort from 1927 to 1948 and was later President of the Council for Scientific and Industrial Research (CSIR).

Chapter Ten

STARTS AND STOPS

I CAME back suddenly, to the early afternoon of the 24th December 1952, to the table in the saloon on the *Dunnottar*, to George Campbell on his feet at my side, his hand on my shoulder, shaking me back to the problem of the Coelacanth, to the reality of the hard times ahead. He squeezed my shoulder, and looking at my troubled face, smiled and said, 'Cheer up, J. L. B., you'll win'. That is how patients are brought back to life.

There was no ordinary telephone for the use of passengers on that ship, the only one was on the bridge for official use, but Captain Smythe had told me that the bridge and telephone were at my service. I realised that I was almost certainly in for some record telephoning, so asked for the head of that Service in the Durban Post Office, gave him a brief outline of the situation, and asked if he would kindly make things as easy as possible. After that the Post Office was marvellous. The willing and efficient service all members of that organisation gave us there is one of our most pleasant recollections, for it certainly lightened our burdens all through that most difficult and exhausting time. In the matter of finding a wanted man I would back the South African Post Office any time against Scotland Yard or even the famous Canadian 'Mounties'.

Practically all the funds for my support, work, and expeditions come from the South African Council for Scientific and Industrial Research, which administers the funds allocated by the Government for scientific work. I was responsible to that Council, so they came first to my mind, and who better than their President, at that time Dr. P. J. du Toit, formerly head of the world-famous institution of Onderstepoort, and he had been one of the most outstanding of all its Directors. Yes, 'P. J.', as he is known, was the man to consult. Would the Post Office kindly find Dr. P. J. du Toit at Pretoria, and say I wished to speak urgently to him?

While I waited on the bridge, my mind was racing, trying to probe the future; would this call never come, half an hour already gone? I called up the Post Office and inquired as to its progress, impatient at the delay—they had not been able to find Dr. du Toit, but the Pretoria office was busy. It is amusing to look back on my early impatience, for I had still much to learn. My wife came up to the bridge and was talking to me when the telephone went. I jumped, but a Quartermaster standing by took the receiver and was apparently having his own conversation, a series of 'Yes, yes, yes', so my attention switched back to our talk, when we heard him say a final 'yes', and 'He's right here', followed by 'Professor, your Secretary wants to speak to you'; and there was Mrs. McMaster, to say that another cable had come from Hunt, addressed to the University authorities this time, wisely covering my possible absence, and it had been sent over to her. It read as follows:

'HAVE SPECIMEN COELACANTH FIVE FEET TREATED FORMALIN STOP ABSENCE SMITH ADVISE OR SEND PLANE REPLY—HUNT DZAOUDZI COMORES.'

She told me she had intended going on a short holiday that afternoon. Would it be in order? I assured her it would, though later, selfishly, certainly wished I hadn't.

My wife left, so I returned to my restless pacing up and down the bridge, covertly watched by sailors polishing brass. The telephone again, I was there in a bound, but it was for the Chief Officer. Up and down, up and down, a cheery greeting from the Chief as he went, then the telephone again. Sorry, there was no trace of Dr. P. J. du Toit, he could not be found anywhere. I was left feeling quite bewildered. I knew that with all his interests he travelled often and widely, but I had just not expected him to be unavailable then.

Well, something had to be done quickly, for if P. J. wasn't there I'd have to get on myself. In my mind I went over the Cabinet Ministers one by one. Donges and Sauer stood out, they were old Stellenbosch University contemporaries, and were personally approachable on that account, but there was nothing to justify it beyond that, their interest could at best be only general. My choice for several reasons settled on Eric Louw,

Eric Hendrik Louw (1890–1968), was South Africa's first envoy to the United States and first representative at the League of Nations. In 1952 he was Minister of Economic Affairs and, from 1955 to 1964, Minister of Foreign Affairs. He was respected by admirers and critics for his unflinching courage and brilliant wit.

Eric Hendrik Louw

The next in the parade of nationalist politicians was Theophilus Ebenhaezer ('Eben') Dönges (1898–1968) who, in his capacity as Minister of Internal Affairs from 1948 to 1961, was one of the main architects of apartheid. After the assassination of Hendrik Verwoerd in 1966 he was briefly appointed as the Acting Prime Minister. He was elected to succeed CR Swart as State President in 1967 but died before he could take office. He was of no help to JLB.

Eben Dönges

Minister of Economic Affairs. The South African Council for Scientific and Industrial Research had not long before been placed under his direct charge; he knew of my work, and I knew him personally. In my dealings with him I had found him prompt, efficient, able, and of keen perception. Yes, he was the man. From the lines on his face his duodenum probably twisted the same way as mine, and I had a suspicion that he might look on Christmas time with the same jaundiced eye as myself. So began my long vigil on the bridge of the *Dunnottar Castle*. Would the Post Office please find Minister Eric Louw and get him on the telephone as soon as possible? Half an hour later the Post Office telephoned to say that Minister Louw was in the U.S.A. on an official visit, and it would take a long time to make contact with him there. In America he was no use in this matter, anyway, so I had to start all over again. Donges? Sauer? Yes. Donges. Would the Post Office kindly find Minister Donges, and tell him I wished to speak to him on an urgent matter.

After some time the telephone authorities told me that they had tracked Minister Donges, but as he was on a train running from Pretoria to Cape Town, I should have to wait until they could catch him on arrival there, some hours ahead. Another agonising wait, then another message, they had got hold of him on his arrival at Cape Town Station, and he said that he would be able to speak to me in half an hour's time from his home. Half an hour. That half-hour seemed like a year. So I spoke to Eben Donges, Minister of Internal Affairs, one of the most able men of our University days, a formidable antagonist in debate. He turned away from one of the most successful and lucrative legal careers for political life, and soon became a character notable in Parliament, where not even the keenest jibes or thrusts, not even those of Smuts himself, could shake his imperturbable calm, a wonderful asset in his position. I rapidly outlined the situation, which he grasped as quickly, for he knew the main essentials of the story of the first Coelacanth. He said that if he had been in Pretoria and in normal times he might have been able to do something, but where he was, in Cape Town, at such a time, the most difficult in all the year, communications almost impossible, he felt that what he could do then and there would probably not help a great deal when time was such a vital factor. Had I tried to approach the Prime

Minister? So I told him something of why I had not, and he agreed with my view. He suggested that Paul Sauer, Minister of Transport, might be able to do something. As far as he knew, he was still in Pretoria. Why not try him? Best of luck.

Back to the starting-post once more. Holidays! Christmas holidays! Would the Post Office please find me Minister Paul Sauer, believed to be in Pretoria. Again I paced up and down that confined space, my brain going over the matter for the thousandth time. It was a long-drawn-out agony. I could not leave that bridge and had no food and no desire for food, even the tea they had brought me remained untouched. The telephone! The Post Office to say that there was no reply from Minister Sauer's office or from his house in Pretoria. Should they go on trying there? Yes! Do anything, only find him. Up and down, my mind as hard at work as my feet. Sauer! My mind went back to our University days, and right back to the first Sauer, 'Red Blanket' Sauer, noted politician of pre-Union times. Paul Sauer the younger soon made his mark at the University as a facile speaker. Some years before the war I was travelling with a former senior member of our University staff, a versatile man whose knowledge covered a wide field, something of an expert on wines, and with an extensive if not undue practical experience of them. We met Paul Sauer in the dining-saloon and lunched together, and those two were soon off on a technical discussion of wines beyond my ken. I noted with some amusement that my colleague had met his match, in knowledge at least. He remarked later that it was surprising to find a South African who knew so much about wines.

The sun went down, the lights came on, and they brought me more tea on a tray. I must have walked many miles on that bridge. It was quite late when they finally succeeded in tracing Minister Sauer and got him on the telephone. So off I started and outlined the matter again to him, but Sauer was immediately emphatic that he could do nothing, he could not authorise any civil plane for such a purpose, he had none to spare, anyway. Why not try to use the French airlines, he believed that they might serve the Comores, they certainly had regular services to Madagascar, but I could rule his Department out. There might possibly be some hope with the French. Had I tried them? Well, see what you can do there and good luck.

The next unlikely player in the ongoing drama, Paul Sauer (1898–1976), was appointed by Prime Minister DF Malan as Minister of Railways (later Transport) from 1948 to 1954. He offered even less help; JLB was now running out of politicians.

Paul Sauer

Velvet worms (*Peripatus*) are an extraordinary group of creatures, transitional between arthropods (joint-legged animals, like crabs and insects) and worms, and can also lay claim to being 'living fossils'. They are caterpillar-like but have many pairs of legs and no segmentation of the body. They trap their prey (small insects) in a sticky fluid and give birth to live young. Three species of velvet worm (in the genus *Peripatopsis*) occur in South Africa.

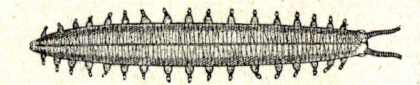

Velvet worm, *Peripatopsis capensis*

Many factors other than size make the coelacanth special. In terms of external anatomy, these include its peculiar lobed fins, unique extra lobe in the tail fin, brilliant blue colour, heavy scales and bony head. On closer examination, the coelacanth's behaviour in the natural environment also proves to be extraordinary.

Back to the start once more, and to telephone the French Consul in Durban just to see. From the Post Office a message, sorry no reply, and too late to do any more. So there I was, baffled, unhappy, in some ways bewildered, my mind not yet settled clearly on any definite course of action.

Lost in a maze of uncertainties I went to the cabin, to the first of a series of restless, virtually sleepless, nights. As I rolled and tossed I could see that fish in the humid heat of the Comores, Hunt trying to turn a responsive nose away from the mounting odour. Was it a Coelacanth, anyway? What a fool I should be troubling Cabinet Ministers if it wasn't. Hunt had been exceptionally interested and persistent, and my wife was quite confident he would know a Coelacanth if he saw one, but, all the same, people far more expert than he had made mistakes of that kind.

'Five-foot Coelacanth.' Size does make a difference, and it certainly was a comfort that this fish was about the same size as the one at East London. Now if Hunt had cabled that he had a Coelacanth 5 inches long, how much less certain it would be, but also how much less trouble, because he could have put it in a bottle and sent it by post, without saying anything to anybody, without my having to think of troubling anyone for a plane. And yet, it was queer how everything has its points, nothing is a hundred per cent. Size is important, very important. An elephant is far more exciting than even a rare *Peripatus*. If you asked the average man if he would like his son to be a hefty 6-footer, with perpetual difficulty in finding clothes to fit him, or a 5-foot skinny runt, you knew the answer. If the original Coelacanth had been only 5 inches long it would not have stirred public imagination a fraction as much. People would have looked at it and said, 'Oh that! Well now, fancy all that fuss about a little bit of a thing like that. H'm!' And they would have gone away with the unspoken thought, 'These scientists!' But a solid great thumping 5-foot fish; now that gives one a feeling of satisfaction. That is at any time something worth looking at, and when it is a living fossil and a phenomenon as well—well!

I woke with a start from an awful confused nightmare about a Coelacanth at the Comores and a dreadful fear that it would probably be putrid before I could get there to save it; and then I realised it was no dream, but real and true.

That long night my tortured and almost fevered brain worked through all the people who might help, and in the end crystallised on the Minister of Defence. Early next morning, Christmas Day, 1952, would the Post Office please try to find the Minister for Defence? And once more I paced the bridge. Late the night before my wife had discovered that the French Consul of Cape Town was on board and she had got from him the private number of the French Consul in Durban. So at a reasonable hour I asked the Post Office to try this and they soon had him on the telephone. I gave him some account of the matter and of the difficulties, but had to confess that at the moment I did not know what could be done, and nor did he; but we parted with the assurance that whatever he could do to assist would be done. It is indeed pleasing to record that every single representative of France in the Union with whom we had any dealings gave our requests the utmost consideration and went to great trouble to assist us throughout.

Christmas day—'Peace on earth and goodwill towards all men.'

Well that might be, but why on earth did Coelacanths want to turn up just before Christmas? Christmas! Well, mine wasn't going to be the conventional Christmas, anyway. I waited for what seemed an endless time, but no message came, so I telephoned the Post Office, but they had nothing to report. For seven agonising incredibly long hours that Christmas Day, 1952, I paced that bridge. The gong went for lunch, but I could not go, no great loss to me because I cannot eat the conventional Christmas meal, anyway; my liver revolts. Even the simple stuff they sent up on a tray remained forgotten, and later I saw a deck-hand polishing off the canned peaches behind a ventilator and tip the rest overboard.

In the late afternoon the Post Office telephoned to say that they had tracked Mr. Erasmus (the Minister of Defence) to his farm at White River, but it was certainly an ideal retreat, for the only available telephone was miles away and worked only in office hours and not on holidays. I dare say this was what anyone outside the inner circle would be told, because it seemed impossible to me that a Minister of Defence could cut himself off like that. What if an emergency arose and the Prime Minister needed to speak to him at once and could not? As will be told that actually happened later, but it was from an 'Act of God' (was it God?), for storms had cut the telephone line.

I wonder whether, in the annals of science, such a series of phone calls has ever been made by a scientist to a parade of senior politicians, all in the pursuit of a single, dead fish? JLB phoned Members of Parliament, senior officials, the President of the CSIR, Ministers of Economic Affairs, Internal Affairs, Transport and Defence, and then even the Prime Minister himself, all during the Christmas holidays!

Only a remarkably determined man who was totally convinced of the importance of science would have dared to follow this course.

In his later years JLB was very selective about what he ate, even under normal, unstressed circumstances. He thrived on fresh, organic foods long before they became fashionable. Furthermore, with his knowledge of organic chemistry, he made his own soap, detergents, washing powder and toothpaste.

According to Keith Hunt, Warden of Smuts Hall when JLB was at Rhodes University, he fed his growing son, William, organic meals that included dandelions in order to keep him healthy. William would then sneak out of the house and visit the Smuts Hall kitchen for a hearty meal!

Anyway, there I was right back at the starting-post again, and now the Post Office very kindly suggested that I might try the Chief of something or other. Yes, thank you, would they get him for me, I didn't know his name or anything about him, and so I went back to my thought-steeped pacing of the narrow bridge, in and out of the wheelhouse.

Soon after our arrival in Durban, before we heard of the Coelacanth, we had arranged an exhibition of the head of the Giant Parrot-fish and other related things on the *Dunnottar Castle*, and notice of this had been featured in the press. My wife, who had taken charge of this, was almost overwhelmed by the crowds who came aboard, so that she could spare only occasional moments to dash to the bridge and see if any hope had dawned, but my face as I paced was enough each time. Eventually the Post Office telephoned to say that they just could not find any trace of the Chief they mentioned, and did not think it was any use going on. After all, it was Christmas Day, and late at that. I thanked them very much and the official suggested that they should try again in the morning, to which I agreed. Christmas! I wonder if I shall ever regain any liking for that time?

Though I had little appetite for food and none for festivity, my wife persuaded me to go down that evening to Christmas dinner in the gaily bedecked saloon, where our steward welcomed me like a lost friend after a long absence. The laden tables seemed to belong to another world. I saw it all through a black cloud of frustration and despair. Soup? Yes, but as the first spoon came to my lips, a petty officer from the bridge leant over my shoulder; there was an urgent telephone call. Who? Don't know, sir. I could not afford not to go, and when our steward found my chair empty, my soup untasted, he stared unbelievingly and said, 'What! Gone again?' It was a quaint meal, if one may call it that, during which I was called away four or five times, mainly the press. There were millions to whom that wonderful range of rich and tasty dishes would have been an unforgettable thrill, but I had no appetite and found no pleasure in that gay scene. There were conventional Christmas crackers, and people laughed and joked and put on paper hats. Great Heaven! and the Coelacanth was probably putrefying. We got out at last, some friends had come, but my mind was so distracted that I cannot recollect anything that was said or done, and quite late I went off to another awful night, more nightmares, more fears, many fears, always worry. So much time had gone and nothing had been accomplished, nothing, just back where I was, at the starting-point.

Boxing Day, 26th December, 1952.

At dawn I went for a quick run along the beach, then back to the bridge and to contact with the Post Office, where I had another wait until they again reported failure to find that Chief of Staff. The Post Office suggested another high official I did not know, and there was another long wait, and again a complete blank, so that I went bang once more right back to the start. Then the Post Office suggested the head of one of the armed services. Yes, please, anyone who might help, and this time they got results quickly, and soon I was having exchanges with a distinctly hostile voice at long range. I introduced myself and started off about the East London Coelacanth. Almost at once he interrupted to ask what on earth a fish had to do with him or with the armed forces of South Africa. I asked him if he would kindly be patient for a moment, I was trying to explain, and went on, but he interrupted sharply with much the same query, so now I answered almost as sharply that if he would listen instead of talking he would learn. It was a conversation that should have been recorded, for at times it almost sizzled. I was compelled to ask him if he really believed that a man in my position would telephone one in his at that time of day and on such a day about any frivolous matter. It was far from that, a matter of national importance. 'What!' he barked. 'A fish! Of national importance.' 'Yes! A fish,' I barked back, so firmly that he listened again. After I had finished my main account, I raised the matter of the plane, and at once he said that there was no plane available that could be sent on such an errand and he could not think of such a thing, anyway. I asked him if I might take that as his final official reply to my approach, when he temporised, so I spoke of the Sunderland flying-boat in Durban and said I had heard that it could be got ready in a short time if its use could be authorised. He asked me if I realised what such a flight to foreign territory meant, it might take a week to organise a thing like that. 'Well, God help South Africa if we should be suddenly attacked!' I could not help flashing back. It was a stimulating conversation, but it became clear that with the best will in

Wars have, of course, been fought over fish stocks before – but never over a single fish! The 'Fish Wars' in the 1960s and 1970s in the United States were a series of civil disobedience protests in which Native American tribes pressured the government to recognize their fishing rights. The 'Cod Wars' were confrontations between the United Kingdom and Iceland regarding fishing rights in the North Atlantic Ocean.

According to aviation expert Willie 'Bomber' Burger, Catalinas had been scrapped by 1952 but Sunderlands could have made the flight, although they were big, four-engined patrol and anti-submarine aircraft requiring a crew of 10–13. The two-engined Dakota was far more economical and had the range to do the job; it also needed a crew of only four (although Smith's flight had a crew of six).

Has there ever been such a strange episode in the annals of bio-aviation as the story of the recovery of the second coelacanth?

Dakota 6832 at the SAAF Museum, Air Force Base Ysterplaat, Cape Town

Today 'bio-aviation' has taken on a new meaning with bio-fuels derived from plants being regarded as the primary means by which the aviation industry can reduce its carbon footprint. The most promising results in the production of bio-fuels have been obtained from the use of algae, camelina oil, vegetable oil and jatropha oil.

the world he really was not in a position to do anything, and he was equally emphatic that even the Commander-in-Chief would not be able to do anything without authority from the Minister, and even he might have to get the approval of the whole Cabinet. I suggested that a flight of this type would provide quite exceptional opportunity for the training of young pilots, but the sounds that came from the receiver in response were not encouraging, and I felt as if I had been beaten flat, face down, into the mud of despair. He concluded by assuring me that I had as much hope as trying to get a plane to the moon. That pulled my head out of the mud a bit, and I told him we should see about that and, after expressing my thanks, returned once more to zero point. I have never met this man, nor do I know his name, and I only hope that subsequent events made him think more kindly of my apparent lunacy.

Our friends were most concerned about our obvious state of tension and distress, and had begged us to take some relaxation despite the urgency of the matter. We had therefore agreed, that Boxing Day, 1952, to go to lunch, to tea, and to dinner at three different homes, it being clearly understood that each appointment might have to be cancelled even at the last moment if any Coelacanth development required it. We dared not risk being away from the ship the whole afternoon, so it was arranged that we should be taken back after lunch and after tea as well.

We were probably deadly guests at that luncheon, but our charming hosts were old friends, considerate and expert at making even corpses feel at home. Even so their tasty food almost nauseated me. There was fish, but I could see only Coelacanth. It was a hot day, and there was a specially delicious cold drink; but I could think only of formalin, and all the time my mind kept on switching over from the Comores to Prime Ministers. It looked as if it would have to be that, after all, and was it going to be the same story over again? Anyway, this was the third day and I was no nearer my goal, still at the starting-point, in fact.

We returned to the ship and had hardly got off the gangway when a distracted young Purser's assistant came shooting out of his office and said, with a distinct flavour of reproof, 'Thank heavens, you've come back, Professor. We've been driven almost crazy by people and the telephone wanting you urgently. Please don't leave the ship again without telling us where you can be called up. The Durban Radio Station asks if you will please telephone them as soon as you can. They say it's very urgent'; and he reeled off a list of others who had sought us, some of whom were still aboard and whom I left to my wife while I went to the telephone.

The Radio Office had received a cable from the Comores addressed to me in Grahamstown, which had been duly sent on. Knowing the situation, however, they had kept a copy, and if I wished they would read it to me as they felt it was of great urgency. I asked for it and they read as follows:

Charter plane immediately authorities trying claim specimen but willing let you have it if in person stop paid fisherman reward to strengthen position stop inspected five kilo formalin no refrigerator stop specimen different yours no front dorsal or tail remnant but definite identification Hunt.

I thanked the Radio Officer for his extremely thoughtful action in telling me of this cable,* as indeed it was, and it is typical of the

---

* Apart from its importance, this cable had an interesting career, which it is worth a diversion to relate. Being Christmas time (Christmas time!), the University offices in Grahamstown were closed and the staffs of all categories were away. An exception was J. Beek, the University electrician, a valued consultant and friend of our Department. He was working at home that day when the telephone called and he found himself speaking to Port Elizabeth. Someone asked him if he would take a radiogram for Professor Smith as they could not track him in Grahamstown. 'H'm,' thought Beek, 'a nice present from somebody for the Professor.' He knew I was far away and did some quick thinking. He replied, 'Yes certainly, send it up and I can keep it in my workshop until he comes.' 'No, no,' said the voice, 'this is a cable, not a machine'; and then asked him to write it down, which he did. He could see at once that it was important, but had no idea where we were. So he set out to find out, and telephoned a member of our staff. No reply. He then had his own taste of Christmas holidays, something like my own, for though he tried the number of every single member of the University staff who had a telephone, it was a complete blank, every one was away. Beek does not give in easily. He had no idea where my Secretary lived; as it happened, it would not have helped him even if he had, for she was away at the sea. He remembered having seen her take part in some play and that she had been associated with a doctor's wife, whom he promptly telephoned, found her in, and told her about the cable. It fortunately happened that she had seen a press report that we were in Durban and knew we were on the *Dunnottar Castle*. So back went Beek, and eventually persuaded the Post Office authorities to relay that cable back to us at Durban. Good for Beek!

Smith's determination to secure the second coelacanth, and his potential rivalry with the French, was a mild confrontation compared to the fabled 'Bone Wars', also known as the 'Great Dinosaur Rush', in the United States in the late 19th century. This scientific conflagration was marked by a heated rivalry between Edward Cope of the Academy of Natural Sciences in Philadelphia and Othniel Marsh of the Peabody Museum of Natural History at Yale University.

ScottRobertAnselmo Wikipedia CC BY-SA 3.0

Tyrannosaurus rex

Both palaeontologists used underhand methods to out-compete the other in the study of dinosaur fossils, even resorting to bribery, blasphemy, theft and the destruction of specimens.

In contrast, after a period of French hegemony in the 1950s and 1960s, international coelacanth research was decidedly collaborative and was characterized by 'coelacanth détente' from the 1970s onwards.

Edward Drinker Cope   *Othniel Charles Marsh*

Madagascar is a large island, and a fabulous evolutionary treasure trove as a result of its long isolation. The meagre corps of colonial scientists in Antananarivo would have had their hands full studying the cornucopia of unique land animals, with little time to focus on the sea, which was bursting with invertebrate and vertebrate life; so it is not surprising that French scientists had not cottoned on to the existence of coelacanths there.

In contrast, the four Comorian islands are relatively small and it is likely that coelacanths were regularly caught off three of them (Grande Comore, Anjouan and, to a lesser extent, Moheli). *Gombessa*, the taboo fish, was well known to Comorians.

thoughtful assistance constantly received from all branches of the public services in South Africa. Many people complain about public servants, but that is not my experience at all.

I read those cabled words again. So that had come into it as well, as if there was not enough already; but it was no very great surprise. My leaflet method had tremendous advantages, but it also had certain drawbacks, one of which was that while so great a sum for a fish, just one fish, would certainly arrest attention everywhere, if one were found by its aid when I was not about, in more enlightened people it might well excite natural human cupidity. If this man was prepared to pay £100 each for these fishes at such a distance, surely they must be worth a lot more? I had certainly never concealed my views about the importance and value of the Coelacanth, the leaflet showed that clearly enough, and if the French had had the remotest belief in my views and in the possibility that Coelacanths might live in their waters, they would have done something about it before if there had been anyone sufficiently interested from the scientific side alone. The very fact that the leaflets taken by Hunt had been new to the Comoran authorities and that they had had no reservations in distributing them was clear indication that the leaflets sent to Madagascar had apparently not really interested the French there, scientifically or otherwise, for if they had been interested in any way, surely my leaflet would have induced them to go into the matter on their own, to hunt and to offer a reward themselves. It could have been done easily enough, and no one could have criticised them for it. Yes, whichever way you looked at it that was my Coelacanth, it had been found as a result of my efforts and ideas, and unless it was just another stray like the one at East London, it did mean, as I had strongly suspected, that it had been under the very noses of the French all along, and they hadn't seen it. I realised full well that one does not value anything until it is desired by others, and I suppose that at the back of my mind there had all along been an unspoken fear that all the fuss and publicity might induce the French authorities to confiscate the fish. In the excitement of the moment they might not be prepared to give full recognition to all that had gone before, to all the years of work that had led to this discovery, to the leaflet that had really tracked this treasure. Hunt must be in a difficult position. He could not afford to antagonise the French, since his trading would almost certainly depend on their good-will.

The French 'authorities' mentioned in Hunt's cable must themselves also be in a difficult position. Willy-nilly, Hunt's excitement and actions and all the circumstances of his relations with me must have forced upon them the importance of the issue. It was true that this fish had been found by no fundamental efforts or ideas of their own, and ethically it belonged to the man whose leaflets they had not only accepted but distributed without any indication that there would be the slightest doubt about his rights of ownership if any fish were found. Now that one had been found, however, it had suddenly been revealed as something of tremendous importance, fantastically more than they had ever supposed. They would naturally be in a quandary as to whether it would be wise to let it go into foreign hands, ethics or no ethics. The cable indicated that they were prepared to concede me certain rights of ownership. That might mean that they really were prepared to concede at least some recognition of my claims, but it might also mean that the fish was in so precarious a stage of preservation, that there might be so much doubt about its being preserved at all, that they were unwilling to take the responsibility for its preservation unless they could say they had been compelled to do so because I would not come. This all left so many uncertainties. If I did not go, nobody could ever question the French authorities taking it from a layman, no matter how interested he might be. I kept on coming back to my uncertainty that they might not know how to preserve it properly? They might not have enough formalin to do it. Like fierce tides all these many complications surged back and forth through my mind, and there emerged from all the obscurity and uncertainty only one clear thought, one certain course—it was like a light that encased my mind and my soul and drove me on. I must go myself, I must go and see that it really was a Coelacanth, and make quite certain that it was properly preserved. Looking back, I realise now that I was in a state of mind that is termed 'Possessed'.

In my half-distracted mind I could see the French standing there over my fish, waiting to seize it unless I came in person, as if they knew that I was beating out my life against an immovable mass of holiday inertia, perhaps of official indifference. This new

In the 1950s there was intense rivalry between the major natural history museums in France, the United Kingdom, Germany and the United States of America. The French would have been less concerned about a coelacanth specimen going to a museum in South Africa than to the British Museum (Natural History) in London.

Later, to initiate the era of coelacanth diplomacy, Jacques Millot and Jean Anthony at the Muséum National d'Histoire Naturelle de Paris donated the 15th coelacanth, a 125 cm female caught off Grande Comore in July 1956, to the British Museum (Natural History), where it is still on display today.

Alexey Fedorenko/Shutterstock.com

British Museum (Natural History) London

The airplane that transported this coelacanth from Grande Comore to Madagascar crashed on landing but the specimen still reached the Scientific Research Institute in Antananarivo intact. Yet another misadventure in the coelacanth saga!

The second specimen, although a recognizable coelacanth, differed markedly from the first in that it had no first dorsal fin and the distinctive middle lobe of the tail fin was missing. No other specimen of *Latimeria chalumnae* caught since then has both these features missing, although some do have parts of their fins missing. It was almost as if this aberrant coelacanth had been sent to test Smith!

The middle lobe of the tail fin in extinct coelacanths varies widely in length, as shown in the attached drawings.

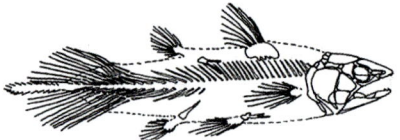

Coelacanths with different length middle lobes on their tail fins

factor had been such a shock that only now did I notice that Hunt's cable read 'inspected five kilos of formalin' (clearly a misprint for 'injected'), about a gallon. That sounded better! But was it concentrated formalin? It had been arranged for Hunt to get a 'bottle' of formalin for that smaller fish, probably about or less than a pint. This cable could mean that he had got about a pint of concentrated formalin which, according to the directions he had been given, would make about five kilos of dilute solution. That was not enough, not nearly enough to preserve a fish as big as that for more than a short time in such heat. If only he had said 'concentrated' formalin; but he hadn't and it was not certain that he realised its significance. So that was still a very big doubt, the fish might even now be going putrid; in any case, because the proper injection of a large fish is not anything a layman can normally do without instruction, especially in places as hot as the Comores. And what in Heaven's name did the statement about the front dorsal or tail remnant mean? All my fears about its not being a Coelacanth at all flared up again in full force. At top speed my mind took me writhing again through all that sea of doubt, though it did help to remember that the fossil record had shown that the small extra tail in Coelacanths gradually became shorter as time passed; in fact, it was apparently absent in some later forms. That in *Latimeria* was about the shortest. Then in the Lung-fishes, which had almost as long a fossil record as the Coelacanths, the first dorsal fin of earlier forms had just disappeared with time. It was therefore within the bounds of possibility that over this past 70 million years such trends had led to a modern type of Coelacanth without extra tail or first dorsal. It was just possible, and beyond that I clung to our faith in Hunt, and tried to feel convinced that even if this was no Coelacanth, it would almost certainly be something just as interesting. If it did prove to be a Coelacanth it must be different from *Latimeria* because that had both first dorsal and extra tail very distinctly. What should I find if I got there? In my exhausted and overwrought condition it seemed indeed as if the dice were being loaded against me with all these uncertainties. Nothing was certain, nothing clear-cut, the identity of the fish, whether there was enough formalin, would I ever get there in time, first before the thing rotted and secondly before the French took it? Through

PREMIO   £ 100   REWARD
RÉCOMPENSE

Examine este peixe com cuidado. Talvez lhe dê sorte. Repare nos dois rabos que possui e nas suas estranhas barbatanas. O único exemplar que a ciência encontrou tinha, de comprimento, 160 centímetros. Mas já houve quem visse outros. Se tiver a sorte de apanhar ou encontrar algum NÃO O CORTE NEM O LIMPE DE QUALQUER MODO — conduza-o imediatamente, inteiro, a um frigorífico ou peça a pessoa competente que dele se ocupe. Solicite, ao mesmo tempo, a essa pessoa, que avise imediatamente, por meio de telgrama, o professor J. L. B. Smith, da Rhodes University, Grahamstown. União Sul-Africana.

Os dois primeiros especimes serão pagos à razão de 10.000$, cada, sendo o pagamento garantido pela Rhodes University e pelo South African Council for Scientific and Industrial Research. Se conseguir obter mais de dois, conserve-os todos, visto terem grande valor, para fins científicos, e as suas canseiras serão bem recompensadas.

COELACANTH

Look carefully at this fish. It may bring you good fortune. Note the peculiar double tail, and the fins. The only one ever saved for science was 5 ft (160 cm.) long. Others have been seen. If you have the good fortune to catch or find one DO NOT CUT OR CLEAN IT ANY WAY but get it whole at once to a cold storage or to some responsible official who can care for it, and ask him to notify Professor J. L. B. Smith of Rhodes University Grahamstown, Union of S. A., immediately by telegraph. For the first 2 specimens £100 (10.000 Esc.) each will be paid, guaranteed by Rhodes University and by the South African Council for Scientific and Industrial Research. If you get more than 2, save them all, as every one is valuable for scientific purposes and you will be well paid.

Veuillez remarquer avec attention ce poisson. Il pourra vous apporter bonne chance, peut être. Regardez les deux queux qu'il possède et ses étranges nageoires. Le seul exemplaire que la science a trouvé avait, de longueur, 160 centimètres. Cependant d'autres ont trouvés quelques exemplaires en plus.

Si jamais vous avez la chance d'en trouver un NE LE DÉCOUPEZ PAS NI NE LE NETTOYEZ D'AUCUNE FAÇON, conduisez-le immediatement, tout entier, a un frigorifique ou glacière en demandat a une personne competante de s'en occuper. Simultanement veuillez prier a cette personne de faire part, telegraphi-quement à Mr. le Professeus J. L. B. Smith, de la Rhodes University. Grahamstown, Union Sud-Africaine.

Le deux premiers exemplaires seront payés à la raison de £ 100 chaque dont le payment est garanti par la Rhodes University et par le South African Council for Scientific and Industrial Research.

Si, jamais il vous est possible d'obtenir plus de deux, nous vous serions très grés de les conserver vu qu'ils sont d'une très grande valeur pour fins scientifiques, et, neanmoins les fatigues pour obtantion seront bien recompensées.

The famous Coelacanth leaflet

Plate 3

JLB's reward leaflet was printed in Portuguese, English and French, the main European languages used in East Africa at the time, but not in Swahili, Kikuyu or any other African language, nor in Arabic, although these latter languages were widely spoken by the indigenous people of the region.

The text in the leaflet is simply written and would have been intelligible to most readers, but the contact details given for Smith seem to be rather thin ('Rhodes University Grahamstown, Union of S.A.'), especially to us in the Internet age.

JLB Smith

Eric Hunt (right) with a coelacanth reward poster

JLB was the eternal optimist. He expected adequate cold storage facilities to be available for a 1.35 m-long fish in tropical African countries in the 1950s! Marjorie Courtenay-Latimer couldn't even find such facilities for the first coelacanth in the city of East London (although that was 14 years earlier).

The various Comorian islands were under some form of French control from 1886 until three of them (Grande Comore, Anjouan and Moheli) declared their independence in 1975; Mayotte, by choice, remains under French administration.

Flag of the Federal Islamic Republic of the Comoros

Sadly, the Comoros holds the world record for attempts to overthrow the government in recent times. Since independence there have been over 20 coups or attempted coups.

Monsieur P. Coudert, Governor of the Comores, near the wharf at Dzaoudzi, 29th December 1952.

Plate 4

"Hunt's trim vessel", at the wharf, Dzaoudzi, showing the Coelacanth box on the left. Only two weeks later this craft was destroyed by a cyclone.

all this there was only one coherent thought in my mind, to which I clung like a drowning man to a life-buoy: I must go myself and see.

Half distracted by this latest complication, we had meanwhile been coping with visitors and the press. The news by now having got to world prominence, you could almost feel the rival reporters growling at one another. Our friends were standing by, and we got away to spend a short time for tea, this latest development naturally almost the sole topic of conversation. Our friends made various wild suggestions for a drastic solution of this problem. They were attractive and even amusing, but while they relieved the tension for a few seconds they were patently impractical. I was nearing a stage of desperation, and it was at that time that I came to veer to my wife's view. It would clearly have to be Malan. In some ways that cable settled it, and troublesome as it was, it was in some ways in my favour in any appeal for help; for it was not only a race against possible putrefaction, but there was clearly a case for my having to go in person. It did help to feel that the French would hardly be prepared to take a step as drastic as confiscating the specimen unless they were almost certainly convinced of its value; but there was clearly no scientist in the Comores, and I could not be certain if the French did seize it that it would be safely preserved for science. Nothing is ever a hundred per cent., and in some ways this cable, worrying as it was, forced the issue.

We got back to the ship again, and the same young officer handed me some telegrams without comment. Among them was the cable that Beek had handled, that the Radio Station had so kindly short-circuited some hours earlier. It was a renewed shock to read the words again. Knowing Grahamstown in holiday time we wondered how it had come back to us so quickly. (See footnote, p. 109.) Nothing of further import had happened, but more visitors and press representatives were waiting. I got away to the cabin for a while and sat, desperately miserable, for time was slipping by and I had still accomplished nothing, just nothing, and here was what seemed an awful blow. I was an experienced and mature man, and yet in this I was failing, not having yet accomplished anything worth-while in this all-important matter. I had beaten my head in vain against this Christmas holiday wall,

o.f.—8

In fact, when JLB did eventually arrive in Mayotte to collect the second coelacanth, no other scientists were present.

This bout of self-flagellation by JLB was hardly merited as he had gone to extraordinary lengths to find a way to fetch the second coelacanth. Not even he could circumvent the total disruption to professional life that a summer holiday causes in South Africa!

The persuasive influence and common-sense approach of Margaret, always urging JLB to do the right thing, is obvious. She was, indeed, the power behind the throne, the rational mind when the sometimes flaky scientist became frenetic, or the strong-willed influence when the man who often questioned his own self-worth needed a shoulder to lean on.

NRF-SAIAB

Margaret and JLB Smith

After JLB's death Margaret became an assertive, sometimes feisty, but always kind leader who mastered the art of team building and created a big, happy family in the Ichthyology Institute in Grahamstown.

a concrete, no! an iron wall. Portuguese friends who had visited the Union had said to us, 'You may talk of Russia and the Iron Curtain, but it is nothing to South Africa on a Sunday or a holiday. That is an Iron Curtain. It shuts down, boom, boom, like that, and everything is dead.' They were right. It had beaten me so far. There was only one hope left—Malan! Malan! 'You will have to go to him in the end,' my wife had said. It certainly looked like it, though my mind still obstinately rebelled against the idea. Prime Ministers and fish just did not mix in my mind.

They called me. Our friends were waiting, so I pulled myself up to go. It was almost as if my body was in bits like my mind, and as in a dream I went up to the lounge where they stood waiting. Malan, yes it would have to be Malan, and when we all sat round before dinner I had nothing else in my mind, and I told them it seemed now that this was all that remained. I had been forced to the very last ditch, with little hope. It was now a question of how to proceed, and someone suggested that Malan might even be at the official residence on the Natal south coast. This would alter the situation very materially for, after all, that was within personal reach, by car; but my first flash of hope soon faded as it was most unlikely, for that was the winter resort, in summer it would probably be the Cape. Whom could we ask? I thought of Desmond Prior, an old acquaintance, sub-editor of the *Daily News* and deeply interested in all this, and asked if they would get him on the telephone for me. Wonder of wonders, I was soon talking to him. Malan? He didn't know for certain, but would soon find out. My number? Right, he would ring back in a few minutes; which he did, to say that the Prime Minister was not in Natal, he was somewhere at the Cape, nobody seemed to know exactly where. It was as I had feared, the Cape, a darned long way off. Prior went on and said that if he might suggest it, would I make contact with Dr. Vernon Shearer, M.P., now in Durban and not far from the Evans's house, where we were. He not only knew all about this affair but was anxious to assist if he could, and in any approach to the Government his position would certainly help. Shearer would be at home that night, and it would be possible to speak to him at once if I wished. So I thanked Prior and told the others what he had said. Frank Evans was enthusiastic, Shearer was a good man, and with my consent he went to the telephone and in a few moments called me to speak to Shearer, whose obvious deep concern and interest were comforting, and though I was still without any real hope that this might rip the curtain of despair that surrounded me, it was arranged to go to his house after dinner.

Vernon Shearer was an influential Member of Parliament from the opposition United South African National Party (known as the United Party), which drew support from different sectors of South African society, including English-speakers, Afrikaners and people of mixed race. Shearer was Mayor of Durban from 1964 to 1966.

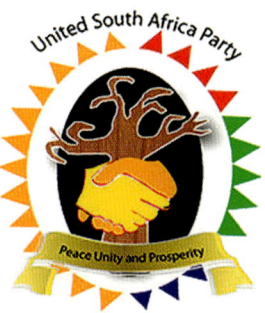

Daniel François ('DF') Malan (1874–1959) was Prime Minister of South Africa from 1948 to 1954. He was a champion of Afrikaner nationalism and his National Party came into power by defeating Smuts in 1948 on the platform of the apartheid policy. The foundations of this racist regime were firmly laid during Malan's term as Prime Minister. Malan's name, which JLB had hoped to honour in his description of the second coelacanth, grew in infamy in later years.

NRF-SAIAB

JLB Smith presents a copy of the book Old Fourlegs *to retired Prime Minister, DF Malan, in 1956.*

Chapter Eleven

I MUST SPEAK TO HIM

MALAN; Malan! Was it to be the same thing as before? This was infinitely worse, for it was Boxing Day, when he would not be in harness, not in his office, but resting, and as this could happen very seldom he was sure to be surrounded by extra concrete and steel walls to guard his privacy. Not only was it pretty certain that those would take some penetrating, but to justify the tremendous request I had to make I could not even say 'This *is* a Coelacanth', only 'I believe it is very likely to be a Coelacanth'.

Dinner was another phantom meal where attractively served and tasty food passed my unresponsive palate and tumbled into a contorted stomach, which pushed it on through a writhing duodenal tunnel. After dinner Frank Evans and I went off and were welcomed at the Shearer home at 8.15 p.m. Shearer is a dentist by profession as well as a Member of Parliament, and is an able and forceful personality as well as a realist.

George Symons, *Daily News* photographer, was there, and we were soon deep in discussion. Shearer knew the main outlines of events, and I gave him a brief account of my attempts to get air transport. I said that I had all along been opposed to approaching the Prime Minister, mainly because he was resting, but that now it seemed the only way. It added to my depression to find Shearer dubious of the wisdom of approaching Malan before it was quite certain that all else had been absolutely exhausted. He said that it would at least be very difficult, for at a time like this there would be a series of extra barriers around the Prime Minister's privacy that only the highest in the Government party would be able to pass, and as an Opposition member, well. . . . It might be best to have one more try for Erasmus, Minister of Defence. So off he went and had a long session on the telephone while we sat dumbly round a table.

My mind was never off that fish in far-away Comores. Time

was passing, so much had gone, and I was yet no nearer my aim except that by then I knew without any shadow of doubt that Malan was the only hope, all else would be a waste of time. I must speak to him. The telephoning now ended in the same way as it had done all along, for the Post Office told Shearer that the line to the station nearest the farm of Erasmus was closed until the morrow, so he booked a call for the earliest possible hour, 9 a.m. next day. So there I was, right back at the starting-point, and we went over it all again, only now I no longer had any doubts or reservations, my mind had veered right over, and I knew that we must stake all and go for Malan. Only Malan could help, and after some discussion I said so, firmly; it was too late to wait for others, I was quite convinced. I must speak to him. Shearer again said it would be difficult; but he was more than willing to do whatever was possible, and went off again to the telephone, which was in an adjoining room. We had decided that it would be best first to try Groote Schuur, the Prime Minister's official residence at the Cape.

Mrs. Shearer had come in, and we sat round the table, silent, concentrating on the conversation at the telephone. 'Dr. Vernon Shearer. Groote Schuur, Cape. Yes, the Prime Minister, a matter of national importance. I must speak to him tonight. Yes, thank you.' And he was back, the Post Office was on the job. I sat steeped in suspense and apprehension, without much hope, for which there was good reason. Prime Minister. H'm! There was Smuts, whom everyone spoke of and regarded as having a real interest in science, he was even almost a scientist himself some people said, and look how he had reacted in my case, a simple one at that? He had not only turned it down flat, he had even re-fused to see me at all, quite probably a following-on of his attitude towards the Coelacanth expedition. I had been told that he regarded the views of overseas scientists as sounder than my own. And now this other Prime Minister, Malan, what would he do in almost similar circumstances? It made me turn cold to realise how very similar they were, for again it was a race against time for fish, and yet this time everything involved was from my side very much worse and very much more difficult. From Smuts I had wanted a plane on what one may term a domestic issue, all within our own borders; but what I wanted from Malan was so much more, not

By any measure, asking a Prime Minister for a military airplane to fetch a dead fish in a distant, foreign country is a very long shot. But somehow, after days of frustration, the pieces of this unlikely puzzle started to fall into place.

Two of the hallmarks of Afrikaner society are respect for elders and deference to experts. In contrast to the arrogant attitude of Smuts, who apparently had greater respect for the opinions of foreign scientists, Malan let JLB have his say, and then followed his instincts. Malan is described by his biographer, Lindie Koorts, as a stubborn man who, once he had made up his mind, rarely changed it, no matter what the odds.

In a biographical article on JLB Smith published in 1996, Peter Jackson states that, 'Smith was a nationalist, with little concern for other countries … he found himself resentful of criticism of South Africa. I suspect his antipathy to Field-Marshal Smuts … had almost as much to do with Smuts' internationalism and pro-Britishness as with his refusal to supply him … with a military aircraft to take him to Walvis Bay. … In discussions on taxonomy he was hostile to the idea of my research staff and myself lodging type specimens of new species overseas, particularly the British Museum (Natural History). "These are African fish and you should keep them in Africa", he told me.'

only very much farther to go, but mostly over and into foreign territory, with international complications as well, which alone might turn the scale against me. Whereas it had been in my favour that Smuts had had the flavour of science, would it not be exactly the opposite with Malan, who was far away from science, very far, a reputedly dour and at one time almost sinister figure in his deeply religious stern Calvinistic righteousness? He was known to be utterly indifferent to the criticism or disapproval of those who opposed him, and once he had weighed a situation and come to a decision he did what he felt was right, not what others wanted. During the war, at a time when the Afrikanerdom he lived for was battling for its very existence and its adherents were rent and torn by schisms, Malan was not prepared to compromise or appease. Indeed, at that difficult time he set out to crush a powerful group in his own camp, a policy that appalled most of his associates of that time; but he succeeded, and later, when the fruits of his policy and victory became apparent, they won him enormous respect for his courage and wisdom. This was indeed one of the main causes underlying Malan's final spectacular and crushing victory over his life-long opponent Smuts. Paradoxically enough he accomplished this in the British way because he won really only the last victory in that long struggle. It was something like the tortoise and the hare over again. Indeed, I reflected, mankind may be divided into the tortoises and the hares, I had myself been like the tortoise this past fourteen years, plodding along steadily, without encouragement, looking for my Coelacanth, and now at the end I was aiming to finish off with a spurt no hare could ever equal, not even a Pegasus. I wondered how Malan . . . the telephone bell crashed through my thoughts and went on ringing furiously. Shearer got up and went out hastily, and there followed a long conversation of which we heard odd fragments.

'Yes, Vernon Shearer, M.P. Member of Parliament. . . . Yes, yes, tonight. Thank you.' Another fruitless effort, for Shearer said that the Post Office had told him that the Prime Minister was not at Groote Schuur, but at his private home at the Strand, where he was resting, and there were strict orders that he was not to be disturbed. Nevertheless, Shearer had stressed the importance and urgency of the matter and we could only hope for the best. He began to speak of Malan in a manner that interested me greatly, since I had for many years not heard anything personal about him, only reports and comments in the English press. The 'Dokter' had mellowed with age, and his political antagonisms had been tempered by achievement and success. My mind drifted away and back many years to the time when this man Malan, the 'Renegade Parson' as many of the more English section stigmatised him, had suddenly emerged as a force in the expression and forging of South African nationalism. Minister of Education in Hertzog's Cabinet, represented in the press as anti-British, a solemn, earnest, and troublesome opponent, he was plainly of such unswerving rectitude that no one could ever find even one foot of clay.

Rhodes University College in Grahamstown was in need of funds and had an Art School to open. A Nationalist Cabinet Minister for such a function would in Grahamstown indeed be a phenomenon, but the invitation was sent and promptly accepted. There were comments and speculations. Would he drown everyone in a flood of Afrikaans? In the morning, formal but affable and correct, he inspected what he was shown and made proper response. Grahamstown turned out in force that day, town as well as gown. I shall never forget the course of the ceremony, the rustle as Malan stood to speak, the relief that it was to be in English, anticipation gradually changing to astonishment that this vaunted anti-British politician should be able to roll out such polished phrases. He stood solid, quiet, and firm, and for well over an hour, without any notes, this amazing man held his audience submerged and quiet beneath an eloquent stream of English, perfect in sense and construction, so smooth that it was only from intonation and accent that one could know that this was not his mother tongue. It was by all standards an impressive achievement. Young as I was then and little as politics meant to me, it was clear that this man could not fail to become a force in South Africa, and that while he might be hated for his strength no one could fail to respect him. He radiated honesty, courage, and determination, and even if these were employed as weapons in a field not to everyone's liking. . . . 'Yes, they call him the Sphinx.' Shearer was speaking again, 'and'—his words were cut off abruptly as the telephone went again, and he was off in a flash. 'Yes, yes, Shearer, yes Shearer.' Silence—then in Afrikaans,

DF Malan obtained a Doctorate in Divinity from the University of Utrecht in 1905 and was an ordained minister of the Dutch Reformed Church. In 1915 he resigned from the church to accept the position of editor of *Die Burger* newspaper, the mouthpiece of the Nationalist Party. He was an ardent advocate for the acceptance of Afrikaans, which was, at the time, an emerging 'African' language battling against Dutch and English.

DF Malan as a young Minister of the Church in about 1914

Margaret loved a cup of tea and, in later years, would often invite Carolynn and me to her home at 37 Oatlands Road, Grahamstown, for a cuppa. She would regale us with stories about her interactions with friend and foe, interspersed with risqué jokes told by her flamboyant elder sister, Flora. Margaret enjoyed reconstructing past conversations about her relationships with people, with the occasional embellishment; and she was committed to developing and upholding the legacy of JLB Smith. One of her favourite expressions, when anyone tried crossing her path, was, 'Over my dead body!'

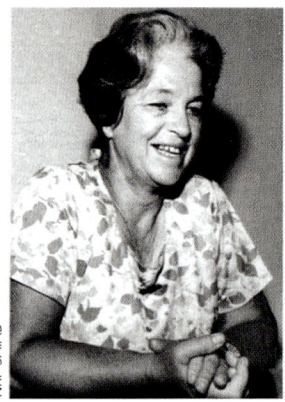

NRF-SAIAB

Margaret Mary Smith in 1972

An important role that Margaret played during this phase of JLB's career was to create situations in which he could relax a little, take a break from his strenuous work schedule, and refresh his mind. It was not an easy task as JLB was driven by a fierce compulsion that few could comprehend.

'Yes, Madam, Vernon Shearer. How are you, Mrs. Malan, is the Doctor well? Yes, I have with me Professor Smith'; and then a brief outline of the main facts. My heart began to pound so madly it almost choked me. Then there was silence, he was clearly listening, it lasted some time, an agonisingly long time, my body was shaking as in a fever. Then Shearer's voice, subdued now, 'You don't think it could be done now?' Another silence. 'Yes, thank you very much indeed, very much obliged. Good night.' Shearer came back slowly, his face despondent. I had no need of his words, my heart fell from my boots through the floor. 'Mrs. Malan,' said Shearer. 'I spoke to her. The Doctor was already abed and she was emphatic that nothing was to disturb him, for he was unusually weary and they were concerned about him. She would promise nothing but would see how he was in the morning, and if she judged it wise would tell him about it. So that's that.'*

10.30 p.m. of the 26th December of the year of our Lord 1952. It was probably the lowest ebb of my life. The sands of time were running out, fate was screwing me down to the dregs, wringing out the last drops of my spirit from the rags of my being. The *Dunnottar Castle* was due to sail at 11 a.m. tomorrow, and what on earth was I to do, for now there seemed no more hope? Even if Malan was told in the morning and took some interest in it, the Strand was an awful long way off, diagonally across the Union, about as far from Durban as Venice from London, and it was Christmas holiday time when public services are difficult or dead. I could hardly hope to have things settled before the ship sailed, and to retain any last shred of hope I should have to pack and prepare to leave the ship and stay on in Durban. Prime Ministers! Well, detachedly, I could sympathise, for I knew what their lives were like. In another sphere I endure something like it myself, but this was hard to take. I sat almost turned to stone with black despair, and was curiously unable to get up and go as I felt I should. I seemed to be paralysed, when Mrs. Shearer jumped up and said brightly, 'No need to be gloomy. One never knows what the morning will bring. You're not to go yet, Professor, I'm going to make you a nice cup of tea.' And off she went.

* Recently (August 1955) when I spoke again of these events to Dr. and Mrs. Malan at their home, he said to me, 'You actually owe a lot to Shearer. But for his help, as things were, it would have been very difficult to get through.'

I once met a cultured foreign woman of great ability whom I admired (without further designs; she was already married and so was I). Once in discussing the British, who annoyed her, she said that when a crisis arose, instead of tackling it in a proper business-like way at once, somebody would be sure to say, 'Well now, let's first have a nice cup of tea', and you would just have to wait until they had all had it before getting down to business. I had often thought of that since, and those words always flash it up on my mind like a screen. Mrs. Shearer bustled back soon afterwards with a tray full of good things, and put tea and cake before me; but misery destroys flavour and I ate and drank what could have been sawdust and muddy water for all it meant to my palate.

Even when everyone had finished and it was already more than time to go, I had a strange reluctance to get up. Although it really seemed that everything was finished, my mind fiercely resisted such a conclusion, for it meant that I was virtually giving up, even though it was beyond my powers to do more. Though I had actually put my hands on the table, I was hesitating even while trying to bring myself to the effort of rising to my feet. They were all looking at me, obviously waiting for me to move, all this in a strained strange silence, when this was suddenly and dramatically shattered by the loud ringing of the telephone, which in that tense atmosphere was like an explosion. It startled us all so much that for a few seconds nobody moved. Then Shearer and I jumped to our feet—he went off, we heard him say a few words, then silence, then a shout: 'Quick Professor, Dr. Malan! Dr. Malan wants to speak to you!'

I found myself at the telephone with the receiver in my hands. 'Dr. Malan himself,' whispered Shearer. I croaked into the receiver, but it was a woman's voice that spoke in Afrikaans: 'Mrs. Malan here, Professor; the Doctor wants to speak to you.' A click, a fumble, then that slow voice easily recognisable over a gap of more than twenty years, in English too. 'Good evening, Professor, I have heard something of your story, but will you please give me as full a summary as possible.' So I began, in Afrikaans, saying that I hoped he would not mind if I stumbled occasionally in technical terms. 'Speak English, man,' he interjected. I said no, I wished to speak in Afrikaans, and taking a

The British certainly do enjoy a cup of tea during moments of stress. John Major's first act after his sudden resignation as British Prime Minister in 1997 was to enjoy a cuppa with his staff and then repair to Lords to watch a game of cricket!

JLB was fluent in Afrikaans but probably struggled to find suitable words to describe the ancient fish, chronicle the saga of its discovery, and expound on its scientific significance. His insistence on speaking Afrikaans may, however, have played a sizeable role in swaying the opinion of the indomitable Prime Minister.

Dr Malan was a deeply religious man, a committed Calvinist, yet he had no qualms about sanctioning such an unusual military mission to bring back a specimen that was critical to the understanding of the concept of evolution. He saw in it no threat to his faith in divine creation.

At about this time (September 1955) Reverend Koot Vorster, a future moderator of the Dutch Reformed Church, wrote in *The Star* newspaper that 'the theory of evolution is in direct conflict with the teaching of the Bible and should not be included in school textbooks'.

Anti-Darwin cartoon

The phone call between Smith and Malan is one of many examples of the extraordinary ability of the coelacanth story to impel people to make inspired decisions. Others who responded to the inspiration – apart from JLB and Margaret themselves – included Marjorie Courtenay-Latimer, Eric Hunt, Hans Fricke, Mark Erdmann, South Africa's then Minister of Science & Technology, Dr Ben Ngubane, who gave the nod for the African Coelacanth Ecosystem Programme (ACEP), and the ACEP leaders – Tony Ribbink, Tommy Bornman and Angus Paterson.

deep breath, plunged straight in. I gave him a brief account of the history of Coelacanths, of the fantastic discovery of the East London fish, the tragedy of the soft parts, my long search, the recent discovery, the heat, the isolation, my fears, and my needs. I emphasised that I could not be certain that this was a Coelacanth, there was a risk for I had only Hunt's word for it, but of all the laymen I knew Hunt was the most likely to know a Coelacanth if he saw one, and I was more than satisfied that it might be the greatest scientific tragedy not to make certain of it. It was a risk and he must clearly realise it. It was a risk for me to take as well, but there was no question in my mind that I at least must take it and was prepared to ask his help to that end. I said that it was not my name that was at stake; in my view with all that had gone before this was a matter of national prestige. While it was of world-wide importance, it was South Africa's responsibility to make sure of that fish, and that was why I was appealing to him for help. I had taken all the cables with me 'just in case', and read him the last one from Hunt, slowly.

He listened all the time and did not interrupt or ask any questions, he was obviously soaking it all in. He was so quiet that at intervals, fearing the line had been cut, I would ask 'Are you there?' and he would say 'Yes, go on'. When I had finished I saw to my amazement that I had been talking a full twelve minutes. There was a short silence, and his first words were, 'I must congratulate you on your Afrikaans. It is excellent.' That was high praise indeed, but what about the Coelacanth? Was this just an unexpected Prime Minister's sugar-coating? I waited in agony, maybe there would be questions, but none came; and after a short silence he said speaking slowly, 'Your story is remarkable, and I can see at once that this is a matter of great importance. It is too late to try to do anything tonight, but first thing in the morning I shall get through to my Minister of Defence to ask him to allocate a suitable aeroplane to take you where you need to go. Where can you be reached by telephone?' I did some quick thinking and said I was on the *Dunnottar Castle* which was due to sail in the morning, but would come to Dr. Shearer's house from 9 a.m. (I looked at Shearer who nodded assent) and wait there for any message. Would that do? Yes, it would. Then I thanked him and we said goodnight.

As I put down the receiver I felt dazed, like a man reprieved on the very scaffold, like somebody suddenly jerked from the hollows of hell to a high hill-top in heaven. A Prime Minister, and one whose educational culture was very far from, some said fundamentally antagonistic to, biological science, and he had roused himself, on a brief holiday, late at night, on his own initiative, to speak to a scientist named Smith from a notoriously English centre like Grahamstown. He had listened patiently, and though not a scientist had been able to grasp, weigh, and assess the importance of the matter straight away. He had not needed to ask any questions, and had come to an immediate decision. Had my name been Van der Westhuizen, and had I come from, say, Potchefstroom, there would have been those who could sneer, but I knew, politics or no politics, that from then on nobody would speak ill of this man with impunity in my presence. My dazed mind flashed back to Smuts, and then I found myself in the dining-room again, where I was unable to utter a word in response to the silent question that was plain on the faces which for some seconds were all I could see. 'My word, you can speak Afrikaans,' someone said; then, as I slowly regained my equilibrium and full comprehension of what had just happened, they questioned me minutely about what Malan had said, which produced such an air of relief all round it hardly seemed like the same place. Even the faces had changed.

I arranged to return at 8.30 next morning, and Mrs. Shearer told me a room would be at my disposal to work while I waited. We went back to Frank Evans's house, and when I saw my wife's anxious questioning gaze, something choked me and I could not speak, but nodded. Frank Evans made good my verbal deficiency, and soon gave them an animated account of the events of the evening.

It was only much later that I heard the outlines of the story of what had happened in the Malan household that night. When Shearer's call was put through, Dr. Malan was abed and as far as they knew asleep. Mrs. Malan answered. She listened to Shearer, but judged the matter should wait at least till the morning; but as she put down the receiver Dr. Malan called out to ask what it was and would not be put off. She gave him a brief outline of the matter. Dr. Malan nodded his head. 'This man Smith is well known. Bring me that fish book.'

Although JLB was described by Peter Jackson as a strong South African 'nationalist', he was not a political animal and judged men (and women) on merit based on how efficiently they performed the tasks allotted to them, irrespective of their ideologies or political persuasions.

Smith was delighted by Malan's response, even though he might have been 'antagonistic to … biological science', but he had no respect for Smuts, the scientist-politician who had scorned him.

Early the next morning, on 27th December 1952, Dr Malan instructed the Chief of the Defence Force, Lt Gen CL de Wet du Toit, to make an aircraft available to fly Smith to the Comoros to collect the coelacanth.

Many years later there was an interesting sequel to this dramatic saga. After the first edition of the revised *Smiths' Sea Fishes* book had been published in 1986, Margaret Smith and I formally presented a copy to the then Minister of National Education (later State President of South Africa), the honourable FW de Klerk, at the Union Buildings in Pretoria.

FW de Klerk

He thanked us for the kind gift and laughingly said, 'I will keep it next to my bed in case I receive an urgent call from you'. He clearly knew about the incident with Malan 34 years earlier. Such is the power of the coelacanth story!

Some time before, since the South African Council for Scientific and Industrial Research, from which most of the funds for my work came, fell directly under the Prime Minister, when Dr. Malan came into power I sent him a copy of this volume, in whose production the C.S.I.R. had played so great a part. (Only three days after it had gone the President of the C.S.I.R. suggested that I should do this very thing.)

Just before Christmas, when they had been preparing to leave for their brief rest at the Strand, Mrs. Malan had assembled a few books, and quite by chance my *Sea Fishes of Southern Africa* caught her eye as eminently suitable for the seaside, despite its size.

So now they fetched him the book and found the Coelacanth pages in it. He read it over slowly and paged through part of the volume. Then he shut it. He called her again, and tapping the book said, 'The man that wrote this book would not ask my help at a time like this unless it was desperately important. I must speak to him.'

Mildly they tried to dissuade him, to wait for the morning at least, but he shook his head and repeated, 'I must speak to him now.' And so the telephones were set going. Thank God for that 'nice cup of tea !' It has helped to win many victories.

Chapter Twelve

DAKOTA DASH

WE went back from the Evans's house to the ship, and I had to try to calm my raging mind to coherent thought. What lay ahead would require the most careful planning, every step, there must be no false moves. This creature that Hunt had was in unknown foreign territory so that I could only guess at what I should find when I got there, and despite the promised plane, the problem still was how to reach the actual spot where the fish was, and quickly. I might have to make a voyage by sea from the nearest landing-ground. To cover that possibility I had to work out all my needs—clothing, food, money, preservatives, medicines, there must be no slip-up with any item. I gave my wife a tablet to make her sleep, but I had much to do and did not get to bed at all. I was hard at work right through that night, and when my wife turned over in the dawn to look hazily across at me, I was pretty tired, but could smile, for my plans were complete.

She was soon up, for we had much to do and discuss. Our baggage and equipment consisted of more than seventy packages, dispersed between the cabin, the hold, the baggage room, the cold storage, on the top deck next to the funnel, and in the magazine. Each bore a number and its contents were listed in a book. Some of the things I needed for the journey ahead had to be dug out of packages lodged in different parts of the ship, and my wife had to be coached so as to be able to take charge of all this diverse material; for in all our expeditions this had always been my own special responsibility, and it was no light task at short notice.

At 7 a.m. I sent a message to Captain Smythe to ask the latest time of departure of the ship and to say I should be leaving her. In a few minutes he was at our cabin eager for news, when I briefly outlined the latest developments. At once he asked if he could help in any way. I asked for the telephone to be kept aboard,

As a result of JLB's relatively frail health, Margaret handled many of the more arduous tasks on expeditions, such as rowing the boat, but the overall organization of trips was always JLB's responsibility.

JLB had a reputation for planning expeditions with military precision, carefully labelling every box or bag in his crates. The flight to the Comoros and back was no exception. Despite the almost overwhelming logistical odds, the journey went surprisingly smoothly, thanks also to the highly efficient crew of the Dakota.

Peter Barnett

Margaret on the 1951 expedition to Mozambique

JLB's Rolleiflex 'Old Standard' twin-lens reflex camera, pith helmet, collecting box and other expedition equipment are still in the possession of the Ichthyology Institute (now SAIAB). Robert Capa used the same model 'Old Standard' to document the Second World War.

Mike Bruton

JLB's faithful Rolleiflex 'Old Standard' twin-lens reflex camera

The use of cigarettes as gifts to curry favour with locals and overcome logistical obstacles is, unfortunately, still standard practice on expeditions in Africa. On our later fish-collecting expeditions we took sweets, small mirrors and a Polaroid camera so that we could give our benefactors a photograph of themselves. 'Selfies' only became popular many years later.

and a gangway open, both until the last possible moment, and for an extra hand to stand by to help my wife when necessary. She would remain in her cabin so as to be immediately available for the telephone. One or two items were needed from packages in the hold and baggage room—could they come up soon? I gave him a short list of food required: cheese, dried figs, biscuits, and Brazil nuts, so much of each. He glanced at it and said it would be prepared at once. When agitated or in action, Smythe has a way of flicking his fingers, which is symbolical of the speed with which he can act in any emergency. All I wanted was done with dispatch, and on his own he had a special breakfast prepared for me which I ate in his cabin soon after. Ours looked like a Customs baggage examination warehouse.

After he left, my wife and I went through the detailed list of my needs: one light suit and a nylon shirt, that I was wearing then. Two khaki shirts, two shorts, compact shaving kit, pyjama shorts, a towel, soap, folding primus, a small aluminium pot, an aluminium water-bottle, six boxes of matches, each in a waterproof bag, a torch, light plastic overcoat, nylon head-net, all in a small waterproof case. Camera (my wonderful Rolleiflex), exposure meter and spare films, Nescafé, sugar, slabs of chocolate, plastic cups, my special silver spoon, two knives, and though I do not smoke, several hundred cigarettes, of course, for without those I never travel in wild parts. They are a wonderful open sesame to primitive hearts. Then last but not least my trusty collecting-box. This is quite a famous box, my inseparable companion on tens of thousands of miles of tropical journeys, by land, sea, and air. It is a neat teak box 18 × 12 × 15 inches high, and it has many compartments and trays with divisions. Its contents are the result of years of experience, and cover a wide range, from almost a full medical outfit to tools, spare parts for pressure stoves and lamps, fishing tackle; there are hundreds of items. Some of them have interesting backgrounds. A tube of rubber solution! We once went on a trip from a tiny isolated coastal settlement in northern Mozambique to the Lurio mouth, about fifty miles of the loneliest track in Africa and through an area alive with wild animals. Lurio lions are notorious. The country there just swarms with them. You can follow the road all right, especially through swamps, for it is made of logs; but you can pass only in the dry season, though there seem to be just as many mosquitoes then as at any other time. If you had a breakdown on that road, your chances of surviving a night in the open were remote, but if you did you would certainly emerge with malaria and worse. We had a Government lorry with a native driver, whom I asked if he had all the necessary tools and spares. He said yes. We set off and how that vehicle survived the jumps and bumps is a miracle; but we got to the Lurio, where there is a small house with netted windows, which the Portuguese always have. Soon after arrival the driver came to report a puncture, and I told him to get it mended. He hesitated, and when I asked sharply what it was, reported that the solution was all dried up in the tube. We tried to dissolve it in petrol, but without avail, it was too hot. Then he tried frying a repair patch on some coals, but that also proved useless. What an idiot I had been to trust a native I had not trained myself. We still had four good tubes, but no spare and no repair outfit, and we had to go back over that awful road. We managed it, but that is why I always carry the rubber solution, and some patching as well.

Spare pump washers ! We once got to a lonely lighthouse with only natives in charge. I asked for lamps and they brought three mantle lanterns, which I told them to light up. Sorry, Senhor, but the pumps won't work; and they showed me the washers stripped to shreds. No spares in your stores ? No, Senhor, nothing. I told them to go and have another search, and when they had left I had a good look, but the lamps really were useless. I had an idea and took out the plunger of our pressure stove, and behold it fitted perfectly. I soon had a lamp going and put the plunger back into the primus, and when the natives came back to report no luck, as it was still light they did not at first notice the lamp. Then one saw it and his eyes bulged; he pointed in amazement and the others stared as well. How had the Senhor done it ? 'I blew it up,' I said, and bulged out my cheeks, and their eyes bulged still more. What a man this must be ! Would I blow up the others, too ? But I said one was enough at a time as it was hard work. We took that secret away with us, but probably left a legend behind. Nowadays I find it less trouble to carry spare washers.

Yes, my 'collecting-box'. I was repeatedly called to the tele-

Playing tricks on gullible people is still standard fare on expeditions. A rather uncouth student with a glass eye accompanied me on one of my field trips. In the evenings he would place his eye on the table near our belongings – we had no problems with theft.

Malaria remains a serious problem in tropical Africa. Unfortunately, one of the methods of combating the disease has had detrimental effects on freshwater fish stocks. Tens of thousands of insecticide-impregnated mosquito nets have been distributed in Africa but many of them are used, not to protect humans from mosquitoes, but as seine nets to catch fish. The insecticide from the nets also dissolves in the water and causes insect and fish mortalities.

The currency in the Comoros at the time was the CFA franc, created in 1945. At the time CFA stood for *Colonies françaises d'Afrique* (French Colonies of Africa). Later, after the establishment of the French Fifth Republic, it became *Communauté française d'Afrique* (French Community of Africa). Today the currency is the Comorian franc (CF). The 50 and 1,000 CF bank notes and the 5 CF coin depict coelacanths.

Coelacanth on a Comorian 5 CF coin

On one of our coelacanth expeditions to the Comoros we carried a bag of white formalin powder in our luggage. Customs officials discovered it at Hahaya Airport and became suspicious, surmizing that we were drug peddlers. We warned them not to sniff the stuff but one of them did, with embarrassing repercussions as the nosy official had to be rushed to hospital!

phone, press and friends seeking information. At 8 a.m. Guy Drummond Sutton came to fetch me, and we were again welcomed to their lovely house by Dr. Shearer and his good lady. Guy Sutton went off with a cheque for £200 to cash and take the money to the ship, where I had arranged with the Purser to change it for East African currency, since that was the nearest I could get to Comoran money, of which there was apparently not a franc available in Durban. I judged that Hunt would use East African currency in his work, and that exchange could be contrived that way.

They brought me the morning paper, and to my amazement and dismay I read there that Professor Smith had last night asked the Prime Minister for help. I was greatly shocked, as the previous night I had particularly asked everyone not to say a word since I had not sought Dr. Malan's permission to do so. I asked for Shearer and showed him the article. He nodded, he had already seen it, but was not at all perturbed. I told him of my dismay as it might upset Dr. Malan if it had leaked out at this end, for I felt that if anything was to be given out it must come from his own office. Shearer said I need not worry about upsetting Dr. Malan. 'He is a tough citizen, believe me, and can take more than most. Besides, this message is marked 'Cape Town' not Durban. It would not surprise me if that was issued by the staff of the Prime Minister's office'; and as it turned out he was probably correct.

I wanted formalin, and what a picnic! It was Saturday morning, 27th December 1952, and the shops were open, but no factories. No chemist had 2 gallons, 2 lb. perhaps. What on earth was I to do? While I was racking my brains, my wife telephoned; George Prior had called and was with her, and I told her my difficulty, which they briefly discussed and suggested Dr. George Campbell. After some time he was located, and undertook to get a friend to go to his factory and have the stuff sent to his own house that day. He would see I got it. Meanwhile, on the *Dunnottar Castle* Prior stood by, for he had undertaken to go and get the formalin by some means himself if all else failed.

Each telephone call saw me jump, but they were all local. How long would it take to get it all arranged? I thought of my Brigadier in Pretoria. Later I learnt that all that enormous amount of organisation for the flight had been completed in less than ten

'he landing-strip at Pamanzi, Comores, clearly once a reef
below the sea.

The landing strip on Anjouan Island in the Comoros juts out into the sea aloft a coral reef, which makes for hazardous landings in adverse wind conditions. The coconut palm grove at the end of the runway of Ouani Airport is littered with the remains of crashed airplanes.

Mozambique Island was a major Arab port and boat-building centre in the years before Vasco da Gama visited it in 1498. The name is derived from Ali Musa Mbiki, the sultan at the time. This name was subsequently taken by the mainland country, modern-day Mozambique. Fort São Sebastião was built in the 16th century.

The Portuguese established a port and naval base on the island in 1507 and built the Chapel of Nossa Senhora de Baluarte in 1522, now considered to be the oldest European building in the southern hemisphere.

"The Beloved Isle" of Mozambique, P.E.A. Note the
famous fortress of St. Sebastian at the near end.

Plate 5

The real coelacanth quartet would have been JLB Smith, Jacques Millot, Jean Anthony and Daniel Robineau. Edgar Barton Worthington (1905–2001) was a British limnologist who had had nothing to do with the coelacanth. His presence demonstrated the dearth of British scientists working in the Western Indian Ocean at the time.

I once met Worthington at Lake Naivasha in Kenya, and also visited him at his historic country home (an ex-highwaymen's den) in England during my sabbatical leave at the British Museum (Natural History) in 1976–1977.

In the latter part of an unending day, 29 December 1952, starting at Lumbo 4.30 a.m. Dzaoudzi 7.10 a.m., now at Durban 9 p.m. telephoning Brigadier Melville in Pretor hemmed in by the South African Broadcastin Corporation and the Press.

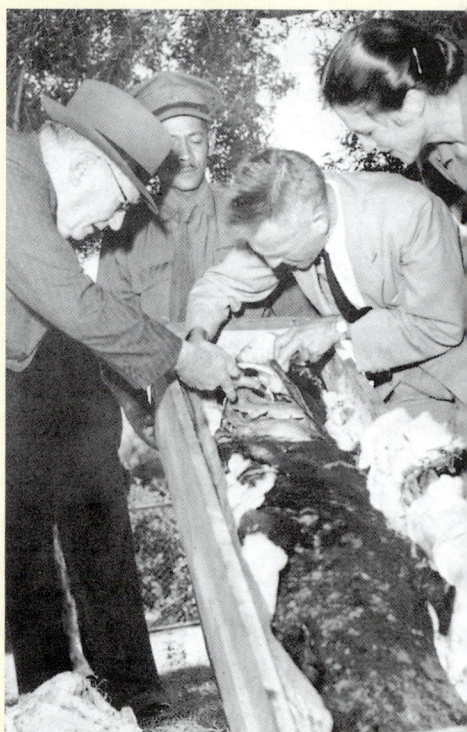

"Do you mean to say we once looked like that?" Dr. D. F. Malan examines the Coelacanth at the Strand, 30th December 1952. Mrs. M. M. Smith is equally interested.

The Coelacanth quartet, Nairc 24th October 1953. *Left to rig* Drs. M. Menache; J. Millot; E. Worthington and J. L. B. Smith.

hours—route, foreign contacts and permits, immigration, Customs, refuelling, the lot. South African efficiency! Again I wondered what my Pretoria friend thought about it all now, but as time passed my anxiety grew. When would that message come? I telephoned Brigadier Daniel, but beyond having heard that the flight was to be arranged, he had received no further news. The Sunderland was all ready and the crew standing by, the moment he had any news he would let me know. A call from my wife— any news yet? Sailing had been put on to 11.30 but could not be delayed a moment longer. Smythe said I had better be there by 11.15 at the latest, which meant leaving at once, so Guy Sutton and I went to the docks, where the *Dunnottar* was ready to move. I raced aboard, did a quick check through the stuff, and my wife and Sutton and I went down the gangway as they were undoing the lashings. A quick good-bye on the quay and my wife went up again, almost yanked aboard by an officer as the gap opened, and she was gone. I saw the telephone wire, held to the last, snaking over the side. Photographed indelibly on my brain is a tiny figure, waving: my wife on the bridge of the *Dunnottar* as she gathered way out through the harbour mouth.

At the request and suggestion of my friend Dr. George Campbell I went to his house, where I might have some peace and rest. I asked Guy not to give me away, but within ten minutes of my arrival the press had found me. No, no news yet. I was worrying about the formalin, but this came soon after. The Campbells kindly left me entirely alone, but I could not rest or sleep, for it seemed impossible that there could still be no news. I was in a fever. Had something gone wrong? There had, but not the way I feared. We heard later that when Dr. Malan tried to make contact with the Minister of Defence, storms had cut the telephone line at several points and eventually a police officer had to go many miles over bad roads to take a message. It was just typical of everything in this whole affair, constant hold-ups.

At 3.30 p.m. Brigadier Daniel telephoned to say that it had been decided not to use the Sunderland, but a Dakota from Swartkops, Pretoria. It might be possible to get it away by about 5 p.m., but we could in any case expect to get away early next morning. Brigadier Melville telephoned from Pretoria to say definitely that it was to be a Dakota, and gave me a résumé of what had been

O.F.—9

The Douglas DC-3 Dakota is a propeller-driven plane with a cruise speed of 333 km/h and a range of 2,400 km. It revolutionized air transport in the 1930s and 1940s and is regarded as one of the most significant transport aircraft ever made.

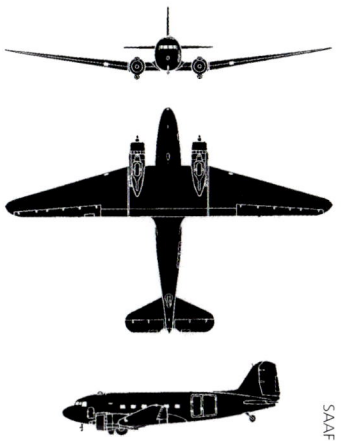

SAAF

Douglas DC-3 Dakota

Dakotas were first flown by the Douglas Aircraft Company in December 1935. About 400 DC-3s (and converted C-47s) are still flying today, many examples being over 70 years old – a testament to the durability of the design. It is possible that the DC-3 will become the first airliner to see over a century of operation. For those interested in buying a flyable DC-3, they cost about US $250,000!

The telephone bill for the calls on Dr Campbell's phone on 27th December 1952 was £31 – an enormous sum in those days. This gives an idea of how complex the negotiations were between JLB, the Prime Minister and Brigadier SA Melville (Acting Chief of Air Staff).

JLB Smith cannot be accused of fear of failure. In fact, he had a view that nothing could be more important than his search for the coelacanth and that no barrier was insurmountable in the fulfilment of this quest. Other scientists gasped in awe at the sheer audacity of his gambit but also secretly admired his *chutzpah*.

Media coverage of the flight by the 'Flying Fishcart', as Dakota C-47A 6832 (KOD) came to be known, was unprecedented, especially after Smith returned successfully with his catch.

Dakota 6832 as she is today at Ysterplaat, in the colours she wore at the time of the trip to the Comoros in 1952

Irene McCulloch

arranged, the course of the flight and other details. It was doubtful whether they could manage to get the plane away that afternoon, but he confirmed that in any case we could leave Durban early in the morning. Later I was told that the Dakota would leave Pretoria before dawn next day, would arrive in Durban, Stamford Hill, at 6 a.m., and get away as soon as we got aboard.

It was a terrific relief to hear this definite news, and I arranged to be fetched by an army car at 5 a.m. Friends were told of these arrangements and I sent the following telegram to my wife on the ship:

'PROBABLY LEAVING DAWN.'

And it was only then that I cabled Hunt as follows:

'HOLD ON STOP GOVERNMENT SENDING PLANE.'

Within a short time I was being subjected to an intensive bombardment by telephone, from near and far. Would I confirm that the Prime Minister had granted a military plane? When did we expect to leave? etc., etc. Then a new phase started. Would I be prepared to take a press representative? I said that would have to be considered, and then there came a succession of more such requests from cinema, newsreel, television, and other agents, until I began to wonder if there was going to be any room for me on the plane. I told them all that the matter would not rest with me. Would I take it further? So I asked the Military Authorities to get in touch with whatever person or department was necessary and to put the matter to them. I gathered it would be the Prime Minister's secretary, but in any case the answer eventually came back a flat 'No, only Professor Smith'.

At the end of an expedition such as ours, financial matters were often troublesome in many ways, and as ready funds would certainly now be of prime importance, I sent the following telegram to the President of the South African Council for Scientific and Industrial Research in Pretoria:

'GOING COMORES IN PLANE GRANTED BY PRIME MINISTER TO FETCH COELACANTH WILL COUNCIL KINDLY GRANT FIVE HUNDRED POUNDS TOWARDS EXPENSES.'

And so, near midnight I went to bed and actually slept a few hours. Despite my extreme weariness I needed no alarm clock; indeed never do need one, for I can wake at any desired moment, but I beat myself that morning, and woke at 3 a.m. and was in the car at five. Mrs. Campbell was up at four and we had coffee and litchis together; George was already away. Knowing my liking for fruit, the Campbells had put piles of all sorts in my room, especially great bunches of litchis, which we rarely see in Grahamstown, and those I had not eaten Mrs. Campbell helped me stuff into my case.

It was misty at the aerodrome, and we heard the plane circling long before she came into view. When the door was opened three huge Air Force Officers emerged and came over towards us. The local Commander called out, 'Commandant Blaauw, let me introduce you to Professor Smith,' and a powerfully built man of about thirty, in whose strong face shone piercing eyes, came up with hand outstretched. To his conventional greeting I answered, 'I bet when you joined the South African Air Force you never expected to command a plane sent to fetch a dead fish.' His face opened a bit at that, and his brief reply left no doubt that he felt that way. He was clearly a powerful personality, and I soon learnt from the others that he was an ace pilot from Korea, one of our best. They were all covertly scanning me closely; what was in this skinny little fellow to get a Prime Minister to send a special plane to look for a fish? I thought again of my Brigadier in Pretoria. I judged Blaauw a tough who would fight to the very end, a wonderful ally, but a dangerous enemy, even as a prisoner he would be a danger and need special care.

They asked if I was ready, I said yes I was, but were they? This startled them, but after a second's thought they said they were. I asked what food they had? They said iron rations, standard on the plane, nothing else, they would not need any. I asked how they knew that, and they said they were confident that those would be all that were necessary. I smiled inwardly, though with what I had brought, plus those iron rations, I felt there would be enough for emergencies, so left it at that. Then I asked how much water was aboard, and was told about three gallons, in the lavatory. Why? I asked if they knew tropical East Africa, and without waiting for any answer went on to say that if all went well that

The contrast between JLB's assertive manner and his skinny frame must have been a surprise to the burly Air Force stalwarts, but they would soon learn that there was more to JLB than met the eye.

Commandant Jan Blaauw was a decorated war hero who flew American P51D Mustang fighter-bombers in the Korean War. He was awarded the American Silver Star for his gallant rescue of Lieutenant Vernon Kruger who had been shot down. Blaauw crash-landed next to him, as he was injured, and stayed with him under enemy fire until they were rescued by an American helicopter.

David Rawlins' art construction of the voyage of the 'Flying Fishcart'

The flight map shown on this page demonstrates clearly that they flew over parts of tropical southeast Africa that had rarely been overflown by South African military personnel since the Second World War. The flight crew took this challenge in their stride and accomplished an immaculate trip.

The map also shows the locations of JLB's 1950, 1951 and 1952 fish-collecting expeditions in northern Mozambique and Tanzania.

would do, but if we had an accident I certainly did not want to fight six men bigger than myself for water; we must have more. Was there any portable tank at the aerodrome? How big? At least five gallons. No! But they had some at the Sunderland base. I asked if we could have two and how long it would take to fetch them?

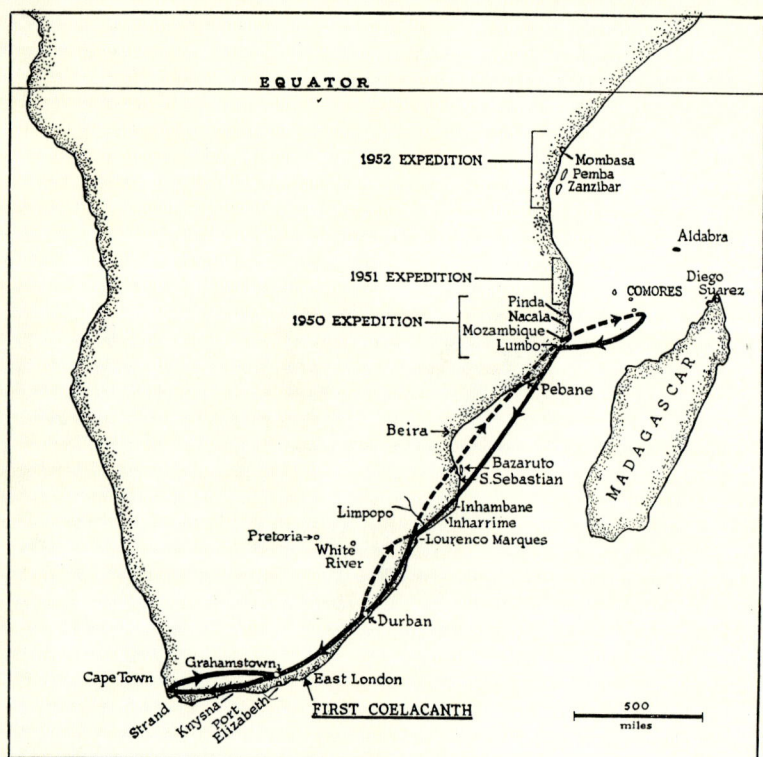

Fig. 4.—Map showing flight of Dakota.

Forty minutes! I refused to leave without that water, and as this flight was for my needs, they did not question this decision and sent a car racing away. That water was never used, and I must record that none of them ever crowed over me on that account. It was not all lost time, for the refueller at the aerodrome gave trouble and needed attention. I thought how everything has its points, even what might seem a disaster at the moment often turns out for the best.

We got away at 7.10 a.m. It was my first experience of a military aircraft. The hull was unlined and she certainly was noisy and not exactly luxurious, while ventilation was provided by 3-inch holes in the sides.

As soon as we had settled on an even keel I went forward and got the names and titles and functions of the crew. They were:

Commandant	.	.	J. P. D. Blaauw
Captain	.	.	P. Letley
Lieutenant	.	.	W. J. Bergh
Lieutenant	.	.	D. M. Ralston
Corporal	.	.	J. W. J. van Niekerk
Corporal	.	.	F. Brink

I asked the course, and was told that we should sleep at Lumbo that night, and go on to Diego Suarez on north Madagascar as early as possible next day. After that? They had been unable to find out if there was any hope of landing at the Comores. The South African Forces had made an air strip on Pamanzi during the war, but it was not known if the French had kept this going or whether a plane like this could land there at all. So even the immediate future was still dark and uncertain, just to add to the black cloud of uncertainty and dread that it might not be a Coelacanth after all. This terrible agony, my life blood, all this fuss and bother— just imagine if it proved to be a common fish. How the world would laugh, and even Dr. Malan would catch it from his opponents in Parliament. There was going to be nothing easy anywhere along the line, that was clear, but at any rate we were on the way at last, we were doing something. I was naturally strung to a high pitch of excitement, but spent the time on the way to Lourenço Marques getting to know these six men, fitting them into the categories every mature man has waiting ready-made. When they thought me unaware, which I never am, they studied me with curiosity and interest, nobody saying very much.

Lourenço Marques ahead. Letley grimaced and gestured to a safety-belt, so I returned to the hold and sat down.

As the doors opened I shot out—hot all right, just think of the Comores! A minor airport official ran out and embraced me effusively—at this tense time my response was not very cordial, and I had seen in the hall the South African Vice-Consul Phillip,

Although the flight never landed in Madagascar, the trip was officially called the 'Madagaskar Coelecanth Expedition' in Lieutenant Bergh's log book.

The flight crew of Dakota 6832 subsequently engaged in many military conflicts and followed distinguished careers in the South African Air Force. Duncan Ralston achieved the rank of Major-General and Peter Letley became a Brigadier.

SAAF

Duncan Ralston

Largely through Margaret's affability the Smiths made friends wherever they went on their expeditions. JLB's main contribution to this cordiality was to share the fish discoveries that he made with officials and local people.

My experience with the Portuguese in Mozambique was very positive, as had been the Smiths' experience. When a coelacanth was caught there in 1991 Dr Augusto Cabral, Director of the Museu da Historia Natural in Maputo, went out of his way to facilitate our access to the specimen, which turned out to be a pregnant female with 26 late-term pups. Hans Fricke and I visited him in January 1992 to examine it and he donated 10 pups to the then JLB Smith Institute of Ichthyology and 10 to the Max Planck Institute in Germany, where Hans worked.

Augusto Cabral

Pregnant coelacanth caught off Mozambique in 1991

who now came out to meet me; the Consul, who had suddenly been recalled to the Union, had sent his regrets and regards, and his Deputy would provide or attend to any needs. Then to my great joy I saw my old friends Comandante Correia de Barros, the Governor-General's right-hand man, and his lovely wife, Senhora Donna Maria Emilia. Phillip had kindly told them I was expected and they had been waiting for an hour, and he one of the busiest men in the world. It was one of the brightest spots in that flight. Then there were greetings from various airport officials, all old acquaintances, who had stood aside until Barros and his wife had greeted me. On the way from Durban I had written a brief account of events in Portuguese, and now sent this to *Noticias*, the Mozambique paper of Lourenço Marques, the staff of which are old friends. I had taken the precaution of having a good supply of Mozambique currency, and soon had the crew filling up on hot coffee, soft drinks, and food. Even apart from our own sense of urgency we did not have to stay long, for the machine was refuelled and all formalities completed in rapid time. Anyone who has ideas that the Portuguese are lackadaisical should speak to airmen or to ships' officers, who will soon disabuse that notion. In all the years I have worked in Mozambique, not a boat or a car or a plane provided by the authorities has ever been late, not one. On this occasion everything was laid on with unusual speed, partly because it was by Government orders and partly as an extra bit for me. We heard later that the Portuguese made no charge for any of the services rendered on this whole flight.

Off we went again straight away north, over the land, on a bee-line for Bazaruto, Inhambane far on the starboard side. In the plane we were now less like stilt-legged dogs walking cautiously around and sniffing one another. At the level of Beira I opened my packages and gave each man a small feed consisting of litchis, biscuits, dried figs, and cheese. Eating together loosens reserve, and there was now less tension and even occasional smiles from the crew. I asked Blaauw if I might light my small primus to make coffee for everyone, but he was plainly shocked to the core. It was against all regulations. I told him I had often lit that very stove in a hold full of T.N.T. without any concern, and would do it in that metal hull and guarantee no danger. After all, man, I pointed out, after all I had been through I wanted to get that Coelacanth. But he would have none of it even when I pointed to matches and lighted cigarettes. I could not help smiling, I knew what I was about. I have sat with that stove alight between my knees, balancing it and atop as well a pot of water, in the well of a madly rolling and pitching boat, surrounded by petrol, paraffin, and explosives, and had no fear. Of all the wonderful things of today I rank a primus stove very high, for without it much of our tropical work would be virtually impossible. It is as wonderful as the fact that boiling water kills amoeba and bacteria. Have you ever thought what life would be like if it did not? Anyway, despite my musings we all did without coffee.

I was back in my seat, tensed and taut, when Blaauw came to me and said with a grin, 'This trip is going to make you rich. A radio message has just come to say that the Council for Scientific Research are giving you an award of five hundred pounds.' I could not help smiling at this layman's view and shouted, 'No, my lad, no riches for me. I asked them for this for expenses, to pay for the Coelacanth and other etceteras.' Good old P. J. du Toit, he had acted quickly and well, and I had no doubt that this would be another tasty item for the press.

I told Blaauw and Letley that it might prove not to be a Coelacanth at all, I had gone on Hunt's briefest words. This shook them, and they whistled in astonishment and some alarm. It was impossible to tell them all that lay behind this. They told me that the plane cost about £40 per hour in flight, so even if it were a Coelacanth I calculated it was going to cost at least £20 per pound landed in South Africa, if it ever was. A costly fish indeed, in more ways than one. No commercial fishing this! They wanted to know if I had told the Prime Minister that there was a degree of uncertainty. I said of course, but I had told him that I was prepared to stake my name and reputation on this venture so as not to risk losing this fish if it really was a Coelacanth. They were obviously surprised, but made no further comment, probably because it would have been difficult to express their thoughts politely, but it also made them realise how much greater was the strain it imposed on me.

These men were patently all so capable that I hesitated to interfere in any plans, but I had to come in, for my knowledge and my needs were both important, and must play their part.

The total duration of the epic return flight was 34 hours and 5 minutes air time, and the distance covered exceeded 7,400 km across three countries. At a cost of £40 per hour in flight, the total cost of the trip to the Air Force, excluding accommodation and other incidental expenses, would have been about £1,360, a very sizeable sum at that time.

The 'landed cost' of the coelacanth specimen turned out to be £16.45 per pound (£36.2 per kilogram), slightly less than Smith's estimate!

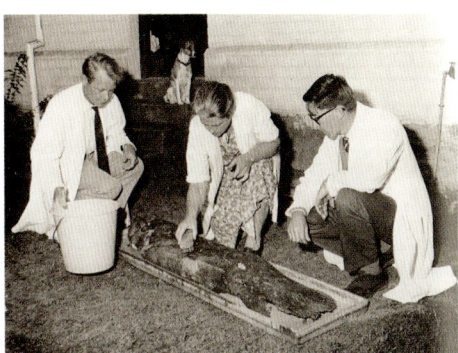

The second coelacanth being examined by Robin Boltt, Margaret Smith and Burke Hill at Rhodes University in 1966

JLB Smith

During the Second World War Madagascar was a French colony and France was Britain's ally. Why then did Britain attack its ally? It was a delicate situation but the decision to invade Madagascar ('Operation Ironclad')

Personal flag of Philippe Pétain, Chief of State of Vichy France

was forced upon Britain by the fact that, with the fall of France and its occupation by German forces, the new Vichy French government was pro-German, and it was feared that it would hand Madagascar over to the Japanese.

The spectacular Japanese successes in the Far East, the loss of the British naval base at Singapore in February 1942, and the advance of the Japanese into the Indian Ocean, presented the British High Command with a conundrum in Madagascar, particularly as it had an excellent naval base at Diego Suarez (later renamed Antsiranana) that was well worth defending. Japanese naval and air bases on Madagascar would have constituted a severe threat to British and South African interests in the region.

Allied soldiers landing in Tamatave, Madagascar, in May 1942, during 'Operation Ironclad'.

The South African Air Force (SAAF) conducted reconnaissance flights over Diego Suarez in early 1942. During the subsequent amphibious/air assault carried out

Emblem of the South African Air Force

by the Royal Navy and Air Force on 5th May 1942, the Vichy French Air Force attacked the Allied fleet but was repelled. Once the main airfield in Diego Suarez had been secured, the SAAF flew a total of 401 sorties from Kenya to Arrachart and elsewhere in Madagascar.

There are few who know that whole area as well as I and few with my experience of tropical East African shores. Besides, I was far from happy about going to Madagascar without first trying for a nearer goal. I do not believe in dodging trouble, and those who know me say I often provoke it to shorten a crisis, and that may be true; but to go to Madagascar when it might not be necessary might well set in train a series of increasingly difficult complications and situations that could easily get out of hand for both sides. I could not help feeling that for this South African military plane on this particular errand to land at Diego Suarez might well irritate the scarcely healed wound caused by the forcible seizure of Madagascar during the war, in which South Africa had played so great a part. There would be officials and others there who had not forgotten, and it would not help that that whole affair merely followed the pattern of the British action at Oran. I did not say this directly, but stressed that we all wanted to get back as soon as possible and sometimes there are unexpected delays at headquarters. I soon sensed that on this matter Blaauw and I saw eye to eye, and I suspected that he probably realised the possible effect of that war-time seizure of Madagascar on this affair as clearly as myself. I pointed out that it was worth a bit of a risk to avoid the certain delay that would ensue if I had to make a voyage by boat from Madagascar to the Comores and back. But even if we had to go to Madagascar, I felt it would be better to fly over the Comores, at least to see Mayotte, and if there was no information, we might just look at that airstrip ourselves. Could we not judge its possibilities by flying low down? Screwed-up faces without words were not encouraging, but I was convinced that this would be the soundest policy and said so, firmly.

By the time the island of Mozambique came into view opinion in the plane had veered definitely north of east—if you look at the map you will see why.

In my own mind I felt no ethical uneasiness about going for this fish, rather the reverse, for even though this Coelacanth had been found in French waters, it was mine by every right. Even though I hoped that there would be no opposition, if it came to a matter of establishing that right in the face of official obstruction, while seven South Africans at Diego Suarez might set feelings running high, at Dzaoudzi they would not have that effect. I did not wish to have to point out to the French that this fish had been under their very noses all the time, and they had not known about it. They had had my leaflets, piles of them, for years, and it seemed to be scarcely possible that they had used them to the best advantage. Indeed, from what I learnt later it seemed that my idea had been decried there as elsewhere. This mad South African! A Coelacanth here, preposterous! But old man Coelacanth would have enjoyed the joke.

I went to the hold and tried to rest—what did tomorrow hold? Where time had slipped by before, now it was oozing past with incredible tardiness. I looked out—Mofamede Island near Pebane—phew! that had been a narrow shave, my mind slipped back. . . . We were at Pebane on one of our trips, and went out from the river in a small and ancient tug to work at Mofamede, a tiny island with a large reef, some miles out. We got some fine stuff from the banks at low tide. . . . Then we went bombing out among the coral heads on the seaward side. Quite a strong wind was blowing and the surf was breaking heavily on the jagged reef. Our bombs brought up very fine stuff, fishes of all kinds, many that none of the crew had ever seen before. They all became very excited, for fish is scarce in Pebane, and many big ones came up that day, which the wind and waves drove towards the reef and we had to go close in to collect them. While I was directing operations at the stern, the man at the wheel forgot all else watching fish being netted at the sides. Suddenly my sixth sense made me look up, and I found we were virtually on the reef. It was an awful moment. I dashed forward, knocked the cox aside, took the wheel, shouted 'A ré!' [astern], and held on. It was a matter of inches, waves and wind driving us on, and a slipping reverse gear barely able to withstand them, let alone take us back. It was one of the nearest shaves we have ever had. The surf was terrible and that jagged coral nearly a mile from the island would have cut us to pieces. . . . I went on dreaming. A place of ill-omen, Pebane. Once in a coaster, as we were on the way in at early dawn, I sat watching my wife writhing in agony with acute food poisoning— she was very near death. The sea was rough, the bar shallow, and we bumped and bumped again, and a huge wave turned the vessel so far over on her side that I held my breath. I could still hear the crashes before she righted herself. What a morning that was! . . .

Coincidentally, 49 years later, the first and only coelacanth known from Mozambique was netted by a side-trawler, *Vega 13*, off Pebane in August 1991. It was caught in an atypical habitat (over a sandy bottom) and in shallow water (40–44 m), potentially in range of JLB's 'bombs' – had it been in the vicinity at the time!

The specimen was a large female (98 kg, 179 cm) and was still alive when it landed. It was immediately placed in an on-board freezer and taken to Maputo, where it was later dry-salted and skinned and donated to the Museu da Historia Natural in Maputo.

Prior to 1898, Mozambique Island was the capital of colonial Portuguese East Africa. With its rich cultural history, high tropical biodiversity and beautiful sandy beaches, it is now a UNESCO World Heritage Site and one of Mozambique's fastest-growing tourist destinations.

Mozambique has tremendous tourism potential and harbours a very rich marine fish fauna along its 2,470 km-long coastline. Mozambique Island, Ibo Island in the Quirimbas archipelago and Bazaruto Island further south are world-class dive sites.

Ralston's head appeared; 'Mozambique,' he yelled. I shot up, and there was the beloved isle, built on the blood and bones of the flower of Portugal. Our wing-tip was near the palms as we banked; Lumbo, 3.30 p.m. As I came out, phew, a blast from an oven wasn't in it, and think of the Comores. . . . There were effusive greetings from the Airport Supervisor, other officials, and the hotel manager. I tackled them at once. Did they know anything about landing on the Comores? No, a complete blank. So I telephoned the Chief of the Radio Station at Lumbo, but he also knew nothing. I asked him to try to make contact with the Comores, but he said it was Sunday, it was impossible. I replied that nothing was impossible really, would he try? Well, he would; but soon after he telephoned to report a complete blank, they would not open till next morning. Shortly afterwards there was a roar and an East African Airways plane from Nairobi touched down. We tackled them, did they know anything about Comoran landing, the possibility of landing anywhere on the Comores? But they knew nothing. We may have looked a mixed and tough lot, but they were restrained and apparently incurious. It was true that we had not been introduced.

I found out that none of our crew had ever been to the island of Mozambique, so by radio telephone I spoke to the Port Captain at the island, five miles across the bay from Lumbo, and asked him to send us a launch. We went to the hotel and had refreshments, and then to the wharf where the launch was waiting. Some of its native crew had been with us on our expeditions, and there were excited greetings for the 'Patrao', and of course as usual he had cigarettes. They shyly gave their news, and I learnt that Salimo's wife's brother had been eaten by a lion. While the airmen went to have a hasty look round the island, I spent a short time with the Port Captain and his family, and gave them a brief outline of what it was all about, the children more interested in some chocolate slabs from my pocket. Then we went back across the bay, a lovely cool trip in the gentle northerly breeze, the stars twinkling.

Dinner as usual was astonishingly good in such a remote place. Iced beer cooled the thirsty crew, Blaauw watching each man's single bottle. The hotel manager whispered quietly in my ear, 'I have the coolest room for the Professor.' I asked Blaauw for a take-off at 4 a.m. There was a general groan, and after a bit of a skirmish we fixed for 4.30, coffee at 3.30 a.m., the cars at 3.45 a.m. Some of them grinned. Would the Portuguese make it? I offered to bet them it would all be on time, as indeed it was—to the minute.

If my room was the coolest, heaven help the others, and if it was like this here, what about that fish at the Comores? I lay and sweated and tossed but could not sleep; for though I had pills with me, I use them as seldom as possible, never when my mind must be at full stretch. At 1 a.m. I got up and smiled as I lit my primus, thinking of Blaauw. I made coffee and almost took him a cup. Once again I tried to sleep, but my mind was too active, so at 2.30 I got up, took my torch, and went to have a look round. I envied the native servants, who were as usual soundly asleep all over the place; indeed, you have to step over them. The night before I had noticed some pineapples in the kitchen, and went to explore the possibilities of the pantry. There were some nice pineapples, bananas, and papaws. I packed a selection in a carton and took them to my room, leaving my card stuck in the brush of one of the remaining pineapples. The owner-manager would smile, for he was an old friend.

At 3.15 a.m., just to be sure, I went round. They were all up, all had coffee, none were effusive.

Lions are not uncommon about Lumbo, and they sometimes ramble round the airport at night; and though for the sake of the crew I hoped we might see one, none were about that morning. It was hot even at that time. Phew! think of the Comores.

We took off at 4.30 to the second.

Mozambique has a tropical climate with two seasons, a wet season from October to March when cyclones are common, and a dry season from April to September. Rainfall is heavy along the coast, with a countrywide average of 590 mm per annum. Average temperatures in Maputo in the south range from 13 to 24°C in winter to 22 to 31°C in summer but it is much warmer further north at Mozambique Island.

Mozambique, a tourist's paradise

The country's economy is based largely on agriculture, with most of the population engaged in subsistence farming. Papayas, pineapples, bananas, guavas, figs, granadillas and other tropical fruits are common. The Portuguese greatly impacted on the cuisine of Mozambique and popular dishes today include *prego* (steak roll), *rissóis* (battered shrimp), *espetada* (kebab) and *inteiro com piripiri* (whole chicken in piri-piri sauce).

Cyclones are common in northern Mozambique and the Comoros during the wet season, from October to March. In late December, JLB and his retinue were visiting at the height of the cyclone season.

'Mae West' was a common nickname for the first inflatable life preserver, invented in 1928 by Peter Markus. The nickname originated because someone wearing the inflated life preserver appeared to be as well-endowed as the actress. Air crew whose lives were saved using a 'Mae West' (or other personal flotation devices) were eligible for membership of the 'Goldfish Club'.

Chapter Thirteen

DZAOUDZI DRAMA

IT was an impressive dawn whose breathtaking beauty only served to emphasise the threat that lay behind its tumbled masses of red-tinted clouds. This was the cyclone season in one of the worst zones in the world, and if one caught us out over that sea, Coelacanths or anything else would not matter any more. As we took off we plunged straight at tier upon tier of gigantic columns that reached up into misty heights, giving the impression that we were approaching an enormous and fantastic tinted marble temple of ever-changing aspect and form. The clouds looked so solid I found myself tensing up each time we plunged into a whirling face, half expecting the plane to be crushed flat. We climbed steadily and I had repeatedly to clear my ears, but could not take my eyes from the wonderful scene that lay before and about us. Never before had I seen such spectacular cloud effects, their size and grandeur were almost beyond description. A tap on my shoulder made me jump, and there was Ralston holding a type of bulging greyish overall. 'It's going to be very cold up here, sir; you'd better get into this Mae West'; which was something new to me, a thick padded overall suit, at which I looked dubiously. 'You float in this,' he added with a grin. That made me think at once of Tiger Sharks; floating in that sea wasn't much use, it just prolonged the end. Twenty minutes on the average was all you had with those brutes about, but Ralston knew nothing about that; yes, and I hoped he would not need to learn, certainly not in my company. He helped me get it on, and it certainly was comfortable, for we had gone up fairly high and it was bitterly cold, the occasional brilliant sunshine deceptive.

I came back to the present and went forward to Blaauw and Letley. Our course was straight for Mayotte. It had been agreed that we should keep on trying to get Dzaoudzi by radio to know if we could land, but so far there had been no response, though

Bergh and the operator van Niekerk were at it all the time. I could see the regular and repeated movement of van Niekerk's arm as he sat hunched over the table in the tiny cubicle.

I went back to the hold and got out my supplies. First I took each man a kind of fruit salad of my Lumbo-foraged pineapple, papaw, and banana, sprinkled with sugar and dried milk. They looked at this with astonishment, and Blaauw at least cocked a speculative eye at me, but nobody asked any questions. Then I gave them a round of biscuits and cheese, and in a loud voice so that all, especially Blaauw, could hear, pictured the delights hot coffee would have added. Blaauw sat like a rock without moving a muscle, but Letley flashed me a quick impish grin.

All this time my brain was buzzing like a machine round the main themes of whether we should be able to land and was it a Coelacanth? As the time grew nearer, I became more and more tortured by doubts and fears. Had I not been an incredible fool to trust a layman's opinion? Hunt had the pictures and all the information we could give him, but those not expert can easily be mistaken, as we know from hundreds of experiences. The situation was typical of most of my life, either hell or heaven, seldom anywhere between. When I asked my wife to marry me, I said I did not know if I could bring her happiness, but I could at least promise that she would never be bored; and she has eased many a tight corner by reminding me of this with a smile, often a very grim one.

Here it was again. Could anything be more ridiculous? In my maturity I had staked virtually my whole life on the identity of a fish I had not seen; but I tried to push these doubts and fears away, for there was nothing to be done until I got to the fish itself, and that was the present problem. Could we land at Pamanzi? Bergh must have got weary of shaking his head in response to the question on my face that there was no need for my lips to utter.

Cyclones were never far from my thoughts. I asked Blaauw what hope we should have if one came on. He said that in the air we might manage as long as we had fuel enough to get up and out of range and reach some safe landing; the greater danger lay in being caught on the ground. Surely the plane could be anchored, I suggested; but they said it just did not work, time and again it had been tried but the wind always won. As the wind

Tropical cyclones are low-pressure systems that form over warm tropical waters and produce gale force winds (63 km/h or greater, with gusts over 90 km/h) near the centre. If the winds around the centre reach 118 km/h (gusts over 165 km/h), then they are called severe tropical cyclones or hurricanes.

The circular eye at the centre of a tropical cyclone is an area of light winds and often clear skies. Eye diameters are typically 40 km but can range from 10 km to over 100 km. The eye is surrounded by a dense ring of cloud about 16 km high known as the eye wall, which marks the belt of strongest winds and heaviest rainfall.

NASA

Tropical cyclone

Tropical cyclones derive their energy from the warm tropical oceans and do not form unless the sea-surface temperature is above 26.5°C, although once formed, they can persist over lower temperatures.

The main island in the Mayotte archipelago, Grande-Terre (or Maore), is geologically the oldest of the Comorian islands. The wide coral reef encircling much of the island ensures protection for ships and provides a rich habitat for fishes and other marine life. In contrast, Grande Comore is very young (only about 130,000 years old) and has no fringing reef.

Coat of Arms of Mayotte

No coelacanths have reliably been recorded from Mayotte, although there are some false records. Coelacanths have been caught off the other three Comorian islands, with most captured off Grande Comore (122) and fewer off Anjouan (52) and Moheli (3).

increased so the anchored planes would first rise from the ground and 'fly', held up by the wind, but when it became really strong the machines were doomed, they were torn and smashed. Well, there were so many worries and anxieties and uncertainties in all this that another didn't matter very much. Compared with this, walking in the dark in a jungle full of wild beasts was just nothing; for I had done it, so I knew. We could not do anything but just go on. We hardly spoke, everyone was tensed up, this was a most critical time for us all. My eyes alternated between the piled-up clouds ahead and that ominous elbow whose hand plied the key. Again and again it went, and my hope and disappointment rose and sank in steady succession like the waves that lay so far below. Suddenly Letley pointed ahead, to what appeared no more than another dense cloud; but my vision is no longer young and I screwed my eyes. 'Mayotte,' he said, and my heart turned right over. So it was there, somewhere there that this fish lay, this fish in whose identity much of my life might be buried. Suddenly the blurred mass turned into a cloud-topped island. Surely now we were so near we should get a reply to our constant battery of signals; but still nothing came.

We were diving down at a steep angle when the clouds parted and drifted aside, giving a clear picture of steep rolling hills and conical peaks whose densely tree-clad slopes rose almost abruptly from the sea. There was obviously little or no flat country here. Could we ever land? We passed at about 3,000 feet over the extensive barrier reef that lies west of Mayotte. It was a marvellous sight, a multi-coloured riot of blues and greens, which for some moments diverted my thoughts. What a place for a fish, a maze of channels and water of all depths, with clearly abundant coral, and even at that height it was possible to see plainly many shoals of fish of all sizes. Whatever the outcome of the Coelacanth, this was a place where we ought to work, for if a fish as large as the Coelacanth had been there all this time unknown, what other treasures might not be hidden there as well? We must come some time, and I was already planning to this end when my thoughts were shattered by a shout from Bergh, who said 'O.K.', and first jerked his hand, thumb up and then down. I suddenly felt queer and shaky. We were actually going to land. By this time we were circling to lose height and had a panoramic view of the whole island, and although it was misty and hazy I took a number of photographs through the small air-holes. We saw Pamanzi and the houses of Dzaoudzi and a small lake, and had a passing glimpse of a tiny toy vessel close to the wharf that something told me was Hunt's. My heart was pounding madly, and they had almost to force me to sit down and tie myself in. I had seen the air-strip. It certainly was small and looked rough, a rounded bit of the island sticking out to the south, obviously not long ago it had been a coral reef under the sea, for it was almost at sea-level now. The waves had showed a north-easter of force less than two; we had to land from the sea and in front was a high hill.

Bump, bump, bump ! It was a wonderful landing on that rough surface, but the instant we came to a stop the heavens opened and the rain just poured down in torrents. Everything was blotted out and visibility reduced to a few yards. The roof of the plane leaked like a sieve, and I had to rush round to get my sleeping-bag and papers to dry spots. I was curiously indignant about this, as I hate leaky roofs; but idiotically I also felt a sudden splash of comfort, for it was as if this was a last effort of fate to try to keep me back from the true end and reality, and I had a queer flash of hope.

The rain stopped as abruptly as if turned off by a tap, the mist parted and figures came running across the flattened coral rag. The door opened and through a blast of hot air I saw Hunt's face looking up at me. For a moment I just could not speak; then with a rush of pent-up emotion the words 'Where's the fish?' burst from my lips like an explosion. With extraordinary intuition Hunt replied, 'Don't worry, it's a Coelacanth all right'; and this had a strangely soothing effect on me, but I was still in a fever. I found myself on the ground, various French officials were introduced to me; but I saw hardly anything, I wanted to see that fish. 'It is on my boat,' said Hunt, and that at least was a terrific relief, for it meant that even in French waters, on Hunt's boat it was not strictly in French territory and technically at least it would be less like taking it from the French. At the same time this was my Coelacanth (if it was one) by every right, and in my almost insane obsession at that moment I knew that those six South Africans who were with me would be prepared to stand by me in support of that right if it became necessary.

The landing on Pamanzi was hazardous. The airstrip, built 10 years before by the South African Defence Force when they occupied the island during the Second World War, was primitive. But a Dakota, in the right hands, is a tough and adaptable airplane.

A peculiar feature of the Comorian islands, more particularly the youngest island of Grande Comore, is the lack of sand, a commodity that we take for granted. Sand is so scarce there that people grind up coral to make the stuff, and sell it in small piles on the roadside.

Coral has many and varied uses on tropical islands, especially as bricks for building and as a source of calcium carbonate to make lime (calcium oxide). Lime mixed with water makes whitewash paint and lime mixed with sand produces a simple mortar. As coral reefs form the foundation on which many marine animal communities are built, the overexploitation of coral threatens their survival.

Adona9 at the English language Wikipedia

Stag corals, which are widespread in the Western Indian Ocean, are very vulnerable to blast fishing.

The forward journey lasted 12 hours and 15 minutes airtime and was extremely taxing. The return journey would prove to be even more demanding, with extra legs from Durban to Grahamstown, then to Cape Town, and back to Grahamstown, added on.

Pamandzi is a commune on the island of Petite-Terre (Pamanzi) off the main island of Mayotte. The Dzaoudzi Pamandzi International Airport, with an asphalt runway 1,929 m long, is now located on Pamandzi, on the same site as the rough air strip on which Dakota 6832 landed. Ewa Air is headquartered there.

Today Pamandzi is a popular destination for kitesurfing, windsurfing, surfing, sailing and paragliding.

I checked that I had my camera and the necessary accessories, and we moved over to the small house on the edge of the field

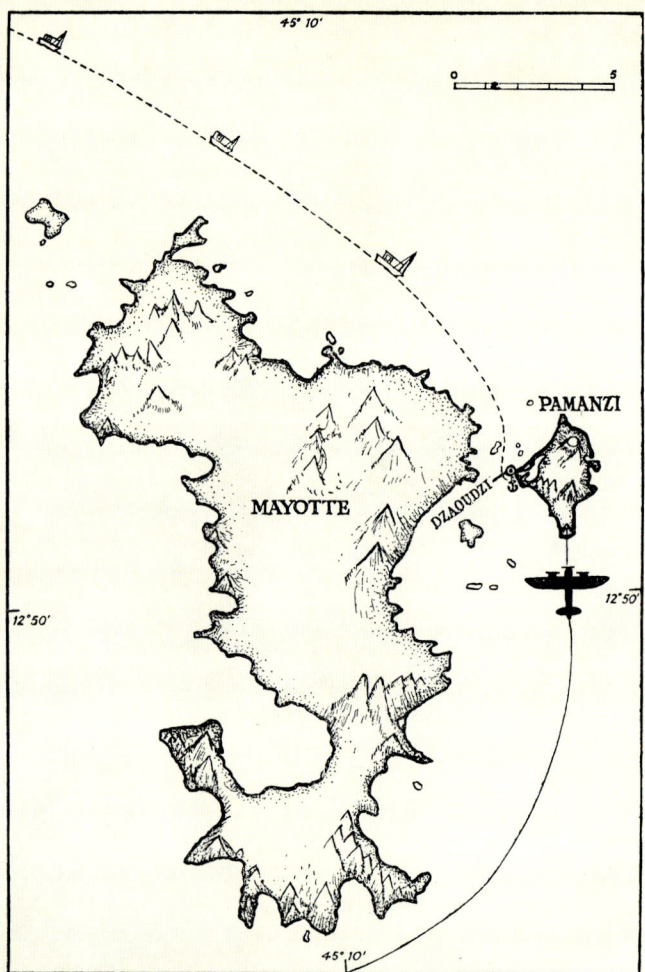

FIG. 5.—The islands of Mayotte and Pamanzi, Comores, showing small projection where the Dakota landed.

where a few cars were parked. 'Where is your boat?' I asked Hunt. 'At the quay,' he said, 'but we shall have to go first to the Governor's house, as he is waiting to meet you. We'll just have to do that,' he added quickly when he saw my reaction. 'For my sake, at least

It won't take long.' I have often suffered from the necessity of paying tribute to officialdom, but this was probably the hardest I have ever endured. It was agony and torture, and I raged inwardly, my mind a searing flame. Blast these formalities! I had not endured all I had been through or come so far to exchange polite words with a Governor at that critical moment. I wanted only one thing, and that was to see that fish, to know if I was a fool or a prophet. But civilisation won, the flames in my mind subsided, and I became *homo sapiens* again. We passed houses and trees and a curved drive, pulled up below a tall, two-storied wooden residence, walked up steep sun-drenched steps, through the front verandah, and into the relative gloom of a large shady room, where dim human figures were waiting. I was formally introduced to the Governor and his lady, and through one of his suite who spoke English I presented our crew. My French is satisfactory for scientific use, but I have little conversation. Most Portuguese speak French, so I wondered if the Governor might perhaps know Portuguese, not very good reasoning I know, but I tried it on him and he did not understand. I thought it better not to try German; so we used Hunt and the official.

They drew our attention to a big table along one end of the room that I now saw was laden with bottles and dainties of all kinds. We were directed towards this, but I could endure no more, and said politely but firmly that while we were more than grateful to His Excellency for all his courtesy and hospitality, I had endured much and come far and wished first to inspect the fish. Would he kindly permit us to return so that we could all the better enjoy his bounteous hospitality a little later, we should not be long? There was a flutter, but my face showed that I was going, so all of us went, the Governor as well, but not Madame. It was only a moment in the car and there below the concrete wall was Hunt's schooner. The place was jammed with idling natives, who delayed our passage down the steps. (See Frontispiece.) Hunt pointed to a large coffin-like box near the mast, and I knew it must be in there. They picked up the box and put it on the hatch-cover, just in front of me, a foot above the deck, and Hunt pulled away the lid. I saw a sea of cotton-wool, the fish was covered by it. My whole life welled up in a terrible flood of fear and agony, and I could not speak or move. They all stood staring at me, but

O.F.—10

The coelacanth does tend to cause emotional reactions in scientists occupied with researching it. I studied dead coelacanths for 14 years, from 1994 until 2008, before I was finally given the opportunity to see one alive in its natural habitat, courtesy of Hans Fricke, Jürgen Schauer and the research submersible *Jago*.

In the Preface to my autobiography I describe this event as follows, 'We shared the cave with five coelacanths for about an hour. They were serenely tranquil, almost spiritual, and stared blankly back at us across space and time. I was transported back to a primeval age when they had wandered the seas and freshwaters before the dinosaurs had evolved. …. Somehow their looming presence was foreboding to me. They are the ultimate survivors, yet we are concerned with their conservation. How arrogant we are!'

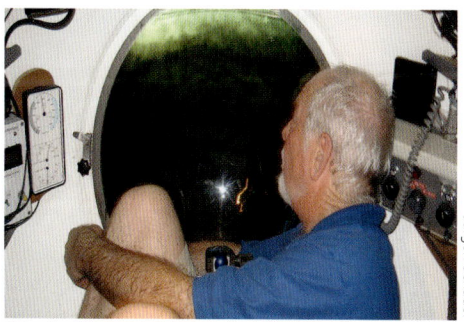

Jürgen Schauer

The author watching coelacanths at a depth of 198 m off Grande Comore in November 2008 in the Jago

It must have been excruciatingly frustrating for JLB to make small talk (which he abhorred at the best of times) while 'his' fish was lying on Hunt's schooner nearby.

Smith's positive identification of the second coelacanth was a landmark event in ichthyology. His long search had ended and his predictions had proved to be right – it was a tropical fish that lived in relatively shallow water over rocky or coral reefs.

JLB was able to recognise the coelacanth immediately due to a combination of unique features, especially the tubercles on the scales, the heavy, bony head and the lobed fins.

Mike Bruton

Coelacanth scale showing the tubercles on the exposed end

This coelacanth differed from the first in that it had no first dorsal fin and no middle lobe in the tail fin (Marjorie's 'puppy-dog tail'). At first JLB thought that it was another species but subsequent research revealed that it was *Latimeria chalumnae*.

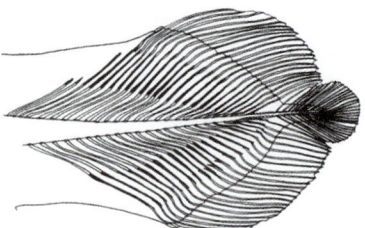

'Puppy-dog tail' of the coelacanth

I could not bring myself to touch it; and, after standing as if stricken, motioned to them to open it, when Hunt and a sailor jumped as if electrified and peeled away that enveloping white shroud.

God, yes! It was true! I saw first the unmistakable tubercles on the large scales, then the bones of the head, the spiny fins. It was true! Malan would not suffer for his action, thank God for that! It was a Coelacanth all right. I knelt down on the deck so as to get a closer view, and as I caressed that fish I found tears splashing on my hands and realised that I was weeping, and was quite without shame. Fourteen of the best years of my life had gone in this search and it was true; it was really true. It had come at last.

Suddenly my mind cleared. Time was passing. Blaauw had told me that we must leave as soon as possible, for in his opinion if the weather shut down we might not be able to get away at all, and I could not but agree with him. I knew those seas. This was something that must go round the world like a tidal wave, and I must work quickly and efficiently. I got them to get the fish out of the box, posed us all and took several photographs. Then I spent about five minutes in rapid inspection of the fish; which was not easy, as my excitement was naturally intense, the sudden release of fear and strain was almost more than the human system could endure. It was certainly staggering to find no first dorsal fin and no extra little tail such as all Coelacanths have had; but it was a Coelacanth. It was probably different, probably a different genus and species from my East London fish, but still a Coelacanth. I must name it quickly—no more risks like the last one. Malan, yes certainly first Malan, genus *Malania* sounded good. *Malania hunti* seemed to be clearly indicated, since each had played so prominent a part in the final act of this long drama. I told them to pack it away. Hunt had been talking animatedly to my airmen, and he now invited Blaauw and myself to his cabin. He produced a bottle of whisky to toast the occasion, and Blaauw who would probably have liked more, took the minimum for politeness, while I put a drop in my water to be matey. I wanted information and kept up such a rapid running fire of questions to Hunt, asking him to answer quickly, that Blaauw, solid as he was, just faded away. Hunt has a keen brain, and it was not long before I had the main outlines of the story from his end. The Governor had several times looked into the cabin, but we did not see him, and Blaauw returned twice to say that we must get off as soon as possible. The clouds were certainly low and threatening, and there was rain at intervals.

I told Hunt that I should be giving this animal a scientific name, and that I proposed *Malania hunti*, the latter to honour him for his important part. He asked if I could not somehow bring in the French, as his relations with them were important, his living depended on that. I said the only way would be for him to drop out, when we might call it *Malania comoroae* or *anjouanae*; and after discussion he said that while it would be very flattering to have his name figured that way, could it be *Malania anjouanae*; to which I agreed only with reluctance and only because he was quite emphatic that he preferred it.

It was clear even then, and it became much more so later, that Hunt was in a difficult position: and in view of all he had done I was prepared to go as far as possible to alleviate this and to aid him. I told him that I was prepared to continue the offer of a reward of £100 for another Coelacanth, through him as my agent; and that if he got another in their waters that way, it was to be offered to the French. I intended to say this to the Governor before I left. Hunt was very appreciative of this suggestion, which he thought would go far to smooth any ruffled pride or hurt feelings.

It was only about 500 yards to the Governor's residence, and I elected to walk as I wanted some photographs. The precious coffin was put on a lorry and Hunt arranged that it would be taken direct to the air-strip to await us there. I hated to let it out of my sight, but felt they knew I would kill anyone who tried any tricks.

At the top of the steps I turned and took one more long look at Hunt's trim schooner, which had played its part, an important part, in this fascinating story. It was his home and his life; he obviously kept it trim and neat and cleaner than most such craft are. I had a queer feeling that had he lived in earlier times Hunt would also have lived on the sea, but he would probably have been an explorer rather than a trader. He and Blaauw had 'clicked' at once, and Blaauw told me, almost wistfully, that he would not mind a voyage with Hunt. I wondered, for I know what it is like to live in a small boat in the tropics, but there would have been something natural in their association. They would make a for-

When new species of animal are named they are sometimes given names that commemorate the person who found the new species. JLB himself had several fish species named after him, including a conger eel, pipefish, flounder, cusk eel and clingfish, and his wife, Margaret, had four fish species (and a worm from a cow's gut) named after her.

Sometimes new fish species are named after fictional people. A requiem shark found in the Red Sea was named *Iago omanensis* after the villain in Shakespeare's play, Othello, as it was fiendishly difficult to classify. Hans Fricke's research submersible, *Jago*, was named after this shark. New species may also be named after institutions, such as the blenny, *Mimoblennius rusi*, named after the 'Rhodes University Smith Institute' ('RUSI', the prefix for all catalogue numbers in its fish collection).

I only managed to have one insignificant beetle, that I collected near Lake Ngami in Botswana, named after me (*Notoxus brutoni*)!

Accounts in the literature of the loss of the schooner *N'duwaro* in a cyclone in mid-January 1953 are confusing, with some even claiming that Hunt lost his life in this tragedy. Hunt survived this sinking but died at sea four years later when his next schooner, *Hairiako*, struck a shallow reef.

We heard several accounts of Comorian fishermen in their dugout canoes being blown far from shore and never returning. One fishermen who survived the ordeal washed ashore on the Kenyan coast over 1,000 km away!

It is odd that JLB refers to Goosen's trawler here as the *Aristea*, and that she was wrecked soon afterwards. In fact, his trawler was the *Nerine* and, within a year, she was refitted as a minesweeper and did sterling service during the Second World War.

midable pair at any time, and in earlier centuries would probably have made history of their own. I looked hard at that schooner, not suspecting that I should never see it again: for a bare two weeks later they were caught in a cyclone and the boat was destroyed and sunk. Our plane did not escape by much. Did Coelacanths bring destruction in their train? I remembered that Goosen's trawler *Aristea* that had caught *Latimeria* was wrecked and destroyed not long afterwards.

I looked about me. Pamanzi is only a tiny island separated by a narrow strait from Mayotte, which is much larger, mostly hills and mountains. I saw canoes and lines and nets, and asked about fish supply. Shoulders were shrugged, and screwed-up faces with few words told their tale. Yes, there was fish, but not much, not nearly enough; the natives were not very industrious fishermen, they got some but were satisfied with very little. During the war there had been plenty, hand-grenades and depth-charges did better than lines, and sometimes brought up a whole shoal. There was little fish close in, those who went out to the big reef did better; but often when they got in the fish was too far gone for European taste.

I told them how impressed I had been by the large reef I had seen from the plane on our way in, and Hunt said he had done some goggling there, and that he had never seen so many fish anywhere, nor so many kinds; the reef was far richer than any about Zanzibar, for example. I said that I was contemplating returning to work on the fishes.

The Governor was enthusiastic. I would be welcome; he had no children there; I could have half his house; they would be delighted. I was touched by this offer, and I got the interpreter to explain how overwhelmed I was at his generosity, but that we could never dream of imposing so uncomfortable a situation on him, as our lives were of course governed by the tides and we were up and about and out at all hours. We had to live by ourselves; was there any other house available? Oh yes, that could be arranged; anything I wanted that was possible would be granted. I found out about the rainfall, the winds, available food supplies, vehicles, and boats. French waters were a big gap in our studies of the fishes of the western Indian Ocean, and I made full use of this opportunity to learn all I could for our work.

The soil was clearly good, and with abundant rain the growth was everywhere luxuriant. I knew there was malaria, dysentery, and hookworm, and other tropical plagues. The houses were not netted in like those of the Portuguese, and the natives were plainly lethargic. The Governor said that this was due not only to climate and disease, but the natural indolence of the natives was much increased by the ever-present cyclones. It was almost impossible to get them to construct decent houses or to carry out long-term planting of palms and trees, when a cyclone could destroy the work of years in a few moments. One result was that whereas fruit in such a clime should be most abundant, the supply was in reality not very good and at times there was little. He indeed spoke prophetically, for only two weeks later a cyclone hit Dzaoudzi and the devastation was terrible.

On the way up they showed me with pride one of my Coelacanth leaflets (Plate 3), with the write-up in English, Portuguese, and French, and the picture, prominently displayed on the public noticeboard of what looked like the equivalent of a magistrate's office. Hunt proved unexpectedly coy about being photographed in its company.

We got back to the Residence. This time I saw it properly, a typical French colonial structure, very high for coolness and shade. Apart from the inconvenience of sharing with others, I would not have cared to live in it; certainly not in the top storey, for anything, for it was all wood and a tinder-box. The furniture was lovely, and there were many antiques and curios. They put us to table, and there were speeches and toasts. A bottle of very precious old brandy of a famous vintage was opened for this occasion, and it was hard to endure their disappointment that I had less than a teaspoonful, while Blaauw's stern eye on his crew reduced them to tasters rather than participants. There was ample wine as well, but no takers on our side. I had coffee. The crew had none of my inner infirmities or restrictions about food, and their inroads on the dainties concealed a good part of my deficiencies, though Madame was concerned at my lack of appetite. Right in front of me was a schoolboy's dream, an enormous cake spread with sticky chocolate icing, the mere sight of which made my liver throb. I had not yet eaten anything that day, but could not risk any upset now, not even for the sake of

My wife, Carolynn, and I experienced a devastating tropical cyclone, *Domoina*, in January 1984 while living on the shores of Lake Sibaya in Zululand. Our wooden yacht was destroyed and all temporary buildings and structures were flattened. Further south, large concrete freeway bridges were demolished by the resulting '100 year' floods. Two hundred and forty-two people died in Madagascar, Mozambique and Zululand and damage totalling hundreds of millions of rands was caused. *Domoina* originally developed off the northeast coast of Madagascar.

Cyclone Domoina making landfall in Zululand in January 1984

Recent research has revealed that one of the reasons why alcohol upsets our sense of balance is because it dilutes the fluid in the cochlea of the inner ear, which causes the tiny hairs that register our balance and orientation to send confusing messages to the brain. This is not a desirable condition for an active air crew!

The third coelacanth was caught about nine months later on 23rd September 1953, north of Mutsamudu on Anjouan Island. The specimen, a male weighing 39.5 kg and measuring 129 cm, was hooked on a hand line by Houmadi Hassani at a depth of about 200 m using fish bait (*roudi*).

The French subsequently prohibited foreign scientists from searching for coelacanths in their waters and proposed instead that an international expedition under French leadership should continue to conduct coelacanth research.

policy or politeness. Although Blaauw shot several short warnings about time across to me in Afrikaans, he had no need to speak, I was as anxious as he to go. The younger members of the crew were enchanted by the place, and would have rejoiced at any, even enforced, stay; but it would have been criminal folly to delay. I had got from these friendly people as much information as could be contrived in less than a stay of days, so as soon as politeness permitted I rose and said we must unfortunately tear ourselves away, and, speaking very slowly, expressed my gratitude to His Excellency for his very kind assistance and co-operation in this matter, which I appreciated very greatly, as I most sincerely did. I appreciated only too well that he was in some ways in a difficult situation, and he had handled it most gracefully. Because of that I went on to say that I would never have come for this fish had I not felt it was mine, because it had been found as a result of my search. Even though I was so interested in Coelacanths, I would not have lifted a finger to fetch one that had come as a result of efforts of someone else. But this one, I told the Governor with a smile, and asked the official to translate carefully, if this one had been found on the steps of your Residency, sir, I would have come for it, for it is mine. However, because you have all been so kind I have asked Captain Hunt to act as my agent here, and he is authorised to offer a further £100 as a reward for another fish, and the very next one he obtains that way in French waters will be given to you as the representative of your nation. It was clear that this was greatly appreciated by them all.

We parted with cordial adieux from the Governor and his lady, with hopes of return at no distant date. I got all their names and titles and occupations. There was still one more thing to do, and I sent off brief cables to my wife, to Dr. Malan, and to the President of the Council for Scientific and Industrial Research, to say that it was a true Coelacanth.

Then we hurried off to the air-strip. The box? Yes, there it was, safe in the lorry in the shade, and soon it was in the Dakota, where I opened the lid and looked inside as well, just to be sure. We had spent barely three hours on Pamanzi, and yet it seemed almost an age. It had been one of the most critical periods of my life; but this whole affair had been just one crisis after another.

It was in many ways really a kind of nightmare, and though I kept on telling myself 'It is true; it is true', my inner self was like an obstinate animal that would keep on stupidly turning back towards the doubts and fears that had torn and tortured me all through those long days and nights.

I took Hunt aboard and paid him the £200 E.A. I had; £100 for the reward to the native and the rest to cover his own expenses. In the few hours before leaving Durban I had got friends to get me as many as possible of the newspapers that covered the affair, as I knew Hunt would value them. I now gave the bundle to him, and they clearly were a treasure.

I reflected that not only he but all these people had shared an event of which they would be able to talk with profit, or at least I hoped so, all their lives.

Twenty-six of the next 29 coelacanths caught in the Comoros went to the Muséum National d'Histoire Naturelle de Paris. Two of the first 29 are at unknown localities and one went to the American Museum of Natural History. A total of 45 coelacanth specimens (the highest number) is now (July 2016) held in museums in France, with the next most in Tanzania (37), Comoros (31), USA (28), Japan (18) and Madagascar (10).

Original logo of the Muséum National d'Histoire Naturelle de Paris

Ordre de l'Etoile de la Grande Comore

In 1953 JLB was appointed as a *Commandeur de l'Etoile de la Grande Comore*, one of the highest civilian honours in the Comoros.

The Dakota DC-3 had many exceptional qualities compared to previous aircraft. Its legendary ruggedness is enshrined in the light-hearted description of the DC-3 as 'a collection of parts flying in loose formation'. Its ability to take off and land on short concrete runways, or even on grass or dirt, makes it very popular in developing countries where runways are not always paved. They are even able to take off from water using amphibious floats.

Current uses of the DC-3 include aerial spraying, freight transport, passenger service, military transport, missionary flying, skydiver shuttling and sightseeing.

Chapter Fourteen

UP IN THE CLOUDS

I LOOKED again to make quite sure. Yes, the box was really there. So Hunt and I got out and, lining everyone up against the plane, had several photographs taken. We said our last adieux to all those present and climbed aboard, the doors were shut and the motors started. I got into that 'Mae West' suit.

I looked hard at the air-strip, as I had seen Blaauw doing several times before we got aboard. It was short, all too short, and the little wind of the early north-west monsoon came directly over the not inconsiderable hill whose steep southerly slopes marked its northern end. We must take off running straight at that hill, and I did not see how a plane of this size could hope to achieve sufficient elevation in so short a distance. However, I was no pilot and Blaauw clearly was one of the very best; so I strapped myself in and reflected that if we did hit the hill it would be over very quickly. It would certainly be a spectacular way to die, and, anyway, I had seen the Coelacanth.

10 a.m. on Monday, 29th December 1952. The engines roared in their test, slowed down, and suddenly we were off in a tearing rush of sound. I found myself gripping the seat and staring through the small air-hole. We left the ground and suddenly the sea and the slopes of the hill tilted so sharply that I caught my breath. Blaauw was banking steeply and the wing-tip was so near the trees I expected it to hit all the time; but we were safely away, and by screwing my eyes I could see the last of the tiny figures on the edge of the strip. One stood out in front waving. I assumed it was Hunt, and hoped that he would not be left to face unexpected repercussions, though he seemed to be very much at home and at ease among the people there. We went westwards and up, up, up, into the towering cumulus clouds whose marbled summits Blaauw estimated to reach up to at least 30,000 feet. They certainly were awe-inspiring and disquieting, great tumbled mountains of

dense whirling vapour with little clear air between, white and grey and black, some with electric storms inside their piled-up masses, so that they were at intervals lit up in sections by concealed discharges. We plunged into almost solid-looking clouds, inside which it was almost dark; then out again into narrow shafts of brilliant sunshine, sudden transitions which were as vivid as the intermittent flashing of a floodlight in a darkened room. We were so high that it got very cold, and I felt it even through my 'Mae West'.

At about 15,000 feet we steadied down. At that height and in such conditions navigation was to some extent guesswork, for there was obviously wind but no means of estimating either force or direction. I could not rest, for the weather looked very bad, and all the crowded events of the recent past hours had keyed me to a pitch of intense excitement; it was almost an intoxication. I went to the cockpit and stood behind the two pilots, who wore earphones. Suddenly Letley made some signs to Blaauw, and started to write down a message, which he showed to Blaauw, who flashed a quick glance at me and then read it again. Letley handed me the slip, and on it I saw: 'Managed to intercept a message stating that a squadron of French fighter planes left Diego Suarez before we took off from Dzaoudzi with orders to intercept us and to compel us to turn back to Madagascar.' My heart missed a beat. The two pilots were staring intently at me, while I did some rapid calculations in my head. 'What speed can they do?' I asked. 'Don't know exactly,' said Letley. 'But they are very much faster than we are.' 'Do you think it possible for them to overhaul us before we get to Lumbo?' I asked. Letley nodded. We were in a clear patch at that time, running between a series of piled-up mountains of cloud, with occasional glimpses of the sea, and where I had disliked those clouds before now there could not be too many for my liking. My mind was racing. 'Any hope of escaping in a cloud?' I asked. 'Radar,' said Letley. 'Well,' I said, speaking slowly, 'I don't know how you chaps feel about this, but I'm not going back. I don't believe they would dare to shoot us down if we refused to turn, but I would be prepared to chance that rather than turn back.' Letley suddenly burst out laughing and Blaauw grinned. So deeply had I been engrossed in weighing every aspect of the situation that it was some seconds

Playing jokes on one another is an effective way of relieving tension on long flights. This was a brave hoax to play on JLB Smith, an intensely serious man who was on an earnest mission.

During and after the Second World War the French in Madagascar used Morane-Saulnier 406 fighter planes with two-pitch propellers. Even a propeller-driven Morane would have chased down a Dakota with consummate ease! The spectacle of a lumbering Dakota trying to outmanoeuvre a squadron of nimble fighter planes in the skies over the Mozambique Channel boggles the mind! Later the Malagasy Air Force acquired MiG-77 jets.

Wikipedia GFDL

Morane-Saulnier MS 406

Malaysians and Indians had some influence on the design of traditional fishing gear in the Western Indian Ocean.

If JLB had seen the efficient tackle made and used by traditional Comorian fishermen to catch oilfish (and, inadvertently, coelacanths), he might not have been so disparaging about it. They use line handmade from twisted coconut fibre (and later cotton) but still manage to catch fishes in water hundreds of metres deep. Today they mainly use heavy-gauge monofilament nylon line (if they can afford it).

Comorian fishermen use large metal hooks and the flesh of fish (especially *roudi*), squid, octopus or crab as bait. A fist-sized stone sinker (*mbize*) is attached loosely to the line so that it falls off at the desired depth when the fisherman jerks the line.

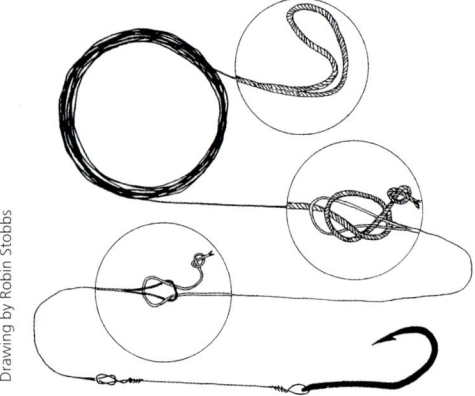

Drawing by Robin Stobbs

Fishing tackle used by traditional Comorian fishermen to catch oilfish (and coelacanths as a bycatch)

before it penetrated my mind that this was a hoax. I made no comment, I was not even angry, it was such a terrific relief, and I went inside, put my sleeping-bag on the icy floor near the box and tried to rest. My eyes would stray to that box. They would not have got me to turn back. This was my Coelacanth. I lay there, going over in my mind again all I had learnt, especially what Hunt had told me, and how it fitted into the background of my life and work all those long years in these parts. It is an amazing story.

In the western Indian Ocean, which embraces the East African coast, Madagascar, and other lesser islands, it is normal to find deep water close to land. There are only a few places where the bottom shelves gently from the shore, in most parts great cliffs beneath the sea plunge abruptly to the depths, and here the clear and lovely green or blue of the water changes to an ominous black. An echo-sounder chart from such parts is most interesting, for it shows the profile of the bottom of the ocean in miniature.

The coastal natives over most of that area have almost everywhere a seafaring and fishing tradition, often derived from the Arabs, whose southerly penetration in exploratory voyages commenced so long ago that its origin is lost in the remote past. Line-fishing of different kinds is common to them all, and partly because the water is too clear for fishing to be successful where it is shallow, in many parts they fish where the water is relatively deep, from 50 to 100 fathoms. With proper tackle and ample fish, the many problems of deep-line fishing can be overcome to make this a commercial possibility, but when you see the clumsy tackle used by most East African natives for this purpose you wonder how it pays; and it continues only because their economic level is so low, their catch per man per unit of time could never be competitive in any efficient civilised community in temperate climates. Those who fish in this way generally use a lump of coral the size of a man's head as a sinker. In some parts they go out with dozens of such lumps, and have a special knot so that when the sinker reaches the bottom or the desired depth a quick jerk shakes it loose from the line, which method at least saves hauling this extra weight several hundred feet to the top. Others find it less trouble to prepare only a few such sinkers, and use them all the time, hauling one laboriously all that way to the surface each time they need to pull up the line; but pulling up one fish and letting the line out again may take twenty minutes or more.

At that depth, 50 to 100 fathoms, the water is colder than at the top, and quite often the fishes down there are different from those commonly found in the surface layers. On parts of the Kenya coast, for example, natives who learnt the trick from visiting Japanese fishermen haul up from this deep, cooler water numbers of fish of kinds that it is astonishing to see in such tropical climes.

The people of the Comores have apparently always practised deep-line fishing. At least that is what you learn on inquiry, but one has to be cautious in accepting statements about the past from natives in those parts, since the average East African native has a much poorer sense of time than any European. Once a thing has happened it is past and done with, and we constantly found that it is exceedingly difficult to establish whether a past event had taken place last week, last month, or last year. In that area the Comores are almost unique in structure, for beneath the sea they apparently slope steeply and uniformly down, in many cases at an angle of 60 to 70 degrees, which is very steep indeed. As a result, there is deep water close to the shore, except only off part of Mayotte, where there is a large reef to the west. It is therefore possible to fish in quite deep water without going far from the shore, which is a great advantage, because it means that on the leeward side of the island such fishing can be carried out during even the windy seasons, which cover most of the year.

The Comoran natives are not distinguished by great energy; indeed, in that respect they fall below the average, already low, and they are not uniform in performance—those on Anjouan being considered the most progessive and energetic, while those on badly disease-ridden Mohilla are notoriously lethargic and hard to move. This is reflected in the proportion who will take the trouble to go fishing and in the time they give to it, so that it is not surprising that the Coelacanth story featured Anjouan, where the most energetic Comorans live.

In this deep-line fishing at the Comores they catch distinctive, well-known fishes, like certain large Rock Cods, some specialised species of the Snapper family, and the cosmopolitan 'Oil-fish' (*Ruvettus*), a rather elongate Snoek-like fish with peculiar scales. This fish is very oily and in some parts has an evil reputation, it

JLB's generalizations about the energy levels of Comorians from one island to the next are interesting, if controversial. I have visited all four islands and would also rate the inhabitants of Anjouan as the most enterprising. This is not surprising as they live in a more congenial environment with more water, timber, natural food and employment opportunities, as well as better infrastructure, than the other islanders.

When I last visited Grande Comore in 2008 there was a marked lethargy among the local populace. Unemployment levels among the youth were very high, infrastructure and services had virtually collapsed, tourists had vanished, and the streets and coastline were littered with rubbish. It is a far cry from the evocative and exuberant place that I had visited 20 years previously.

Litter along the shore of Moroni on Grande Comore, November 2008

Mike Bruton

Traditional hand-line fishermen in the Comoros mainly target oilfish (locally known as *nessa*). They eat the flesh and use the oil for a variety of purposes (it is also known as the castor-oil fish). Comorians occasionally eat salted coelacanth (locally known as *gombessa*) but, in Madagascar and Tanzania, the fresh fish is cooked and eaten; Malagasies also use coelacanth flesh as bait. *Nessa* and *gombessa* live in a similar habitat and at about the same depth.

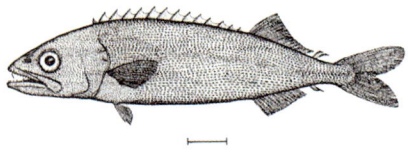

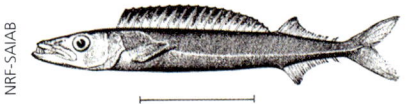

Oilfish locally known as nessa *(top), which are targeted by Comorian fishermen, and* roudi *(above), which are used as bait*

In the Comorian language *Shikomori* (the 'language of islands'), *gombessa* means 'taboo', in reference to the fish's oily flesh and inedibility but possibly also due to its sacred status. Like sharks, coelacanths retain urea in their flesh, which contributes to their foul taste.

Technically, Comorians do not use long lines, which are typically attached in series to a floating surface line, but hand lines, which are deployed singly and held by the fisherman in his wooden outrigger canoe (*galawa*).

being said that the oil is strongly purgative and even poisonous, and there are many records of ill-effects and even deaths following the consumption of its flesh. In other parts, however, as apparently in the Comores, it is a valued food fish and commonly eaten without harm. It is interesting to note that this fish is oily, for so is the Coelacanth, very oily.

By late December 1952, the leaflets that Hunt had taken in the previous October had, by orders of the Governor, been distributed to all the islands and by special runners about them, so that the more intelligent natives at least were aware, even if they could not believe, that the enormous sum of 50,000 Colonial francs would be paid for one special fish.

According to what has been told, some at least of the European officials read the leaflet with scepticism if not with amusement, and I was told that though a visiting scientific officer from Madagascar was shown the leaflets Hunt had brought (he may have seen them before on Madagascar), he attached so little importance to them that he apparently did not indicate anything special to the local authorities nor did he mention the matter on his return.

On the night of the 20th December a native, one Ahmed Hussein, of Domoni, a small village on the south-east coast of Anjouan, with another fisherman, went out in his canoe to fish. First they went to his palm-leaf-strip traps, close to the shore, from which they took small reef fishes that are used for bait. Then they went somewhat farther out, and let out their long lines. Hunt had heard that the depth was about 20 fathoms.*

In the night Hussein hooked a large fish, which he eventually subdued at the canoe by battering its head, a merciful way of killing a fish, but scientifically a shocking tragedy. Nobody seems to know if he caught anything else, but from what I gathered of the Comorans, one such large fish would have been more than enough to satisfy any of them, so they went back to shore and to bed.

* The French later gave this depth as considerably more; indeed, stating that they had been able to find the exact spot and take soundings, and they quoted the depth to a metre. This indicates work of unusual precision, for it would depend on the memory of a native to find the exact spot where he was when he hooked the Coelacanth, on a dark night in a drifting canoe without any anchor. With the bottom sloping as steeply as in this area, even a few yards out would make a considerable difference to the depth. However, a few fathoms more or less are not important. What is important is that where the Coelacanth was hooked was emphatically not any 'inaccessible depth' of the sea.

NRF-SAIAB

This is in many ways an immoral story, for what was bad proved good. Even in our colder climate I have taught my sons that when angling it is a crime not to clean any fish you have caught before taking it home, even if you get in cold, wet, and sleepy, and no matter what hour. Thank heaven, these Comorans had not had a training of that kind; for they apparently threw the fish down at

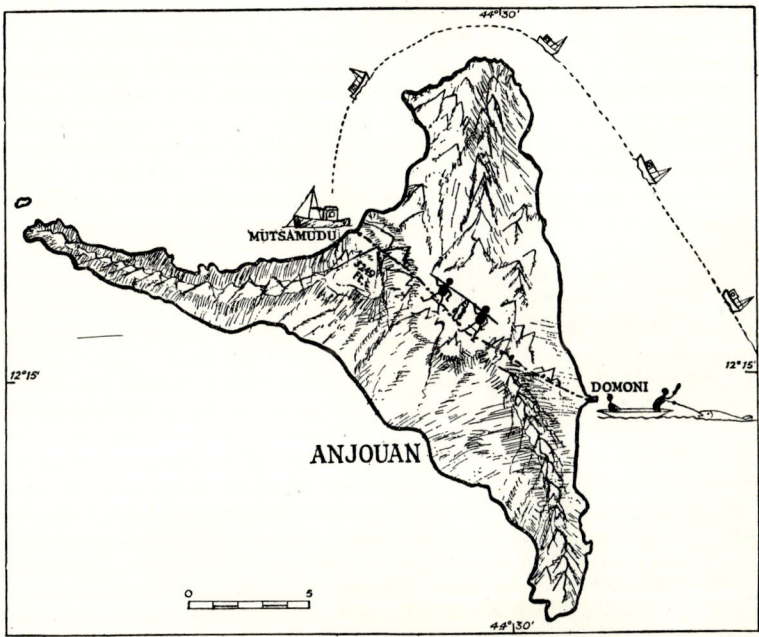

Fig. 6.—The island of Anjouan, Comores, formerly known as 'Johanna', showing where *Malania* was caught and position of Hunt's schooner.

Hussein's hut, just as it was, without scaling or gutting it, despite that oven-like climate.

Next morning Hussein took it to the local market for sale, and it was about to be cut up when a native who looked at it advised them not to do so because it looked like the fish on the leaflet, Hunt's paper.* While the French authorities did a tremendous amount, it is clear that Hunt was always hunting, so that the natives came to connect him with the fish. One could imagine the

---

* It later became apparent that this surprisingly intelligent behaviour on the part of the native was due to his being of a higher class, a trained teacher.

Fishermen who have previously caught coelacanths claim to know immediately when they have hooked one by the reactions of the fish. Unlike oilfish or sharks, which tug hard and fast, *gombessas* are comparatively slow but dogged, like groupers (locally known as *sahali*), and may take up to two hours to bring to boat. The fisherman who caught the eighth *gombessa* off Anjouan in November 1954 knew his trade and eventually caught four more coelacanths.

Fishing boats off Anjouan Island depicted on a Comorian postage stamp

Once on the surface the *gombessa* (or other large fish) is gaffed through the jaw or gills with an *ulowo*; this keeps the fish captive and fresh (even alive) until first light the next morning when the fisherman paddles ashore.

Several large wrasse species in the Comoros are also called *gombessa* by traditional fishermen.

The trek from Domoni to Mutsamudu is through very mountainous terrain. Eugene and Christine Balon and I travelled this route via Bambao and Patsy in April 1987. Along the way we encountered a fragrant ylang-ylang factory using coconut husks as furnace fuel in the jungle at Tsimbaeo. Then, Indiana Jones-style, we clambered down cliffs and ancient ladders to reach the Tatinga River deep in a steep-sided valley, where we electro-fished for unusual migratory gobies.

The second coelacanth was not, in fact, carried from Domoni to Mutsamudu but travelled in a public works department truck that was transporting the Anjouan soccer team to a soccer tournament!

Mountainous terrain on Anjouan Island depicted on a Comorian postage stamp

debate that ensued. The instructions on the leaflet were clear: 'Do not cut it or clean it or scale it, but take it at once to some responsible person'; and who more so than Hunt, whose schooner, by a miracle, was known, as natives do know such things, to be anchored at that moment at Mutsamudu on the other side of the island, about twenty-five miles away. But what a twenty-five miles, the path mostly through deep, densely bushed valleys and over high mountains.

It still remains in my mind as one of the most astonishing things of this whole story that anything could stir people like the Comoroans to the stage of even considering carrying that 90-lb. fish that terrific distance over such hard country in that blazing, tropical heat. That they actually came to do it shows the tremendous power of money. It was only a fabulous sum like 50,000 francs that could have got them to do it, but it would have been an achievement in even a temperate country, in any country. They did it. My blood ran cold when I heard the story, the uncleaned fish first in the close tropical night, and then that whole long day in such torrid heat. It is a miracle that it had not liquified in putrefaction long before it reached Hunt. It may be the preservative nature of its oil that saved it.

According to Hunt, when those men reached Mutsamudu they came straight to him, and he recognised the fish at once as a Coelacanth. It was already putrefying and he had no formalin; so he went posthaste to the office of the local doctor, but found he was away. Remembering Mrs. Smith's instructions, Hunt told his crew to cut it for salting, in a hurry, and unfortunately while he was getting the salt they cut it as they were accustomed to do in salting such fish, hacking it open along the back, through the body and head like a kipper. Next time you see a kipper, look at it. That is how my precious *Malania* was cut, but of course by Hunt's instructions the insides were left intact, most of them.

Hunt realised the full importance of this find, and wisely questioned the natives. Was this fish known to them? Oh yes, they knew it well, they were rare but caught regularly. They called them '*Kombessa*', and they were not much valued as food when fresh, but were good salted. When cooked fresh the flesh became soggy and jelly-like, and was not very good to eat; but they were eaten. They were nearly always caught with the Oilfish in the deeper water and on flesh bait, using Squid or any kind of fish. They fought hard on a line, were difficult to kill, and some were lost after being hooked and pulled up. Hunt found out afterwards that a good many people there knew these curiously rough scales, for they were used to roughen bicycle tubes in mending punctures. These fish were usually caught in cyclone time; that is, towards the end of our year. Most of them were big fish, more than 30 kilograms, some very big, but there was another and smaller kind occasionally seen. Some of this might be doubtful, but Hunt was satisfied that they really knew the Coelacanth, and that though it was rare, it turned up regularly. When he told me all this, I was worried by the name '*Kombessa*', because in East Africa the rather rare large Kingfishes (*Caranx*) are called 'Kambesi', and to the undiscerning eye they might well appear not so very unlike the Coelacanth. They have the same ferocious appearance and a large, powerful mouth.

Hunt had to act, and he did so with speed. It was not clear if a radio message was sent or could be sent from Anjouan to Dzaoudzi, but Hunt set out in his schooner and arrived at Pamanzi on the following day, the 22nd December 1952. He at once informed the Governor of his find, the local doctor willingly gave all the formalin available, and Hunt himself injected this into all the parts of the fish, and from what I saw he did it well. Hunt had a metallined box made to hold the animal. He sent me the cable mentioned before, expecting me to be at Grahamstown, but realised from my reply that I must still be on the *Dunnottar* at Durban.

The French can scarcely be blamed if they were at first somewhat sceptical of the great importance Hunt attached to this fish; but his intense excitement had its effect, and a cable about it was sent to the Scientific Institute of Madagascar at Tananarive. Not only was this cable so mutilated in transmission by the native operators as to be undecipherable there, but the official to whom it was addressed was absent at that time. Christmas time! (See p. 88.)

If I had endured many difficulties and uncertainties, those at Dzaoudzi had had their own. Hunt was shrewd enough to realise that the importance he attached to the find was having an effect on the French authorities, and he eventually had to face the situation that, despite their initial scepticism, he might well have

Based on catch statistics from 1952 to 2010, most coelacanths (for which the month of capture is known) were caught in the Comoros from November to March. This pattern is partly due to the seasonal habits of the Comorian fishermen, whose fishing effort is greatest during the northerly monsoons (*kaskazi*) and inter-monsoon calms, when sea conditions are favourable. Little fishing activity is possible from July to October during the southerly monsoons (*kusi*), when sea conditions are far less favourable.

Comorian fishermen have a separate name for small coelacanths – *mamme* – but they are the same species as the adults. They claim to have seen them swimming over a sandy bottom in shallow water during the day but this seems highly unlikely. The preferred habitat of young coelacanths is still unknown but is likely to be over deep rocky reefs and in caves.

JUVENILE COELACANTH

Eugene Balon

Juvenile coelacanths have occasionally been caught. In 1974 Said Ahamada caught an 800-g, 42.5-cm juvenile, less than one year old, off Iconi on Grande Comore. In 1976 a prematurely-born juvenile about 25 cm long (the birth size is about 33 cm) was found swimming slowly on the water surface off Iconi.

The stonefish is the most venomous of all fishes. JLB was stabbed in the thumb by a stonefish at Pinda and Margaret's remedy was to pour very hot water on the wound so as to denature the protein in the neurotoxic poison. Stonefish stabs can be fatal but fortunately Smith survived. Stonefishes are not aggressive; rather, they are sedentary and use their poisonous spines to deter predators.

Peter Barnett

JLB Smith with a stonefish in Mozambique

The wreck on Baixo do Pinda is that of a large, steel cargo ship ('*BP*') that foundered there and has now created a cosy artificial reef, which is inhabited by corals, sponges, sea squirts, crabs, crayfish, batfish, potato bass, scorpionfish, sweepers, soldierfish, lionfish and snappers, as well as hawksbill and green turtles.

to surrender the fish to them. He was in a most awkward situation, as his trading depended on the good-will of the French authorities, and he could not afford to offend them. Not one of them would apparently believe that anyone would come from so far off as South Africa just to fetch a fish, certainly not in a special flight such as Hunt visualised and obviously expected. Hunt got the Governor to promise that if I did come to fetch it in person that way, no difficulties would be raised about its being handed over to me. Even when my final cable came there was apparently still scepticism; they were not finally convinced until the roar of the Dakota brought them all running from their homes to look at the skies. It must indeed have been sweet music to Hunt's ears. . . .

I must have dozed off, for I woke with a start to a shout in my ears. Ralston's head was poked through the door and he was gesturing ahead. 'We can see land,' he shouted and withdrew. I was up in an instant and went forward. Yes, there was the African coast and north of us was a bay, but it was not the bay of Mozambique. We all stared. It proved to be Mokambo, south of Mozambique. The wind up top had been northerly after all, and we had been driven southwards. Swinging round in a wide arc, we soon picked up my beloved isle of Mozambique, and I could see right beyond Fernao Veloso, which is the northern point of Nacala Bay, to the wreck on the end of the Pinda Reef.

Pinda! I have told you something of our life there and of the lions. Pinda! Where I endured unspeakable agony and nearly died from the stabs of the dreaded Stone-fish; but all that is a story by itself.

As the air was fairly clear just then and the light good (11.55 a.m.), I asked Blaauw to circle close north of Mazambique which enabled me to take the photograph reproduced on Plate 5, showing the whole island, with the famous fort of St. Sebastian at the near, north, end. At 12.05 p.m. we landed on Lumbo. The 'Chefe' of the airport ran out and eagerly sought my news. I told him we had got it and he stared at the box. 'Could he see the fish?' I shook my head, for on the way over I had resolved that not a single person was going to see that fish until I had shown it to Dr. Malan.

It was baking hot, the wind from the land almost scorched the skin. I asked the crew their needs, and they voted for iced shandies.

So I telephoned the hotel and asked the manager to send the necessities, 'muito muito depressa', with plenty of ice, and, as is usual with the Portuguese, it was all there in record time. We did not stay long and, after hasty but cordial adieux, took off at 12.55 p.m. into a mizzling north-wester. Although weather reports stated that it was cloudy with rain most of the way south, Blaauw said he hoped to reach Lourenço Marques by 6 p.m. and to get to Durban that night. It seemed ambitious, but he was the expert. I calculated that my wife, due at Port Elizabeth early that morning, should already be at home, though I naturally knew nothing of the road breakdown she had in reality endured on the way.

This was almost the worst part of the flight. We went up to 15,000 feet again, and there was nothing to see but cloud. It was beastly cold in that unlined metal shell, so I lay down on the floor wrapped in my bag and tried to sleep, but despite two restless, wakeful nights my mind was still racing madly, weighing up all that had happened and planning all that still had to be done. I must not let my exultation lead me astray, for I knew only too well that enthusiasm can end in a long and painful walk back, and this was a peak of achievement I had not before attained. Virtually alone in the scientific world, I had held to my conviction that the Coelacanth was to be found in the reefs of the East African region, and if what I had learned at Dzaoudzi was correct, the fish I now had with me was not just another stray, like the one at East London, but a homebird at home. It should be only a matter of time before others were found. Yes, I was exulting. It had not been pleasant to have one's deductions just pushed aside. I thought again of Smuts, with his ready ear for the overseas experts.

It was strange to look back on how almost all other scientists had been united on that issue. It was almost a kind of conspiracy between them. There was the opinion of the British Museum that the East London Coelacanth was obviously a stray from the deeper parts of the ocean. . . . The Danish deep-sea expedition had gone hunting hoping to find the Coelacanths in the great depths; palaeontologists in America and other countries! They were satisfied that Coelacanths lived in the 'Inaccessible depths of the ocean'. All of them knew it was ludicrous to go fishing for

O.F.—11

Research since 1952 has revealed that the Comoros has one of the largest concentrations of coelacanths, with the main population occurring along the west and south coasts of Grande Comore. Substantial numbers have also been caught off Tanzania.

Grande Comore depicted on a postage stamp

The true abundance and distribution of coelacanths is still poorly known. They have lived off the coast of northern Zululand since time immemorial yet they were only discovered there by mixed-gas divers in 2000.

The 'British captain' who settled there and made a famous garden is almost certainly a Frenchman, Léon Humblot (1852–1914), an entrepreneurial botanist who arrived in the Comoros in 1884. Like many island dwellers, he arrived with the intention of leaving soon, but stayed for years. He collected plant specimens for the Muséum National d'Histoire Naturelle de Paris and played a key role in ensuring that France gained dominion over the Comoros. He introduced spices and perfume plants into the Comoros and established the foundation for lucrative industries based on these crops.

Property of M Dulac

Léon Humblot

Today the Comoros supplies over 80% of the world's demand for ylang-ylang and also exports other perfume essences and spices such as vanilla, basil, jasmine, bitter orange, ginger, frangipani and lemongrass. Coconuts, coffee, jackfruits, cashew nuts, sisal and tobacco complete the potpourri of botanical exports.

Coelacanths with hooks and lines. It could not be done. And now it appeared that malaria-soaked, worm-ridden, bone-headed blacks of the Comores had been doing exactly that for centuries past, and not all the learned deductions of museum scientists could unsay it. What is more, Coelacanths were no strange food on the Comores. It was not even impossible that ancestors of these very scientists may have eaten Coelacanth, for in the early days of sail many British ships engaged in the spice trade with the East called regularly at 'Johanna', as Anjouan was called in those days. One British captain liked the place so much, he settled there and had a famous garden. At Mutsamudu, deep water close to land gave complete shelter from the hellish wind, and here the ships revictualled and filled their tanks, while the crew banished incipient scurvy with vitamin-rich tropical fruits. Slabs of salted Coelacanth may well have been among the stores they took aboard.

Thinking of food brought me back to the present, and I got up and gave the crew a snack of biscuits, cheese, and dried figs, but could not eat myself. The cloud was still so dense that neither land nor sea was visible. Though I was restless and overstrung, I compelled myself to lie down again, and made a stern effort to sleep; but a sudden stab of intense pain brought me to a sitting position. From before we left Durban my ears had been troublesome, and now, after several severe spasms, this settled down to a steady toothache in my right ear. This was a predicament, for before the advent of penicillin I had once had an abcess there, and it had not been a pleasant experience; and with all that lay ahead another now would be a disaster of the first magnitude. With anything of that kind treatment is a matter of hours, so I went forward and told Blaauw of my predicament, and asked to light the primus to sterilise a syringe for a penicillin injection, for which everything was in my 'collecting-box'. Blaauw was shaken, but I would not press the matter too strongly, as the plane and our lives were his ultimate responsibility, and of course the Coelacanth was with us now. It probably would not have helped even if I had gone on, it was clear that I would have to stick it. The agony was considerable, and I could scarcely sit, let alone lie down. So I turned to a resort I have often employed in such circumstances, which is to keep the mind so busy that it does not register pain, and decided to use the remaining time to write notes on the events of the past few days while they were still fresh in my mind, especially what I had learnt at Dzaoudzi. Soon I became so engrossed in getting the many points into their correct order that I forgot all else, and was brought, not to earth fortunately, but to look at it, by Ralston, who shouted 'Bazaruto', and there it was far beneath us to starboard, the clouds at that time being wispy and thin. We cut across inside Cape Sebastian, when the cloud over the land shut down again, and the plane became once more just a noisy box encased in cotton-wool. I went on with my notes.

My ears suddenly told me we were at a lower altitude, and looking out I saw the lakes about Inharrime, with glimpses of the sea between the ragged clouds. We dropped to only a few hundred feet above the dunes, and in quick succession passed the Limpopo mouth, then along the chain of coastal lakes about San Martinha, Chefina Island, and the Incomati mouth, and there was the Polana beach and the bay of Lourenço Marques, all old friends.

We touched down at 6.20 p.m. Our Vice-Consul Phillip was waiting at the airport, and there were greetings and inquiries from the various Portuguese officials. My chief concern was refreshments for the crew, who again alternated between coffee and cool drinks with sandwiches.

Despite my great stress, I had always kept in mind the Brigadier's request that the crew should not be driven too hard. This had been a long and trying day, and eager as I was to return I put this to Blaauw and suggested a night in Lourenço Marques; but like horses scenting home they were all eager to get on and to reach Durban that night. I asked Phillip to telephone my friend Dr. George Campbell in Durban to find me an absolutely trustworthy photographer to meet the plane, for I wanted to have my precious films developed at once.

It was raining as we took off at 6.45 p.m., and in the gathering gloom it seemed a flight that would never end. I had discovered that the crew had all been on holiday when they were abruptly recalled for this flight, and while they had not offered any account of their reactions, it was only natural that there should have been some degree of resentment. I spoke of this, and said that some compensation for them would be to have participated in what

JLB's 'toothache in my right ear' (page 162) was almost certainly caused by the unequal air pressures that would have developed on either side of his eardrum, especially if his eustachian tube was blocked. As a Dakota does not have a pressurized cabin he would have been subjected to widely changing air pressures as they changed altitude. This condition might have been exacerbated by an abscess.

The chain of coastal lakes near San Martinha, and those further south such as Lake Piti, remain largely unexplored as far as the fishes are concerned. We mounted an expedition to Piti in the summer of 1967/1968 and found estuarine and marine fishes living in this land-locked lake that had last been open to the sea in the 1920s.

The epic flight to fetch a dead fish has remained etched in the memories of the flight crew for many years. Today, over 60 years later, websites on the history of South African aviation, and of the Dakota, carry detailed accounts of the flight of the 'Flying Fishcart'.

In December 1992, to commemorate the 40th anniversary of the flight, SAAF C-47 Dakota 6832 flew from Cape Town to Grahamstown with three of the original crew on board – Duncan Ralston, Willem Bergh and 'Vanski' van Niekerk. Major Vic Fouche of the South African Air Force Museum announced that 6832 would be restored to its 1952 livery and donated by the South African Air Force to the SAAF Museum in Pretoria.

Commemorative cover issued in 1992 to celebrate the 40th anniversary of the flight of the 'Flying Fishcart'

On this occasion, organized by Robin Stobbs and the SA Dakota Association, the Ichthyology Institute issued a commemorative envelope bearing the four South African coelacanth stamps that had been issued in 1989 and a logo showing the coelacanth and a Dakota.

would remain an historic flight. Being curious about this matter, at about 8 p.m. I wrote on a sheet:

'Will you kindly write here what you said, when you heard you had to come on this unusual flight to go hunting a dead fish near Madagascar. (Leave out unprintable words.)' This was passed around in turn, and it is interesting to record their comments, which are given below:

1. *Commandant J. P. D. Blaauw*

It must be a pretty important fish if the Prime Minister is prepared to give an aircraft and crew to some hare-brained scientist to fetch it.

2. *Captain P. Letley*

The first time that I knew we were going to fetch a fish (DEAD) was when the Orderly Officer told me. My reply, as you requested, cannot be written down. Anyway, I enjoyed the trip.

3. *Lieutenant D. M. Ralston*

Not very impressed at first and was doubtful whether it was the correct fish. Professor Smith's enthusiasm is infectious, and I have found this an extremely enlightening trip.

4. *Lieutenant W. J. Bergh*

As I was all set to go on a special visit (my girl friend) for the week-end, I did not like the idea very much at first. I had to cancel all arrangements by phone and said, 'Somebody caught a fish that should long since have been dead!' The trip was, however, so enjoyable that I was all for staying at Dzaoudzi.

5. *Corporal J. W. J. van Niekerk*

When I heard about the fish story, I thought that we were going to have chips with it, too, but I enjoyed even the smell.

6. *Corporal F. Brink*

Although I made arrangements to visit some friends on Sunday for a swell dinner, I had no idea that I would have to come on a fishing trip for a fish that was dead already. But the trip was very enjoyable, and the Professor was like a nurse to us, feeding us all the time.

I suddenly realised that the pain had gone from my ears and the enforced inaction had produced relief from strain; but only to merge into an incredible weariness, and I could think of nothing more wonderful than getting between cool sheets in a quiet room. Though what I had already planned meant a very early start next day, I might manage to snatch some sleep at Durban; for the few friends who knew of my return would soon leave when they saw how weary I was. It is quite amusing to look back on my ignorance of what lay ahead.

'The lights of Durban,' Ralston again, it would not be long now, but it was a full half-hour before we could land. Round and round we buzzed, because a stubborn aerial refused to come in and eventually a plate of the hull had to be unscrewed and opened before it could be retracted. We went down. As the door opened I was first at the steps, and was immediately blinded by a battery of flash-bulbs. In that fraction of a second's vision I was appalled to see a seething crowd whose dimensions enlarged as further flashes flared. How on earth had all these people got there? My bewilderment and dismay increased when that human dynamo George Moore of the S.A.B.C. at Durban grabbed me by the arm and pushed a microphone with trailing wires under my nose and started to question me. The sounds from my lips were more frog-like than human, and, indeed, later that night when my son in Grahamstown heard those words from our radio set, he said to my wife, 'But, Mom, that's not Dad.' His mother's sole retort was a terse 'Shut-up', as she sat tense before the machine.

In this confused milling crowd I was moved across the ground towards the office. I needed a drink badly and asked for coffee. A Customs officer pushed a form under my nose, put a pen in my hand, and asked me to sign on a line. The officer in charge said the Commander-in-Chief in Pretoria was waiting on the telephone, would I speak to him, and so I made my formal report and received his congratulations (Plate 6, facing p. 129). He asked if I was now finished with the plane or should it take me to Grahamstown? So I told him my plans, could the plane take me to Cape Town in the morning as I wished to show the animal to Dr. Malan? Would he kindly make contact with the necessary authorities as soon as possible to say that I wished to bring it? All this he promised to do. I then told Blaauw that we might have to go to Cape Town in the morning and why. He was by then clearly resigned to anything and told his crew. Only later were they able to get my wife on the telephone at Grahamstown, when I gave her a brief account of events and told her my plans.

My coffee had got cold, so they fetched more, and it was scarcely easing my parched throat when George Moore brought me to

The interview that JLB did with the SABC on the tarmac at Durban Airport on that evening of 30th December 1952 is one of the most famous in the annals of South African science. The Ichthyology Institute subsequently obtained a vinyl recording of the speech and it has since been digitized. I use extracts from it regularly in my talks on the coelacanth. It never fails to move the audience.

JLB was so moved by the events that he was re-enacting that he wept again during the interview, and many of his listeners wept with him. I doubt that any scientific event in South African history has had such an emotional impact on people.

East London Museum

SMITH RETURNS WITH COELACANTH

Historic Find for Science

100-lb. FISH "ALMOST PERFECT"

FISH NAMED IN HONOUR OF MALAN

Cape Times Correspondent
DURBAN—As the coelacanth is of a slightly different species from that found near East London, 14 years ago, Professor Smith has named it provisionally *malania exlorinae*, in honour of the Prime Minister, Dr. Malan, who placed a military Dakota aircraft at his disposal and to commemorate the locality in which it was found.

Coelacanth newspaper headlines

The reason why the story of 'old fourlegs' has gripped the public's imagination is that the coelacanth has a rich natural history as well as a fascinating cultural history. Biologically, it is one of the most extraordinary creatures on the planet today and the history of its discovery is populated by an astonishing range of captivating people, with the passionate Smith at centre stage.

Pity the poor lungfish! It too belongs to an ancient group of fishes, close to the origin of four-legged animals on land (closer, it seems, than the coelacanth). It too has a bizarre anatomy and some truly astounding adaptations. Yet hardly anyone knows (or cares) about it. What it lacks is a swank cultural history and a good PR agent!

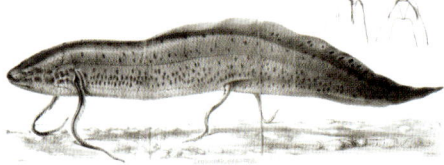

The much-ignored lungfish

earth with a bump, saying that they wanted me to say something over the microphone, as the whole of South Africa was waiting for me. It was only much later that I discovered that all programmes had been dislocated that evening for this purpose. This was terrible. It was already 9 p.m., we had provisionally arranged to take off at 3.30 a.m., and I was so weary that it was an effort to utter any word at all. I looked at him in dismay, and said, 'What must I say, man?' 'Anything you like,' he said. 'Even just a message.' I sat back and thought. If people were waiting and had been waiting, a mere message was a poor return; they would want to know what it was all about, and it was emphatically not a story to be told in a few words. Suddenly I remembered my notes, and told Moore I had made them and could use them to give an outline of events, but that it would take at least twenty minutes. Instead of saying, as I hoped, that that was far too long, to my dismay he shouted, 'That will be fine, Professor, you take as long as you like'; and there was no doubt that he meant it. The notes were still in the plane, and with my directions one of the crew went to seek them while I drank more hot coffee and tried to compose my mind to the ordeal ahead. I must not make a hash of this. Those notes were merely a sketchy scrawl. The man came back, the sheets were pushed into my hand, and I sorted them out while those who filled the room shuffled and pressed tight around me like a wall, though I could scarcely see them through the black mists of weariness that enclosed me. I began to speak, or rather to croak at the microphone, my throat muscles almost refusing to function, and then suddenly I was no longer aware of my surroundings but almost miraculously back in it all, living it over again, the strain and agony and suspense. It was so real that when I told how I had found myself weeping at my first sight of the creature on Hunt's vessel, I was there again and even though I knew that tears had again run from my eyes, I did not care. Once from sheer exhaustion my voice failed completely, and I had to stop and gesture for more coffee, then went on. At the end I felt like a pricked balloon, and just sat hunched up while whirling black mists shot with points of light submerged me, but Moore brought me up by asking if I had any idea of the name I intended to give this creature. I said, yes, to honour Dr. Malan, the genus *Malania*, and from the island Anjouan, where it was found, so came provisionally the name *Malania anjouanea*. There was dead silence, then a sudden burst of applause. Someone seized my notes and said 'Let me see those notes'; I was too weary to care what anyone thought, it was enough that it was over at last. Perhaps now they would leave me alone; I wondered if I would ever again know what it was like not to be unutterably weary. But that merciless Moore pushed the microphone nearer and said, 'Now just something in Afrikaans, Professor. You must not disappoint our Afrikaans listeners.' Only half conscious now, I tried to rouse myself, and from far off, speaking of this same event in Afrikaans, heard a strange voice that it was hard to realise was my own. The words came without volition. I have never had the courage to ask to listen to that part.

JLB's Afrikaans broadcast was equally moving. It was clear to listeners that he was totally exhausted, almost shell-shocked, and was under considerable strain; yet he agreed to give the interview, which made the audience even more appreciative.

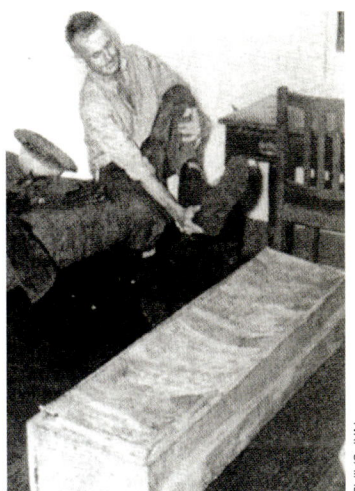

NRF-SAIAB

Coelacanth coffin with JLB Smith in his bedroom in Durban (December 1952)

In 1965 JLB delivered a series of popular science talks (in Afrikaans) on SABC radio and stated, 'Knowing the most wonderful thing in the world is useless if you do not share it with someone'.

JLB's almost maniacal determination to show the fish to the Prime Minister, and to him before anyone else, seems almost comical today but there is no doubt that, without Malan's intervention, Smith would never have got his fish.

NRF-SAIAB

The second coelacanth, with Margaret, JLB and William Smith and the crew of the 'Flying Fishcart' standing behind it, at Ysterplaat Airforce Base in Cape Town in December 1952

In his exhausted and agitated state Smith assumed that Malan would want to see the fish. As it turned out, Malan and his wife welcomed their visit and laid out the red carpet for JLB, Margaret and the beast. Perhaps Malan saw the fish as a novel political opportunity?

Chapter Fifteen

MALAN AND MALANIA

As I lay back, my friends swam into my consciousness and took hold. I could not speak, but tried to smile thanks for all they said. Photographer George Symons took my two precious rolls of films, and swore to let me have them before 3 a.m. 'When can we see the Coelacanth?' my friends asked, and this was drowned in a chorus from the press, who asked to photograph it. There was near consternation when I shook my head, then explained that Dr. Malan was to be the first to see it; until then, no one else. They tried all they could, but I was adamant. Then came a telephone call from Pretoria, to say that in the matter of taking the Coelacanth to Cape Town to show Dr. Malan, it was felt that as he was resting and on holiday, and in view of his age, having need to conserve his strength, it would be too much for him to come all that way in from the Strand. They had obviously mistaken my intentions. I asked if the matter had been put to Dr. Malan himself, and the answer was 'No,' they did not want to disturb him, but he had been told of our safe return. So I said that I appreciated all their kind concern for Dr. Malan, but I certainly had as much myself and had never expected him to come to Cape Town and certainly did not want that. I wished to take the Coelacanth to him wherever he might be, and unless he expressly forbade it, would do so. I added firmly that if all else failed I would push the thing there on a wheel-barrow myself. At any rate, that could be settled later. Would the Commander-in-Chief kindly authorise the flight to Cape Town, and the issue could be put to Dr. Malan in the morning; but it must, please, be put to Dr. Malan himself, for whatever else might happen I intended to go and find out myself if it had been. He very amiably said go ahead, and he would pass on my message. Then they put through a call to my wife in Grahamstown, when I told her what had happened and asked her to bring all the necessaries for coffee and food for seven to the

Grahamstown aerodrome for 6 a.m., also four copies of my *Sea Fishes of Southern Africa*, each inscribed with the name of the officer concerned, and as follows:

WITH MY COMPLIMENTS, IN MEMORY OF THE 'COELACANTH' FLIGHT TO PAMANZI, COMORES, 29TH–31ST DECEMBER 1952.

Would she also find and bring along a large square of white cloth, a piece of clean plank about a foot long and 10 yards of string? My wife knows me and confined her curiosity, merely saying 'Right.' I realised that all this meant no sleep for her. Could she cope? 'Yes.' Then she asked, as we intended calling at Grahamstown, could not she and our son William come to Cape Town as well? Phew! That was a poser. I said I would see what could be done, and after ringing off spoke to Blaauw, who at once said it was quite impossible as women were never permitted on military aircraft. So I telephoned Brigadier Melville in Pretoria again, 12.20 a.m. A weary voice spoke and I said, 'Smith here again. Are you married?' when weariness gave place to surprise. 'Then you will perhaps understand better why I am disturbing you again at this unholy hour'; and put my request. He whistled, there was silence for some seconds, and when he spoke again he seemed to be pulling reluctant words from his lips: 'It would be highly irregular, of course, highly irregular, and I do not remember that it has ever been done before; but everything about this flight has been highly irregular, so I suppose you can do it. O.K.' And so at what had now become an early hour of the morning, they made another call to Grahamstown to warn my wife to be ready to leave at six. Then I thought I might get a little sleep. They asked where I wished the Coelacanth put for safety, and when I said in my bedroom of course, they were not surprised. I was escorted to my room by the press still asking to photograph the fish, and even when I started to undress, that did not discourage them; they went on firing questions even when my shirt concealed my head. Had I any photographs? I nodded. Any hope of getting any? 'No.' One by one they faded, but Natalie Roberts of the *Daily News*, an old press friend, still stayed on. Though I was clad only in shorts, she sat on the bed and talked, and when I pointed out how scandalous it was, she just laughed.

The book, *The Sea Fishes of Southern Africa*, is regarded as one of the best taxonomic accounts of a regional fish fauna produced anywhere in the world and was, of course, a very useful diplomatic gift to have available. Margaret Smith and I presented many copies to glitterati of various kinds.

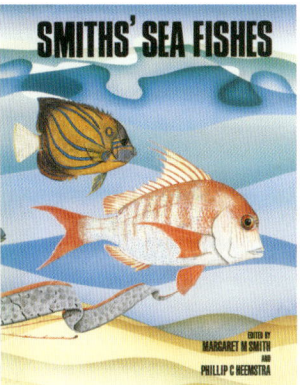

The Sea Fishes of Southern Africa

On the 'Flying Fishcart' flights JLB broke nearly every rule in the Air Force's books (except being allowed to light his primus stove!). Margaret and William did accompany him and the fish to Cape Town and everyone turned a blind eye.

It was a mistake to release a photograph of the fish and the proposed name of the new species before JLB had formally examined the specimen. His error is understandable, however, as he was under tremendous pressure from the media.

Once again, JLB persuaded the crew to break all the rules, fly low over a residential suburb and drop an object from a military plane. It is surprising that Captain Blaauw retained his commission!

Yet another prank played on JLB, but he took it in good spirit. Having gone to so much trouble to secure the second coelacanth, Smith was determined to carry the task through to completion. He was also skeptical that this message had come from the Prime Minister himself.

At 2 a.m. George Symons came in with the negatives, which were a real thrill, for they were perfect. Mrs. Roberts begged hard for one, and I told her what had happened in the matter of the pictures of the first Coelacanth; but eventually relented to her persistence and handed over one negative with the condition that the scientific name *Malania anjouanae* had to be printed with it, and that it remained my copyright. Within twenty-four hours that photograph had appeared in every newspaper of any rank in South Africa, and within forty-eight hours in most others throughout the world.

We took off at 4.50 a.m., and for once it was nice and clear; but above East London we ran into low cloud, and Blaauw doubted if we should be able to land at Grahamstown. It was only by flying below the cloud in a long valley from near the sea that we managed it, and we were mighty close to the trees much of the way. We touched down at 7.05 a.m. Aided by Mrs. Hester Locke, one of our earlier 'Book' artists, my wife, a sleepless night's ravages skilfully concealed, served coffee and food, and we all got aboard and were off at 7.40 a.m., emerging from the clouds only west of Port Elizabeth.

My eldest son was on holiday at Knysna, where I have a riverside house and a laboratory. On the piece of board I had asked my wife to bring I wrote a message, and attached this to the cloth shaped to a parachute and asked Blaauw to fly down over the place, which I pointed out to him the first time we passed. Then we made a long circle out to sea to lose height and zoomed down low over the house, and Blaauw dropped the 'bomb', with perfect judgment, for we heard later that it fell only a few yards from the fence and safely reached my son; but the huge Dakota coming back to pass so low down caused a near panic among the coloured folk who live in the valley behind. They expected bombs.

Over Bredasdorp I was in front with Blaauw and Letley, discussing our possible return flight for that day. Letley suddenly interrupted by pointing to his earphones and wrote down a message which he showed to Blaauw, then handed to me, 'Message from Dr. Malan, he thanks you very much for having taken the trouble to come so far, but he does not wish to see the fish and wishes you a safe return to Grahamstown.' This was a shock. Was it the old business of evolution? There was continual trouble about it in America, for example; and all those letters I had received! I did not let them see a thing, but nodded, put the paper into my pocket and said, 'Oh well, now we are so far we can lunch in Cape Town and go back early this afternoon.' Privately I decided that I was going to make quite certain there that the message really came from Dr. Malan. I looked down to see them both grinning somewhat impishly, and this time knew at once what they were up to and could cheerfully have killed them. They knew full well that only Dr. Malan himself could have turned me back.

10.35 a.m., Ysterplaats Airport. The Cape Commander-in-Chief, Colonel Louis du Toit, and other officers were there to meet us and a lorry with guards was ready for the Coelacanth. Dr. Malan's compliments, and would we spend the day and lunch with them, and a function had been arranged for tea-time with numerous guests. So there was to be no return that day.

To his partial disgust, William was left behind, but the promise of looking at jet planes consoled him. We were soon at the Strand, where we were warmly welcomed at their holiday home by Dr. and Mrs. Malan. He can best be described as a grand old man, greatly respected, admired, and loved by his people. No man can ask more. She is a gracious and dignified woman whose life is clearly centred on the care and service of this man. The bond between them is almost tangible, and their home breathes serenity and peace. Once formalities were over, refreshments were offered; what would I have? Prominent on the tray was a bottle of whisky. Whisky for the Professor? When I said we did not use alcohol except to preserve fish, there was something of a silence, then laughter. The Malans take no alcohol, but when I was on the way some discussion doubtless took place. The life I lead, going to many remote and wild parts and enduring adventures I would avoid if I could, has gained for me the reputation of being a pretty tough person, and those who do not know my ways would almost certainly credit me with a taste for whisky and strong tobacco. I use neither, except cigarettes, and then only in our work as bribes for fish or as presents, which is probably why my relatively frail body is tough. Yes, whisky was clearly indicated for this English Professor from Grahamstown, so as it is no light undertaking even for a Prime Minister to get a bottle of whisky early in the morning in South Africa in a place like the Strand,

There continues to be 'trouble' in America with evolution. A recent Gallup poll revealed that 51% of Americans in some states believe that humans co-existed with the dinosaurs, and over one-third believe that humans have existed in their current form since the beginning of time. According to a recent Eurobarometer poll, 46% of Belgians are also creationists.

Over 99% of people in Muslim Middle East countries do not, or are not allowed to, believe in evolution by natural selection, the most rational explanation for the diversity of plants and animals we have today.

JLB and Margaret were known in Grahamstown and beyond as the 'Fishy Smiths'.

The misconception that we should look like creatures that are ancestral to us is common. We share certain (usually specialized) characters with our ancestors but there are so many intermediary stages between us and them, most or all of which have subsequently died out or have not been preserved in the fossil record, that it is highly unlikely that we would resemble them at all.

The photographs of JLB and the second coelacanth taken in the Comoros, Durban, Grahamstown and Strand were soon published in major newspapers worldwide. I doubt whether any undescribed fish species has ever been so widely publicized before being examined.

NRF-SAIAB

JLB and his son William (right and centre) with the coelacanth in Grahamstown

a real hunt took place. I have often wondered what happened to that whisky, and must ask Dr. Malan some time. Their young adopted daughter Marietjie was intrigued by all the fuss and followed me around. She called me 'Oom Vis' (Uncle Fish).

The box containing the Coelacanth was put under a tree on the lawn. I had it opened and showed the fish to Dr. and Mrs. Malan, and to my wife as well, for she had not seen it before (Plate 6). Dr. Malan looked for some moments, especially at the head, then turned to me, and with a twinkle in his eye said, 'My, it is ugly. Do you mean to say we once looked like that?' I replied, 'H'm! I have seen people that are uglier.' Then I gave him a scale from the animal, which he handed to his wife with instructions that it was to be put in the family archives. The press photographers had a good time with us all, and I noticed that Mrs. Malan was always on the watch to see that they did not worry 'The Doctor'.

Word of our arrival soon got round, and crowds of curious sightseers were constantly passing. Despite the guard under the tree, I kept a wary eye on the coffin myself, and soon noticed a secretive-looking man moving about the drive and along the fence. I watched him for some time, and eventually told Colonel du Toit I couldn't help feeling suspicious of that chap. He was greatly amused and said there was no need to worry, he was the Prime Minister's special guard! After lunch Mrs. Malan, aware of my exhaustion, insisted on our resting awhile in their own room for comfort and quiet. But sleep would not come and I wondered whether I should not after all have had some of that whisky.

A request came to Dr. Malan that the Coelacanth should be put out on exhibition in a public place for everyone to see. He passed this on to me and I considered the matter. I had brought this fish so far for the special purpose of showing it to Dr. Malan. He had already selected those he considered would have exceptional interest in seeing it. I had refused to show it to the crowds in Durban, and felt it would not be fitting to do more now. From the scientific point of view, it seemed wiser to refuse at that special time. Any public exhibition needs careful planning, and with little time for that would then have involved a degree of risk I was not prepared to undertake, and I did not want the animal exposed more than was necessary. I told Dr. Malan I would prefer not to have any public exhibition then, and gave my reasons; but added that if he wanted it done we could go into the matter. He at once said that the decision must be mine alone, and I trust that those who were at the Strand at that time will understand my reluctance.

In the afternoon some hundreds of guests from all over the south-western Cape, invited by telephone only that morning, assembled at tea-time. Dr. Malan had asked me privately if I would address them, to which I agreed, but said also that I was so exhausted it would be a terrible strain to speak in Afrikaans. He patted my arm, and took it on himself to announce that he wished me to speak in English. After several speeches the company, composed of leading scientific and public personalities of the Cape, filed slowly past the Coelacanth bier, while crowds assembled outside in the street, and their cameras constantly clicked and buzzed.

After that day of intimate contact, we left the Strand knowing that behind the solemn, stony face of the newspapers lay a warm humanity and an active if dry sense of humour. It left us with more. Back at Ysterplaats my wife put into words what had been filling my mind; she said, 'That man could do no wrong to anyone.'

We learnt that night that the *Dunnottar Castle* was due at 7 a.m. next day. Early transport was arranged, and we arrived at the docks as the ship was being made fast; but as the Port Officials had not yet arrived, there was no entry. Time was precious, and I was raging at the gangway when a door opened in the side of the ship near by and Captain Smythe jumped out and hailed us, naturally dumbfounded that we could be there, as may be understood. One of the officers had seen us on the quay and had hastened to tell him. In spite of the guards, we all went aboard through the same hole.

On our voyage in that ship from Mombasa southwards I had found 8.30 a.m. very late for breakfast, and had had occasional passages with the head waiter, for I much preferred having my simple meal at children's time, 7.30 a.m. I left my wife and Smythe and went to the dining-saloon, now just 7.30, and seeing this head waiter coming from the kitchen, stood at the side of the door, and as he passed said, 'Would you mind if I had breakfast with the children?' He started, and with bulging eyes stared

In hindsight, how wrong Margaret was in her assessment of Dr Malan! Most would say that he did irreparable harm to millions of South Africans through the leading role that he played in the implementation of racist apartheid policies.

JLB's speeches on the coelacanth were always charged with passion and emotion. He left little doubt in his audience's minds that the study of fishes, and science in general, is a vital part of human endeavour and essential for the progress of society.

JLB's little joke with the head waiter reminds us that he still had a sense of humour, even in times of stress.

Flying low over a harbour in a military airplane, and 'bombing' the Prime Minister's residence! What next?

GG Smith, an amateur botanist with a particular interest in endemic succulents, was the Chairman of the Board of Trustees of the East London Museum from 1941 to 1973. He oversaw the design and construction of a new building for the museum that was 10 times larger than the previous building and was very supportive of Marjorie Courtenay-Latimer's work.

East London Museum

Marjorie Courtenay-Latimer with GG Smith at the coelacanth display in the East London Museum ca 1941

Many years later, in 1989, Marjorie told Hans Fricke and me that she had been jealous of the amount of publicity that the Ichthyology Institute in Grahamstown had received for its coelacanth research and felt that the East London Museum had not been given enough credit.

speechless at me and almost dropped his load. He obviously thought I was a ghost. It could not possibly be me, in the flesh, for the last he had heard was that I was careering about somewhere off Madagascar.

When we got off the ship we discovered that our car had vanished, so we had to find a taxi, no light task in that area, and eventually discovered one, far from new, with a coloured driver, also an older model. I told him to go to Ysterplaats, and as our speed did not satisfy my haste, after a while asked him to hurry as a plane was waiting. A second or two later the car swerved alarmingly as he jerked round to look, and he said, 'My God, sir, you aren't the genilman with the fish are you?' I said, 'Yes, I am.' He said, 'Oh, what a honour, what a honour for me and my taxi.' There was silence for a while, then with agony in his voice; 'But the only trouble is, sir, no one will believe me. Can't you give me something to prove it, sir?' I found all my cards had gone, but my wife had hers so I autographed one of those and he left a very happy man.

We took off at 8.30 a.m. First we went north, low over the *Dunnottar* at the docks to see Captain Smythe waving his arms from the bridge. Then we circled and cut across the Cape Flats to the Strand, where I got Blaauw to circle the Malans' house twice low down. Judging by the effect on the hastily emerging populace, they also expected bombs, but we saw the Malans come out on the lawn and wave. We 'bombed' them with several copies of the early morning papers, then turned east and up, over the high escarpment that soon blotted out the line of surf where their house lay.

There was more bad weather with low cloud-ceiling all along the coast; once again there was grave doubt as to whether we should be able to land at Grahamstown. But Blaauw was equal to the emergency, and we got in to find Mayor McGahey and numerous citizens there to meet us. It was a great joy to see Miss Latimer there, too, with G. G. Smith, the present Chairman of the Board of Trustees of the East London Museum, who had come over from East London to welcome our return. Her presence brought back poignant memories of the first Coelacanth. Then she was a young girl, finding her feet. Now she was a mature woman of established position, her Museum famous, her Chairman and Board well aware of her value to the Institution. She kissed me before them all, and nobody was embarrassed, not even myself. The crew of the Dakota had got the box out and were standing round politely, but clearly very conscious of the compass needle pointing northwards. In a brief ceremony each of the four officers was presented with one of the inscribed volumes my wife had brought, while the others each received a special memento. They were soon aboard and off, and we watched that huge machine fade to a minute speck, then she was gone. As she vanished it was as if a curtain had suddenly been dropped at the close of a tense act.

It was fitting that Malania was loaded with Miss Latimer into G. G. Smith's van, and we all went off to the laboratory. Before the cars had stopped we heard the telephone ringing madly, and as I got out Mrs. Locke's head came through the window, calling 'Trunk-call, Professor.' It was prophetic.

The epic adventure that had gripped South Africa and the world was finally over after eight intense days.

NRF-SAIAB

JLB Smith (right) and the flight officers on Dakota 6832 KOD: Commandant JPD Blaauw, Captain P Letley, Lieutenant DM Ralston and Lieutenant WJ Berg, with William Smith in the middle

The second coelacanth is still relatively intact and on permanent display in the South African Institute for Aquatic Biodiversity (SAIAB), as the JLB Smith Institute of Ichthyology is now called.

BOOK III

THE WAVE RECEDES

Chapter Sixteen

FLOTSAM AND JETSAM

A CHAOTIC account of a chaotic time.
The days and nights that followed our return to Grahamstown on the last day of 1952 were something of a nightmare, of which it is difficult to give a coherent account. My Secretary had gone on holiday with her family and left no indication of her whereabouts, nor should I have worried her had I known them, for the Christmas–New Year period in South Africa is dedicated to release from work. She was due to be away for another two weeks. At that time even the Prime Minister rests, at least he tries to, but of course unexpected events like wars or Coelacanths may intrude.

Despite efficient handling in my absence, there was the usual accumulation of troublesome matters, reports and queries and inescapable financial commitments and payments. I find myself constantly grateful that while I have enough to eat and a comfortable dry bed to sleep in, I am free from those cares that great wealth brings. There is more than enough to do besides.

From the moment we arrived, telephone calls came constantly from far and near. Would I confirm this or that? Were we going to remain in Grahamstown? Could such and such a representative, press, radio, television, cinema, publishing, and others, come for an interview?

In normal times pressure of this kind would have been hard to bear, but now it was almost unendurable, for I had been virtually without sleep for a whole week. Fortunately even in normal times I can do with less sleep than most, and my body and brain are rather like a worn engine whose minor deficiencies are concealed when it functions at high speed. My wife was equally overwhelmed, and tore herself into bits, acting all at once as secretary, buffer, and general assistant. Even the laboratory boy was away enjoying a restful holiday!

It is typical of JLB to equate a war with the discovery of a fish. Although his comment was probably meant tongue-in-cheek, it is clear where his priorities lay. He was always cynical about humankind's tendency to go to war to resolve conflicts, just as he was cynical about the inconvenience of being rich, or frittering away one's life watching television. He expected everyone to subscribe to his spare, highly-focused lifestyle.

Until the discovery of the first coelacanth, and then, 14 years later, the second, Grahamstown was not known at all as a fish research centre, but that soon changed. Today it is one of the leading centres for the study of fishes in the southern hemisphere, and a vital link in the international fish research and conservation network.

JLB's appreciative comment on 'the wonder of modern communications' (p. 180) and 'Such are modern times!' (p. 181), referring to telephones, telegrams, ciné and airmail postage, is, of course, amusing to us today in the age of Internet, emails and Skype. I suspect, however, that the availability of additional communication media would have made JLB's life even less bearable.

The last day of 1952! My weary brain shot occasional probes of worry into the future. Could we stand this pace and pressure?

A trunk-call. The London *Times* wanted an article. When? At once, please, its point would otherwise be less, and if I would write it, it could come by cable. I said it was doubtful if it could be done, as life was more than full, and would not promise, but said I would see how things went and if at all possible would do it. I would let them know by next morning at the latest. They asked if they could make provisional arrangement meanwhile? Yes, but it must be understood that this did not bind me, and at the moment there seemed no hope. This was the afternoon of our return, the 31st December 1952.

This story is told partly to show the wonder of modern communications and partly because this article played an important part in countering the unfortunate effects of that 'Missing Link' appellation that had been tacked on to the Coelacanth.

My brain often annoys and sometimes alarms me, for it has a part that works secretly, even when I am hard at work at something else. Although exhausted that night, I knew I would have to write that infernal article, but put it from me then. Telephone calls pursued us to our house, through a meal, and until we had to ask the Post Office to draw the curtain. I slept like a log for five hours, and at 3 a.m. found myself fully wide awake with that confounded article clear-cut in my brain. I slid out of bed, though probably not even a bomb would have wakened my wife, crept downstairs, and had a cup of Nescafé. I can never condemn those who are addicts to alcohol or tobacco, for I am just as bad with that stuff and turn to it in every crisis. 'A nice cup of tea!'

I wrote steadily until daylight dimmed the electric lamps, and by then most had been got down on paper. This was only the first part. It is doubtful whether those who hear a good broadcast or read an interesting article know how much labour lies behind it. I put a call through to say that the article would be ready by noon, and learnt that though it was a public holiday, 1st January 1953, when all public services are dead, provisional arrangements had been made with the Post Office at Grahamstown, and a telegraphist would be waiting to send this off by cable to London. This would now be confirmed: he would be on duty from noon, and would I make contact and send it to the Post Office?

Typists! They are normally scarce in Grahamstown, and this was New Year's Day. At the aerodrome Mayor Patrick McGahey had said that if he could assist us, privately or officially, would we ask. I had that in mind and telephoned him now, 5.30 a.m., and he answered, clearly without ill-feeling, from his bed. Two typists at 9 a.m., please? Without hesitation, yes! he would see to it, he believed two of their staff were still in town. More Nescafé, and to the laboratory at 6 a.m. First a quick look at the Coelacanth; yes, it was still there, a good solid fish, there was nothing imaginary about that. It was real.

At 8 a.m. Mayor McGahey telephoned to say that he had managed to find two expert typists,* and that they were prepared to forgo the holiday to help us, and would accept no payment for their services. They arrived at 8.45, and worked all the time until 11.30, when it was complete, the final clean copies. At noon it was at the Post Office, and next day I got a cable from London thanking me for the article and saying that it had been published that morning, the 2nd January 1953. Only four days later a copy of that paper was in my hands. Such are modern times! (This article is reproduced in Appendix *D*, p. 243.)

During this same morning, 2nd January 1953, there was a trunk call from the press. Had I heard that the night before Dr. White of the British Museum had broadcast over the B.B.C. to say that it was nonsense to speak of the Coelacanth as a 'missing link', and it certainly was not the ancestor of man, and did I agree with him? It was the same Dr. White who in 1939 had relegated the Coelacanth to the abyss and reproved me for naming that first one after Miss Latimer. That 'Missing Link' caption was a great trial, especially as some overseas reports had virtually tacked it on to me (see p. 87). I knew that White would already have been answered by my article which had appeared in the London *Times* that morning, but told the press now that while it was doubtful if the Coelacanth would ever be proved a direct ancestor of man, it must be pretty close to the main evolutionary stem. It is interesting to note that a day or two later, Julian Huxley, the famous British biologist, took up the whole matter in a broadcast over

* Miss R. M. Koen and Miss M. Goetsch—of Grahamstown.

When I was Director of the JLB Smith Institute of Ichthyology (1982–1996) I was fortunate to have one of the most efficient and knowledgeable personal assistants imaginable – Jean Pote. She had previously worked with JLB and Margaret and typed the manuscripts for many of their fish publications. She was not only an expert shorthand typist but also had an extensive knowledge of the common and scientific names of fishes.

NRF-SAIAB

Jean Pote with Margaret Smith in February 1986

JLB was right to say that the coelacanth 'must be pretty close to the main evolutionary stem' leading to humans, although the lungfish is closer.

The British evolutionary biologist Sir Julian Huxley (1887–1975) was a staunch supporter of Darwin's theory of evolution by natural selection. It is therefore not surprising, considering his clear understanding of the process of evolution, that he agreed with Smith but disagreed with the rigid views of Errol White.

Charles Darwin

JLB must have felt very frustrated by all the intrusions on his research time as he finally had a nearly complete coelacanth specimen with most of the soft anatomy intact, yet he did not have the opportunity to study and describe it.

the B.B.C., and in this he expressed views rather like my own, saying indeed that the ultimate vertebrate ancestor of man was probably something like a Coelacanth. He said that when Dr. Malan asked his rather whimsical question about the Coelacanth, 'Do you mean to say we once looked like that?'—the answer really was 'Broadly speaking, yes.'

We could barely find time to open the constant series of telegrams and cables from all over the world, and the sheaves of letters were piled up and treated in rotation. Post Office revenue for that period must have been very considerably augmented by the Coelacanth. There was a constant succession of visitors from far and near, all eager to see the Coelacanth, and it was difficult to cope with them, so an illustrative exhibition was hastily prepared. Everyone was so full of questions that, as we all had only one voice each, I dictated a brief outline of the main points of the whole affair, and had this posted up as well.

Press representatives and photographers of papers and journals from all over the world kept on coming, new arrivals being viewed with disfavour by those earlier on the spot, and at times I wondered what we should do if the almost open general hostility broke its bounds. They reminded us of a pack of bristling dogs waiting for one to start it off.

Mayor McGahey came to ask if he could arrange an official welcome, and as the public would like to see the Coelacanth, could we have it exhibited in the City Hall, and combine the two? We agreed to do all this on the 9th January 1953, and this was duly carried out with a record attendance.

At the request of the authorities of the East London Museum, we later took *Malania* there, and with *Latimeria* alongside it was exhibited for two days, which were a repetition of the earlier historic times, for thousands of people in lengthy queues pressed in constantly all the time. The fish was also exhibited by request at Port Elizabeth, and attracted great attention. Many further requests for exhibition of the fish came from far and wide, but these had to be refused, as was also the generous offer of a Port Elizabeth business magnate who offered to fly the fish and myself to the Rhodesian Centenary Exhibition: neither the C.S.I.R. nor I felt the risk justifiable.

A Coelacanth can do strange things to scientists. My wife and

I posed for photographs and became ciné and television stars. I would leave broadcasting engineers fixing a tape machine in my office to face more flash-bulbs or to wave the Coelacanth's fin for a ciné. We were told that within three days a television record from our laboratory had been shown all over America, and later we got letters from scientist friends in remote places like Japan, Alaska, and Timor to say they had seen us on the screen. We were, in fact, carried along by a kind of tidal-wave we could scarcely control, a wave that went right round the world many times, to its uttermost corners, and the backwash still keeps on coming back to us even after this long time. In this process an obscure and highly technical scientific term became part of the common speech of mankind.

When I snatched a moment for a closer look, it was a most unpleasant shock to discover that the first cutting of the fish to save it, as told by Hunt, had been a crude hacking open as the natives do when they wish to salt any big fish. So far from its being a complete animal, the whole brain was missing and much of the viscera had been badly hacked and torn. This was indeed a bitter blow. Once again it was not a complete fish.

On the morning of the 3rd January 1953 my Secretary appeared, ready for work. Hearing of my predicament only the day before, she had summarily cut short their holiday and swept the whole family back home. It was a fine gesture, for she came at a critical time when we were almost desperate with all that was besetting us.

No sooner had the movie and television people gone than there came a flood of requests for articles, popular, informative, and scientific, about the Coelacanth, of which only a small part could even be considered.

During this time we estimated that not less than twenty thousand people came to look at old man Coelacanth. Not many humans have achieved that in death. It reminded one of Lenin in his glass coffin, or dead royalty in state.

A bright spot in that difficult time was the arrival of three young coloured men* at the laboratory. They stood for some time quietly watching and, when asked if they wanted anything, said, rather shyly, that they were graduates of Fort Hare, the non-

* F. Backman, B.Sc., L. Backman, B.Sc., and N. Dennis, B.Sc.

I also felt the backwash from the coelacanth saga when I took over from Margaret Smith as the second Director of the JLB Smith Institute of Ichthyology in 1982. At my first staff meeting I boldly announced that we would be leaving the past behind and entering a new era of post-coelacanth research. How wrong I was!

The constant stream of communications that I received about the coelacanth convinced me that it would be a permanent part of the Institute's DNA. Instead of fighting the 'backwash', we went with the current and launched into a new era of coelacanth research and conservation. Secretly, I was delighted as the coelacanth had been part of my own life from an early age.

The University of Fort Hare in Alice has launched the careers of many African leaders, including Nelson Mandela, Thabo Mbeki, Oliver Tambo, Mangosuthu Buthelezi, Robert Sobukwe, Julius Nyerere, Kenneth Kaunda, Robert Mugabe, Govan Mbeki, Seretse Khama, Desmond Tutu, Chris Hani, Khotso Mokhele and Sizwe Mabizela.

The 'one or two of the more primitive types that still survive' that have spiral valves include the bichirs, which are ancient, ray-finned fishes that are endemic to tropical Africa. They are very unusual in that they have jaws similar to those of four-legged animals, two lungs, and fleshy pectoral fins like coelacanths.

Stan Shebs, Wikipedia CC BY-SA 3.0

Bichirs are ancient ray-finned fishes that are endemic to Africa.

European University College, they had heard we needed assistance, they had taken Zoology and could they help in any way?

Our enormous collection of fishes from East Africa stood untouched, in sealed bottles and tins, packed in crates that were piled up in the laboratory, an angular tower that reproached us every time we passed. So we set the three young men to unpacking these treasures and putting them into more suitable containers. It was no pleasant task, formalin is stifling, and it was hot. They laboured for many days and would accept no reward. We remain grateful for their kind thoughtfulness and service.

Even though the Coelacanth had been so badly mutilated, most of the soft parts were there, and so, as before, I made a thorough but more or less general investigation so as to be able to give information that was eagerly awaited by scientists everywhere. As was mentioned previously, the gills of the first Coelacanth were lost, but these were intact and they were remarkable. Most fishes have gills of cartilage (gristle), relatively soft, with 'gill-rakers', soft finger-like projections, above. The gills of the Coelacanth proved to be bony and hard, and instead of soft gill-rakers there are teeth. In fact, stripped of the soft filaments below, they looked almost like jawbones; they could easily be mistaken for them. They showed at least that jaws and gill-arches had the same origin.

I found in the intestines a structure known as a 'spiral valve'. If ever you open a Shark or a Ray, run the intestines through your hand and you will come to a peculiar, rather hard, purplish part, which you will see clearly has a spiral structure. Cut it open longways and you will see that it is a device to make a short bit of gut do the work of a much longer straight part. The digested foods must go round and round, and so are exposed to longer absorption. This structure is characteristic of Sharks and Rays and not of modern fishes, though it is found in one or two of the more primitive types that still survive. The study of fossils had progressed so far that some workers had come to suspect that Coelacanths might have a spiral valve, and here it was. (Some other things suspected to be present, like the internal nostrils, were not found.) This type of intestine clearly carries one back to the very earliest beginnings of vertebrate life or even farther. In the intestines I found fish remains, a few scales and the eyeballs of a fish. Fishes' eyes vary in size, but, taking an average, those were from a fish of about 15 lb. in weight. A 15-lb. fish is quite a size. This confirmed my views about the way a Coelacanth fed. Clearly he pounces and grabs. His powerful jaw muscles would enable him to hold the struggling prey, and it seems likely that the toothed gills would come into play and rasp that struggling fish's flesh and life away. Then it could be pulped and mashed and broken up, and swallowed just as a big Rock Cod does it. Degenerate! Deep-sea refuge to escape competition! Not this fish.

To my great sorrow, this was a male. I had, of course, hoped for a female with young. We are still waiting for that.* There were many other points observed and noted, but they are rather technical.

Scientific publications normally take some time to prepare and appear. Even in 1939 it took a month before my first account of the Coelacanth appeared in *Nature*, and another four weeks before I saw a copy. In 1953, on the 2nd January, two days after our return, my wife and I set to and worked all that week-end, furiously, and a seven-page manuscript and photographs were sent by air on the 6th January 1953. This appeared in the issue of *Nature* of the 16th January, and a copy was on my desk in Grahamstown on the 19th, by which time it was in the hands of scientists all over the world as well. Modern times!

Several letters came from Hunt giving further news. Some days after we had left the Comores, the ceremony of presentation of the 50,000 francs took place. I have no record of events, only that bare statement. Hunt wrote to say that some photographs had been taken and he hoped to send prints later, but the cyclone finished that. This presentation payment of so vast a sum must have had a profound effect, the kind of effect we had hoped for, and we waited and waited, hoping to hear of more and other Coelacanths.

After these events, Hunt returned to Africa, and soon after set out again on his Comoran round. By a strange trick of fate, when he returned to Pamanzi, only two weeks after we had left, a cyclone caught him there, and after what he describes tersely as 'a terrifying experience', he and his crew escaped with their lives; but the

* See p. 237, Appendix C.

The coelacanth diet consists of cartilaginous fishes such as swell sharks, bony fishes (lantern fish, beard fish, cardinal fish, snapper, witch-eel and beryx) as well as cuttlefish and octopus. Coelacanths live in a complex and cut-throat marine community where they have to compete with large sharks and other predators. They are certainly not degenerate!

In the past, coelacanths were preyed on by fish-eating dinosaurs with long, crocodile-like snouts. We know that *Spinosaurus aegypticus*, which lived about 95 million years ago, ate 'old fourlegs', as coelacanth fossils have been found in fossilized *S. aegypticus* stomachs!

Luis Rey

Spinosaurus aegypticus preying on a coelacanth

Today coelacanths are likely to fall prey to large sharks. Divers in the *Jago* research submersible observed bumpy-tail ragged tooth and sand sharks in coelacanth habitats off the Comoros, and shark expert Len Compagno has predicted that six-gill and frilled sharks may also occur there.

NRF-SAIAB

Bumpy-tail ragged tooth shark

JLB did not know that Capt. EE Hunt had died at sea in 1956 when his second schooner, *Hairiako*, sank after striking the shallow Geyser Bank between the Comoros and Madagascar. Although five of the 26 crew and passengers eventually reached Grande Comore safely, after drifting for 17 days in a rowing boat, Hunt and the other passengers and crew were lost at sea.

When the *Hairiako* ran aground Hunt had apparently tried to tow both the rowing boat and a raft using a small, outboard-powered dinghy, but this was unsuccessful. He then left, Shackleton-style, with two others in the dinghy to seek help. The occupants of the rowing boat admitted later that they had cut loose the raft with 18 survivors on board in order to prevent them from climbing into the boat.

Hunt's overturned dinghy was later recovered by a Dutch frigate. It still contained £300 in cash, Hunt's personal papers and those of his African cook and a French passenger. Near the dinghy were three shredded life-vests. All three men had apparently been eaten by sharks.

NRF-SAIAB

Eric Hunt on board his first schooner, N'duwaro

schooner was smashed and sunk, beyond any salvaging, a tragic sequel to its eventful participation in the Coelacanth adventure. Hunt apparently lost everything, though, providentially covered by insurance, he has since found another vessel. For a time he stopped trading about the Comores or anywhere in French waters, but recently he has written from there again.

Regrets for Hunt's misfortune are of added poignancy, for had such a cyclone caught us at Pamanzi, we might have had more of the Comores than we bargained for. It was a close enough shave.

Dzaoudzi and the whole of Pamanzi and Mayotte from the air were pleasant to the eye, but the damage wrought by this cyclone to buildings, plantations, gardens, and the vegetation was, according to reports, both devastating and heartbreaking, and will take many years of labour, and many years of persuasion and encouragement by the officials, to repair. In such parts you see men labouring earnestly to administer the territories they control, living in discomfort, far from their real homes and congenial company, in constant danger of deadly or crippling, often almost incurable, diseases, and one wonders why on earth they do it, quite often when the peoples for whom they labour clearly do not appreciate this or desire their presence. They are the last of this passing phase of 'Colonial' administration, the condescending gesture of White superiority that is arousing increasing resentment in the awakening consciousness of existence that is stirring in the backward ebony mind. Where will it all end?

I have told previously of my voluminous monograph on what remained of the first Coelacanth, and which would have been even more ample had the specimen not been recalled for exhibition. Even with what had been lost in this second specimen, the full investigation of the various parts of the body would mean many years of careful investigation. Such work is slow, for it involves the most delicate manipulation. Our knowledge of structure and life comes from many years of intensive study by a series of leading experts. It has progressed so far that the proper study of a whole organism has really got beyond the powers of any single man. Scientists have therefore specialised, one man often devoting his whole life to the study of only one organ such as the eye, the kidney, the pituitary, or the liver.

It will therefore be understood that to reap the greatest advantage from this wonderful opportunity of an almost complete Coelacanth, it would be desirable for each organ to be examined and studied by an expert who already had in his brain a full knowledge of that organ in most or all other creatures. This therefore ruled out an examination only by myself, for, despite all I know about fishes, it would take years before I could master all the existing knowledge of each organ in turn; and besides that, it would have been selfish and unjustifiable. My chief aim was to extract the utmost scientific value from the Coelacanth. In addition to this, now that there was a real hope that the true home of the Coelacanths had been found, I wanted to get on with what had become my greatest work—the investigation of the fishes of the whole western Indian Ocean.

It is a curious paradox that while the fishes of most seas have been almost fully investigated, before my work there commenced, those of East Africa had hardly been touched. Our series of expeditions had indeed shown this to be one of the richest and most interesting areas in the world, and our discoveries had astonished ourselves as well as the world of science.

I put my view about the study of the Coelacanth to the South African Council for Scientific and Industrial Research, and suggested that we should invite specialists to apply to be included in a panel of workers to carry out the whole investigation. This was approved by the Council.

There was another aspect to be considered. This was the only reasonably complete specimen of a Coelacanth. What if no others were found? Full study of the fish meant that it would have to be cut up; dismembered. I wished to put that off until there was no further hope of more specimens. In my own mind I decided to wait for another eighteen months, which would cover a full intermonsoon period, and if the Comores proved to be the real home, there was every hope that others would turn up or be found within that time. Another important point soon emerged, which was that for the proper study of certain organs, formalin-preserved material was useless, quite fresh tissues, treated in various special ways, were necessary. So the full study was, in any case, going to depend on finding more specimens.

Since the pioneering work of the Smiths in the Western Indian Ocean, the fishes and other aquatic life of this region have become far better known through interdisciplinary research by the South African Institute for Aquatic Biodiversity (SAIAB), the Oceanographic Research Institute in Durban, the Western Indian Ocean Marine Science Association, the African Coelacanth Ecosystem Programme (ACEP) and its associated Global Environment Facility (GEF)-funded project, the Agulhas and Somali Current Large Marine Ecosystems (ASCLME) project, and by East African universities and museums.

ASCLME
Agulhas and Somali Current
Large Marine Ecosystem Project

Multi-national collaboration has since become a feature of coelacanth research and several multi-authored books and journals have been published on their evolution, genetics, anatomy, physiology, demography, feeding, breeding, locomotion, behaviour, habitat preferences and conservation.

Eugene Balon

Coelacanth on display in the President's palace on Grande Comore

The coelacanth, after surviving unknown to science for so many years, is now one of the best studied fishes in the sea!

As a result of this, the following letter was published in *Nature*, on the 28th February 1953, over my name:

INVESTIGATION OF THE COELACANTH

It was my privilege to carry out detailed investigations on the first Coelacanth, and to have discovered what appears to be the area where those fishes still live.

The recent Comoran Coelacanth, while mutilated more than was at first realised, nevertheless retains most of the soft parts, including the abdominal viscera. This extends enormously the scope of the investigational work that may be carried out on the specimen. There will be still more that can be done only on parts, exudates, and secretions from an untreated fresh specimen, which it is hoped to seek before very long.

It is in keeping with the importance and scope of the investigations on all parts of this fish that they should be assigned to leading experts in the field in which they fall. I have advised the South African Council for Scientific and Industrial Research, and have requested the Council's approval of, and co-operation in, this matter.

Application to be included in this scheme should be sent either to the South African Council for Scientific and Industrial Research, P.O. Box 395, Pretoria, or to me personally at Rhodes University, Grahamstown. While every possible facility will be granted to selected visiting specialists, it should be noted that there is no possibility of financial aid from this end.

The ownership of the next specimen or specimens is of less importance than their proper preservation for scientific purposes. As certain organs and body fluids require special treatment and preservation, it is intended to compile a set of special instructions to be issued to those in areas where it is possible that a fresh Coelacanth may be obtained. It will be appreciated if those interested will kindly furnish detailed special instructions composed in language as simple as possible, giving full directions, and not only the names, but also the actual composition, of any materials to be employed.

Since there is a hope that more Coelacanths may be found at the Comoro Islands, it is desirable that all such materials should be available there.

This elicited a widespread response from scientists all over the world, and as a result my wife was able to publish a composite account giving detailed instructions for the care and preservation of a Coelacanth, so as to enable the maximum to be obtained from investigations.

Among the general letters we received in all this time were many different types. The clerical staff had a fat file of letters which they privately and irreverently termed the 'Crackpot' file (which I did not see until later), in which they housed letters to which that term could be applied. As in the case of the first Coelacanth, there were many letters from apparently religious persons of a certain type which came from virtually all English-speaking countries. Some of these correspondents were really shocked that a man like myself should so misuse his position as to mislead the public by talking of millions of years. One man, whose attacks editors had refused to print, published a denunciatory pamphlet about my views. In some missives my worldly end and my future in another sphere were vividly portrayed; I was told that it would be better for mankind if people of my type had never been born, and there were even threats of personal violence. One good lady in Grahamstown called in some indignation with an issue of a well-known paper from England, in which somebody quite scurrilously 'Gave me blazes'. She expected I would write to pulverise the author, and was surprised when I laughed and said that I never replied to anything like that, nor would I permit her to do so on my behalf. A day or two later she delightedly brought me a further issue of the same journal in which somebody else (in England) had pulverised the author. It generally happens that way.

An R.A.F. officer who had been at the Comores during the war wrote to say that he hoped I would not be shocked at the tale, but as food had been a problem at that time, depth-charges were used to get fish. Many queer creatures came up that way, and after seeing the picture of the Coelacanth he was almost certain that there had been several in the catches they had made about those islands. I need hardly record that I was not shocked by this disclosure; in fact, if he felt that way I wondered what his reactions would be to what I could tell him.

The descendants of a missionary who had lived near Mount Kilimanjaro wrote from Germany giving a good deal of information about flying-dragons they believed still to live in those parts.

I have also had my fair share of 'crackpot' letters and encounters.

When I taught science at the Bahrain Science Centre in the Middle East, the integrity of modern science was constantly questioned. What visible proof is there that life has existed for millions of years? How do we know that fossils are that old? Why are we certain that life evolved? Did continental drift really take place?

Charles Darwin

If Charles Darwin were alive today he would be shocked to learn that one of the most basic tenets of biology, that plants and animals have evolved over time, is still not believed in many parts of the world.

In fact, fundamentalist ideas on creationism and so-called 'intelligent design' are on the rise in both the Islamic and Western worlds.

Stories of mythical dragons and other monsters still abound in Africa. The discovery of the coelacanth, and other 'Lazarus' species, has done nothing to quell those myths.

Lake Sibaya in northern Zululand, for example, has a mythical 'Fiery Draco' that devours people and boats when a storm sweeps across the lake. Such myths are by no means limited to Africa, though: Lake Windermere in England, for instance, has a legendary monster, 'Bownessie'. The incomparable 'Loch Ness Monster', one of Scotland's main tourist attractions, even has a scientific name, *Nessiteras rhombopteryx*, coined by no less a scientific luminary than Sir Peter Scott!

The Loch Ness Monster

The family had repeatedly heard of them from the natives, and one man had actually seen such a creature in flight close by at night. I did not and do not dispute at least the possibility that some such creature may still exist. A man of foreign birth reported having seen a dragon at a place on our own south coast. It had left clear tracks on the ground before it vanished in dense bush, and though he had told the police, nobody had succeeded in tracking it. I suggested a Leguaan (a big lizard of South Africa). People from many countries wrote to tell of Coelacanths they had seen there. An American soldier stated that they were common in the fish-markets of Korea. A woman in Bermuda was positive one had been offered to her by a fisherman there. One somewhat politically minded person wrote to reprove me for naming the fish after Dr. Malan, and said that it would have been much more fitting to have honoured in that way the native who had caught it. Several natives did the same. An American who wrote about the Coelacanth, concluded by sympathising with me for having to live in such a dreadful country as South Africa, a visiting native professor had told them all about it. I concluded my reply by saying that many years ago I had heard a talk by a visiting American about life in his country which had left us all very thankful that we lived in South Africa, and that what they had been told was probably as accurate as the story we had heard. An American ichthyologist wrote: 'Now I can die happy for I have lived to see the great American public excited about fish.'

The broadcast from Durban about the whole matter had apparently been greatly appreciated, though friends laughingly reproved me for having made many others weep from my emotion. My young son certainly disapproved of that part. Anonymous letters are normally despised, but we received some that do not fall in that category, and one is reproduced on p. 254, at the end of that broadcast.

The whole affair had some peculiar consequences. All over the world it led to greatly increased sales of books about fishes; in Britain especially of one by a late member of the staff of the British Museum. I had sent a scale of the first Coelacanth to an American museum, and this had been kept for safety in the strong-room. Now it was brought out for exhibition, and thousands of people filed past to see it. A prominent member of the British

Parliament, in attacking an opponent, called him a 'Coelacanth' on the grounds that from his long silence in that august assembly it was a surprise to find him still alive. The able retort was that the Coelacanth lived long, had great endurance, and never spoke unless it had something to say. At least a thousand people told us personally that while it had been very good of Dr. Malan to do what he did, it was only right that he should have, and of course everyone knew that General Smuts would have done at least the same if he had been there. We just smiled! It is significant that literally a flood of letters and telegrams of congratulations and thanks had been sent to Dr. Malan, not only from South Africa; and among them were many from persons, from angling and other clubs, societies, and institutions, composed mainly of those opposed to his political principles. Many of those who had done this told us personally or wrote to say so.

When the general election early in 1953 showed the trend of public opinion by returning Malan to power with a greatly increased majority, one close but very sore old friend, to whom this represented almost the end of the world, wrote in accusation that 'you and your darned old Coelacanth helped this on', and this view was very widely held. But in that form it was certainly not correct. If Dr. Malan gained any advantage from the Coelacanth affair, it was because he had earned it himself; it was entirely his own doing. Not only did he put himself out on my behalf when no one could have blamed him if he had not, but he took the decision alone and the very considerable risk that this entailed entirely alone. It was a risk, especially at that time. It was very different with Smuts, who would not have had to take any risk to help me, but who would not even spare a few moments to hear me.

The series of events which had compelled me to seek assistance from both Smuts and Malan had been almost fantastic in their similarity, not only in the subject of my appeal, but in each case by the most curious coincidence the climax had come not long before a critical general election. It was even more fantastic that each of these two men had, almost dramatically, within a very short time afterwards been himself treated by his own people almost exactly as he had treated me—the one spurned, the other supported. One could not escape the conclusion that this indicated

In 1973, eight years after Winston Churchill's death, George Lichtheim said of him: 'In one of his aspects he is a political coelacanth, a prehistoric monster fished up from the depths of the past (thus he appears to British left-wingers, who are nonetheless secretly proud of him)'. For other examples of coelacanth symbolism, see Chapter 5.

Winston Churchill

Jan Smuts was defeated by DF Malan of the Herenigde Nasionale Party (Re-United National Party) in the general election of 1948 and died of a heart attack on his farm 'Doornfontein' in Irene near Pretoria some two years later on 11th September 1950, aged 80 years.

DF Malan died in Stellenbosch on 7th February 1959, aged 84 years.

that the manner in which each man had behaved in my case was symbolical of the way he had treated his own people, and once again it showed clearly that, though this may take a long time, in the end, if he lives long enough every man reaps what he sows.

Chapter Seventeen

FALLING THROUGH

EVEN though the discovery of this second Coelacanth at the Comores, and the information gleaned by Hunt, appeared to pin-point that area as the home of those animals, it was clear that they could scarcely be abundant there. Although it was true that this Comoran Coelacanth had been found in the type of environment which satisfied every condition I had deduced and predicted, I knew only too well that a Coelacanth in any place does not necessarily mean that it is at home.

Most of Hunt's information was got from natives, and South Africans know from bitter experience that the average native attaches little importance to factual accuracy or to veracity. Their approach in such matters is very different from ours, and when questioned a native will almost invariably tell you what he thinks you want to know, rather than what he knows about what you ask, or even if he knows nothing at all.

At the same time, however, the evidence that Coelacanths were caught occasionally indicated that if their true home was not at the Comores themselves, it would not be so far off this time, and there was all the more likelihood that this might be found more easily. The field of search would certainly be greatly narrowed down.

It had long been one of the main objectives of my existence to establish or to see established the certain home of the Coelacanth. Even before my return to the Union with the animal, and on the flight itself, my mind was busy revolving this matter and exploring possibilities. I hoped that the discovery of the Coelacanth would abate the scepticism of the French and stir them to action, in the Comores at least, but at the same time nobody had studied this matter as deeply as I and nobody else knew as much. While I know only too well that nobody is unique and nobody is irre-

O.F.—13

Now, 60 years after JLB wrote these words, we know that the Indian Ocean coelacanth lives not only off the Comoros but also off South Africa, Mozambique, Tanzania, Kenya and Madagascar, with another species living off Indonesia.

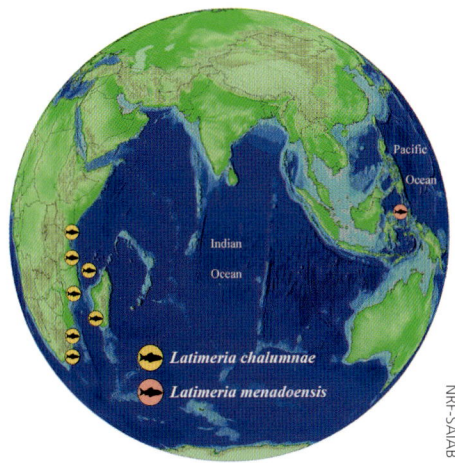

NRF-SAIAB

Distribution of the African and Indonesian coelacanths

The largest population of coelacanths known lives off the Comoros, and especially off Grande Comore. In the 1990s, Hans Fricke and his team estimated that the population may reach about 500 adults.

The capture of many coelacanths off Tanzania, and the suitable habitats identified by ACEP off northern Mozambique and eastern Madagascar, suggest that we should keep an open mind as to where their main population lives.

Smith died before he could realize his ambition to 'know the certain home of the coelacanth'.

JLB's appeal for a coelacanth to be seen by 'the ordinary man' in an aquarium has been echoed by both Marjorie Courtenay-Latimer and the author.

JLB's 'evolutionary' thinking was different from that of many others at the time who regarded species as relatively unchanging entities. The American evolutionist Ernst Mayr suggested that this is partly due to our 'pattern-forming' minds that

Ernst Mayr

University of Konstanz, Wikipedia CC BY-SA 2.5

like to classify things in discrete chunks. We have difficulty understanding that something as 'definite' as a species (a human construct, after all) can (and must) change over time.

placeable, this had become a matter of personal honour, and before I died I wanted to see it completely tied up and tidied away in its proper niche in the halls of science. This desire was greatly increased when I discovered that, after all, we had not yet got a complete animal. That also I was determined to get or see made available for science, and though tired of the tropics, I was quite ready to go again, and as often as need be, now that this more than encouraging discovery had been made.

It had long been my ambition to catch a Coelacanth alive so that the ordinary man could see it in an aquarium, and be given the opportunity to look back to the kind of creature that lived hundreds of millions of years ago. There is probably no other true scientific story which has given the ordinary man so clear a vision of what is meant by time, and to have a live Coelacanth on view would round it off in a way that H. G. Wells would certainly have appreciated. It was in one sense his idea of a 'Time Machine' come true.

There were many other things as well. Was there still another species, the small one Hunt had mentioned?* In addition, the startling difference from *Latimeria* observed in *Malania* raised difficult problems. Was *Malania* really different, or was it just an exceptional, perhaps extreme, variant?

Any species reproduces itself with comparatively small variation between individuals, but I had long held the view that this 'mass production' would tend to weaken or slacken in the course of time, that in a relatively broadly unchanging type like the Coelacanth, while the general form remained constant, after long ages there would be increasing variation in characters of lesser importance, like the position or size of the fins and other parts. There is one curious, rather primitive, rare type of fish (*Tetragonurus*), which is like this. Though it has been known for centuries and odd specimens have turned up over almost the whole globe, comparatively few, less than thirty adults, have been found in all that time. What is amazing is that hardly any two specimens agree in minor characters, like fin counts, and to this day scientists do not really know if there are a dozen species or only one. It will need hundreds of specimens before this can be settled.

* I have since come to believe that this is actually the Oil-fish (*Ruvettus*) that the Comorans catch in the same way as the Coelacanth.

In the case of the Coelacanth it was clear that problems like these could be solved only with more, probably only with many, specimens. There was also the strong possibility of mutilation or of deformity, or both, to account for these differences, but one needs to be cautious in seeking to explain things that way. We had one striking lesson.

When working in the northern part of Mozambique, in a canoe-catch one day I saw one of the peculiar Unicorn-fishes, a type of fish that has a long horn on its head, and this one even more peculiar because it had a hump on its back as well. You often see humped-back fish, a result of being either deformed or mutilated, usually when young. I looked at this one closely and decided it was probably a deformed specimen of the large Unicorn-fish that is quite common up there; but at the lighthouse that night my wife remarked that she had seen a deformed '*Naso*' that morning, it had a humped back. That made me sit up, and I questioned her. Next day I watched for that 'deformed' fish, and within a week we had got a dozen; and what is more, while all had humps some had no horns. It did not take long to discover that this was no 'deformed' fish at all, but a true species, and I found that while the males had horns the females had not. I described that fish in scientific literature and named it '*Naso rigoletto*'. Following the publication, from scientists over a great part of the world came a series of letters, many quite amusing, some chagrined, to say that they had had that species in their collections many, many years, and had just dismissed it as an abnormal or deformed fish. So I was wary about 'deformed' fish. It is risky to decide anything like that on one specimen, and in the case of the Coelacanth it all pointed clearly to the need for more, for many more, specimens.

I wanted to clear up all these points; it was something I wanted to see settled before I died, and I was prepared to go on myself. The Governor had given a most cordial invitation to return to the Comores, where it was clear that no scientific work like mine was in progress; nor had it probably ever been done at all. Coelacanths were apparently caught near the end of the year; the best time, with little wind, for any work on fishes. We could do our normal work and hunt Coelacanths as well, as we had been doing for years. We must get an expedition going.

Even when I composed that early article (p. 243) for *The Times*,

---

In 1957 JLB predicted that 'the coelacanth doubtless sheds its eggs inside a special case, quite possibly like those produced by some sharks and rays. Who will be the first to find one?' In one of the biggest *faux pas* of his career, he did not accept the offer of the 29th coelacanth, a 65 kg female caught in 1962 off Mutsamudu, on the basis that, 'My coelacanth work is done'.

This same specimen was sent to the American Museum of Natural History (AMNH) in New York and was subsequently found (when it was eventually dissected 14 years later) to contain five embryos, the first evidence that the coelacanth is a live-bearer. The discovery was announced by James Atz in a 1976 article entitled '*Latimeria* babies are born, not hatched'. In 1979 the AMNH donated a cast of one of the embryos to the Ichthyology Institute in Grahamstown.

NRF-SAIAB

Late-term coelacanth pup from the American Museum of Natural History

Wealthy yacht owners have often provided invaluable assistance to marine fish researchers. In 2009 Hans Fricke continued his research on the coelacanth off Grande Comore in *My Octopus*, the motorized yacht owned by Microsoft co-founder, Paul Allen, and he has also searched for coelacanths in the Pacific Ocean aboard a vessel owned by a Japanese billionaire.

most of this was in my mind, and it led me to insert the concluding paragraph as a venture, in which I asked if some yacht owner would make his vessel available for this search. There was little possibility of the large and perpetually busy South African Government Fisheries research vessel being seconded for such a project, but there have been a number of wealthy men, owners of sea-going yachts, who have rendered great service to science by placing these at the disposal of biologists over wide areas of the ocean. One of them might be attracted by this venture.

In the nightmare weeks that followed our return to Grahamstown this matter had to be relegated to a temporary background, but it pushed itself right to the front with the receipt of a most interesting letter from Jersey, Channel Islands. Dated 16th January 1953, it came from W. J. Stuttard, who represented himself as owner-master of a 150-ton twin-screw yacht, *La Contenta*. He offered this vessel, himself, and crew. He had a photographer, and stated that their expedition was not a profit-making concern other than what they might make out of travel-films, and he mentioned their thirst for knowledge and adventure.

The Channel Islands! The place, we are told, to which people of means or those with enough to live on without working go to escape British income-tax. This looked like the answer, but even though with our resources this seemed the only way, it is almost always cheaper in the long run to pay in cash for what you get rather than in other ways, and anything like this would need the most careful handling, and even then endless difficulties and complications can arise that had never been foreseen. No matter what terms or conditions might be agreed upon in advance, should any dispute arise in a case like this my party would be under most severe handicaps. The possibility of trouble is much greater when arrangements and agreements have to be made by correspondence.

So I replied with gratitude but without committing myself, and immediately put the issue to the Prime Minister's Department, as a matter of policy, and because this gave a sounder approach to the possibility of a Government vessel of some kind being seconded for this purpose, one which despite its global interest had acquired so clearly a South African national character. The reply indicated that there appeared to be no valid reason for refusing the offer of the *La Contenta*. The issue was then in turn submitted to the South African Council for Scientific and Industrial Research, and from them was received a recommendation to accept what appeared a generous offer. The Council also generously voted a thousand pounds towards the expedition, and I set about raising funds from the public. Thanks mainly to handsome contributions from a group of Johannesburg business magnates, this was rapidly achieved.

After having taken what seemed satisfactory precautions, I wrote on the 7th March 1953 to accept this offer of Stuttard, and it was settled that the sole direction of the scientific work should be in my hands. It was my intention at that time to manage by some means to make a short visit to Jersey, to meet Stuttard and his crew and to inspect the vessel, but this proved impossible, as there was too much that always needed my personal attention.

Arrangements for any such expedition as was now visualised to hunt further Coelacanths involved negotiations in widely different fields. First there was the critical matter of a vessel, now apparently solved by Stuttard's offer of *La Contenta*. There were all the special equipment and stores necessary for our work and maintenance. Thirdly, there was the matter of participation by other scientists and contribution by other institutions. Since at least part of our work would be in French waters, it would be necessary to seek permission and possibly some type of co-operation from the French.

Of all these, transport was the most troublesome, and once that was settled it seemed at least possible that no matter how many other obstacles might arise, they could probably be overcome. Even if the French were not co-operative, there were other parts not far off where we might have as good a chance of finding Coelacanths, and there are many things one can get done on the spot that volumes of correspondence will not even loosen. So transport was undoubtedly number one.

Museums and other institutions in various parts of the world had been asking if I intended to go to seek more Coelacanths, and some were eager to have at least some part in any such venture. Some were anxious to have Coelacanths. The press published a statement by a member of the staff of the British Museum that they were prepared to support an expedition for this purpose if I got it going. Although Stuttard was apparently

JLB Smith was a remarkably successful fund-raiser at a time when scientists rarely engaged in raising funds to support their research. In the late 1940s he mounted a successful campaign to cover the substantial publication costs of his book *Sea Fishes of Southern Africa*, one of the first books to be produced in South Africa with colour illustrations. He had also established an excellent working relationship with the CSIR, which had great respect for his pioneering work.

Several unsuccessful attempts were made to find living coelacanths off the Comoros in the 1950s. In 1954 and 1955 Jacques Millot collaborated with Jacques-Yves Cousteau and his dive team to investigate coelacanths off these islands using deep-diving techniques, but failed to find any.

Miniature sheet of the Calypso and postage stamp of Jacques-Yves Cousteau issued by Palau to commemorate Cousteau's death in 1997

In the 1950s and 1960s coelacanth research expeditions were launched to the Comoros from Britain, France and the USA, some of which tried to catch live coelacanths, but without luck.

prepared to offer his vessel and crew without payment, and I had sufficient to cover fuel for his vessel, as well as for food and equipment, as I well knew there is always an element of uncertainty in the cost of such ventures. We had some financial reserves, but there was no harm in having more, and no harm in permitting others, if they were eager to do so, to make contributions, provided no uncomfortable terms were attached to them. I wanted to find Coelacanths, not for myself but for science, and if science would help financially, all the better.

After weighing the matter carefully, I decided to select a limited number of the leading institutions in the world and to send to each the following proposal, namely that the payment of the sum of £500 towards the expedition would secure the right to purchase a Coelacanth for a further £500. This meant that a specimen would cost £1,000, which was very considerably less than it might be expected to cost any individual institution if it conducted the search alone, while it limited liability to £500 if none were found. In the event of there being fewer Coelacanths available than participants, they would be left to settle the allocation between themselves. It was to be accepted, however, that the first Coelacanth taken in French waters was to be given to France. The money subscribed in this way was to go to Stuttard for expenses.

The British Museum and several others replied that they did not wish to participate, a large American institution apparently found £500 too large a sum to risk, and indeed only two of those approached hoped to be able to do so. Fortunately, however, as far as it was possible to judge at that time, we apparently had enough for our purpose without any such extra aid.

We shall now follow the course of negotiations with Stuttard.

Even before I accepted his offer, he had informed me that he intended to come via the Cape, giving sound reasons for that course. In letters dated the 17th February 1953 and the 12th March 1953 respectively, he visualised leaving in good time, stating in the latter his intention to leave in the first week in May (1953).

Knowing what I did about preparing for expeditions, I wondered whether he would find it possible to do this, and was not surprised to receive a letter, dated the 15th April 1953, in which he reported difficulty in getting equipment delivered in time.

This difficulty evidently increased, for in a letter dated the 30th April 1953 Stuttard reported that the delay in getting equipment might mean that he would have to go via the Mediterranean.

In my reply to this, dated the 5th May 1953, was the following:

> If you should go through the Mediterranean it will I think cause the very greatest disappointment out here as there will be great numbers of people, especially those who are supporting us, who are hoping to see the vessel. Pictures of the vessel have appeared in all the leading papers. I would not let them have any of you without your special permission, and in any case felt that it would be better when you are actually in South African waters. Another great drawback of your going through the Mediterranean would be that you would be compelled to battle southwards against the monsoon, which I can assure you is really appalling, and the run from the Red Sea downwards would be something of a nightmare at that time. In addition, I am planning to provision the ship very largely here, but that is no insuperable matter if you can get sufficient foodstuffs in England. You are likely to encounter difficulty in getting things in Kenya. If the foodstuffs we are planning are to be taken, then I think it will be advisable for you to load them here, as to get them round to Port Amelia in time will be more than a nightmare. With regard to all this, I should stress that we shall be more than content if we get to the Comores by September. There is no hurry, and we get the best tides then. If you leave England even as late as the end of June, assuming that it takes you 25 or 30 days to the Cape, you should easily get to Mozambique Harbour well before the end of August. I think it would be advisable for you to make Mozambique Harbour our point of departure from the African mainland, as it gives us a real northing in our crossing which we may be very thankful to have.

Stuttard considered this matter and wrote several letters about it. Eventually, in a letter dated the 13th May 1953, he stated that he would sail not later than the first week in July and come via the Cape.

That seemed definite and satisfactory, and in the meantime Dr. Eigil Nielsen of Copenhagen, one of the foremost palaeontologists of Europe, had written to say that he was waiting for a final decision from his University about participating in our expedition, and that he was, in any case, planning to go and hunt for Coelacanth fossils on Madagascar.

It is interesting that JLB goes to such exhaustive lengths to chronicle the negotiations surrounding the abortive *La Contenta* expedition yet he makes hardly any mention of the successful arrangements he made for the use of a variety of other ships from South Africa, Mozambique and the Seychelles to support his expeditions.

It seems that he was more preoccupied with recording the almost insurmountable hardships that he had to overcome and less interested in documenting arrangements that went smoothly as they, presumably, have less dramatic appeal and are less revealing about his indomitable character.

Fossils of extinct coelacanths from the genera *Coelacanthus*, *Piveteauia* and *Whiteia* have been found in Triassic deposits in Madagascar. In addition, 13 living *Latimeria chalumnae* have been caught off Madagascar, which may still prove to be one of their primary habitats.

Coelacanth fossil from Madagascar

Harun Yahya

With this in view I wrote to Stuttard and suggested that if Nielsen came in he might perhaps travel on the *La Contenta*, which would be of benefit to all. Stuttard welcomed this, and stated that he could easily be accommodated.

In a letter dated the 23rd May 1953 Stuttard stated that I could take it that he would definitely be sailing within the first few days of July, and that such small items as could not be got aboard in time would be sent on to Durban.

On the 5th June 1953 came a letter from Nielsen dated the 28th May 1953:

Today I got the welcome news that the money had been granted, which means that I now am in the position definitely to accept your most kind invitation to contribute to the Coelacanth venture as well as to carry out my work on Madagascar.

I am extremely glad to know that it seems possible to arrange for my actual partaking in part of your expedition, and as to eventual inconveniences I am accustomed to rather much in that way from the life in our small crowded Greenland vessels, which often spend more than a month in covering the distance between Denmark and Greenland. The further possibility that an arrangement could be made so that I could join the yacht already in England sounds of course perfect to me. My equipment does certainly not take up much space, as my first plan was to go by airplane. I beg you ask the owner of the yacht to be so kind as to inform me as to time and place for the departure, and to thank him on my behalf for his invitation.

As to my contribution to the expedition, please instruct me as to the deliverance. I can send the money to England or to you or deliver them personally, I don't know which method you prefer?

I shall be very grateful for your advice as to what sort of camping outfit, clothing, and photographic equipment is necessary for work in Madagascar. I suppose that it is possible to buy much of the things on the spot, Leica-films (i.e. Kodachrome films), etc., probably can be had out there?

On the 5th June 1953 I wrote to Stuttard as follows:

I have this moment received a letter dated 28th May 1953 from Dr. Nielsen of Copenhagen, to say that he has been granted the necessary funds and that he wishes to be a contributor to the expedition, also that he would be very pleased indeed to travel with the *La Contenta*. He had originally planned to go out to Madagascar by air. He is delighted at the opportunity of coming out with the *La Contenta*, and says that he has not got very much equipment. He asks that you should make the earliest possible contact with him, especially to notify him of the date of departure. I am enclosing a copy of my letter to him, which goes by this same air post, and you can see that I have told him that the payment of the original £500 can be by arrangement between you and him, because I feel that should go to you. Please make your own arrangements. In addition, he should pay for his keep. I suggest that you work out some reasonable sum like 10s. per day to cover the cost of food, etc., even more if you think it wise, as he would otherwise certainly have to pay quite a considerable amount to get out. He will doubtless have to make his own arrangements for the return journey.

To avoid any possiblity of misunderstanding on that point, I had quite early sent Stuttard an explicit account of my resources and had said that this was the limit from my side. The financial aspect had in the later correspondence begun to obtrude itself, and in a letter dated 4th June 1953 Stuttard wrote that as he feared in the beginning for the financial success of this venture and could not himself afford to purchase all the equipment required, he considered it necessary to seek further funds on my behalf. He had indeed approached the Nuffield Trust.

From this and other matters, on the 9th June 1953 I sent the following cable to Stuttard: 'If financial burden too heavy you better cancel everything writing unable assist further.' To this Stuttard cabled in reply that he intended sailing as planned and that he had written to Nielsen.

On the 10th June 1953 the President and Vice-President of the South African Council for Scientific and Industrial Research visited Grahamstown and called at my Department. I gave them a résumé of all that had occurred, showed them the correspondence, and told them that in my opinion a distinct element of uncertainty about the whole project had arisen.

We had a full discussion and it was their opinion, as it was mine, that it would be inadvisable to spend money on the project unless all the arrangements were quite sound and satisfactory to all parties concerned.

Stuttard wrote again on the 11th June 1953, enclosing a copy of a letter of the same date from himself to Nielsen, in which he

Dr Eigil Nielsen, a prominent palaeontologist from the Museum of Mineralogy and Geology at the University of Copenhagen, researched fossil vertebrates, including coelacanths, in Devonian and Triassic deposits in Greenland and Triassic deposits in Madagascar. The extinct coelacanth *Whiteia nielseni* from Greenland is named after him.

PierreSelim, Wikipedia
CC BY-SA 3.0

Whiteia nielseni fossil from Greenland

JLB Smith's keen knowledge of currents, wind and sailing conditions in the Mozambique Channel was helpful in the planning of the expedition. It is a treacherous stretch of ocean with many surprises for the uninitiated. The large number of shipwrecks along the East African coast and around coastal islands bears testimony to its hazardous nature.

informed Nielsen that he was planning to sail about mid-July, the date being dependent upon the delivery of a large deep-freeze essential to my requirements. He stated that the route would be via the Cape.

Although the date of sailing had been moved from the first week in July to mid-July, this still fell within my time-limits, but all my anxieties were aroused by a letter from Stuttard dated the 2nd July 1953, in which he stated that he had reluctantly come to the conclusion that it would be out of the question to come via the Cape if he could not leave before the 15th July. He undertook to try to decide early the following week and to advise me by cable, so that I could get all my stuff up to Mozambique if he had to come via the Mediterranean. When this issue had arisen earlier, on the 12th May 1953, I had written to Stuttard as follows:

> I am indeed pleased that you have abandoned the Mediterranean approach. Not only would it have caused me endless difficulties at this end, but for you to have gone battling into a four- and sometimes five-knot current against a southerly monsoon as well would certainly not have been pleasant. It will be quite bad enough going up the Madagascar Channel against the current, though you will have the wind with you in the southerly part, and proper use of eddies can avoid a good deal of trouble (north of Madagascar the current runs north). All that we can go into here, and it may be possible to arrange for you to have a discussion with the Captain or Mate of one of the Portuguese Coasters at Lourenço Marques. Regarding the explosives, I have heard from the Portuguese Authorities, who state that it will be almost impossible to arrange for them to be sent to Port Amelia. I think it will be quite clear for them to go to Lourenço Marques and be stored in the Government magazine there, and the Port Captain has promised every facility and aid in this matter. So after what you say I am going ahead.
>
> I wish to emphasise again that as far as we are concerned, it will suit us quite well if we get to the Comores by the beginning of September. The four best tides of the year for reef work are the 24 August, 8th and 23rd September, and the 9th October. If we can contrive things so that we do the circumnavigation of Madagascar in November, that will probably be best of all, as we can expect the least wind and slowest current at that time.

In South Africa both persons and firms had donated a wide variety of goods and supplies for the expedition, and it had been arranged for them to be sent to the nearest port. At Cape Town, Port Elizabeth, East London, and Durban there would be generous consignments ready for *La Contenta*. At Lourenço Marques there was more equipment and that ton of explosives, whose part as essential to the expedition had been established between us before I accepted Stuttard's offer. In addition to all this, our Government had kindly conceded a number of special concessions for *La Contenta's* visit that would mean a great deal to Stuttard as well as to the expedition.

At that stage, 8th July 1953, it was clear that even to attempt the by then enormous task of getting all this dispersed stuff to Port Amelia within six or seven weeks to meet this abrupt change of plans was too much. The Portuguese authorities had been quite explicit that it was virtually impossible to arrange for the transport of the ton of explosives from the Government magazine in Lourenço Marques to a port in the north and to store them there. I knew what transport along the East African coast was like at that time, and short of chartering a special vessel or by other expenditure on a quite unwarranted scale, it would have been virtually impossible to do so. I should have to consult those who had provided the money, but on general principles I was opposed to the risk of spending in that way so much of the funds that had been donated, for even if I succeeded in this formidable and apparently impossible task and got the essential goods there by the stated date, sea travel in a small vessel is so uncertain that there could be no guarantee that Stuttard would arrive in time to carry out the expedition, or indeed at all. With the best will in the world he could never guarantee to do so. I could picture our party perched on top of a pile of partly perishable baggage in the humid torridity of a remote East African port, while this small vessel battered its bows southwards against the powerful monsoon-impelled waves and the current of the long stretch down the East African coast. We should have no redress if Stuttard did not arrive, it might prove beyond his powers or control, and all that tremendous effort and expenditure would be wasted. From my experience and knowledge, I was quite firm in my opinion that what Stuttard now asked would be almost impossible without unjustifiable expenditure, that it lacked certainty as justification, and in short to attempt it at all would be folly.

With his excellent knowledge of chemistry and strong conservation ethic, JLB would have used explosives carefully and effectively, but one wonders what impression he created among local fishermen.

Today, blast fishing with explosives, together with deep-set gillnets, represent two of the major threats to marine communities in the Western Indian Ocean. Blasting is arguably the most senseless form of fishing as, used incorrectly, it instantly and permanently destroys coral and rocky reef communities that took millennia to develop.

Branching corals are particularly vulnerable to blast fishing as they are reduced to useless rubble. In addition, this fishing method is indiscriminate and wasteful as many fish and invertebrates that humans do not use are killed.

It must have been a nightmare for Smith to attempt to organize a complicated international expedition with someone in another hemisphere whom he had never met, especially considering the military precision with which he organized his own expeditions. It was almost inevitable that the planned trip would fall through.

Serious as it was, this was not the only complication that arose from this proposed abrupt change of plans. Stuttard had apparently not succeeded in obtaining all the items he had hoped free of charge. Notable among these was a deep-freeze unit, to which he referred in almost every letter. According to his letters, he could not leave without it. Owing to the unexpected but obvious concern Stuttard had revealed about finance, I had got interested friends to agree to cover the cost of the deep-freeze unit (the intention being to sell it on their behalf after the expedition), and wrote to Stuttard as follows on the 22nd June 1953:

> Following on our recent correspondence I have taken further action with regard to finance, going to certain friends who are in some ways a reserve I do not easily care to tap. At any rate, the result is that they are prepared to pay for the deep-freeze, on condition that at the conclusion of our association it shall either pass to my possession or be disposed of and the amount returned. This should relieve you, especially as you have indicated that the deep-freeze is not a piece of equipment which you will greatly value for your vessel.

In the letter dated the 2nd July 1953 Stuttard wrote that he was pleased to note that I was accepting the responsibility of the deep-freeze, and as there was a twenty-five per cent. discount I should not experience difficulty in disposing of it at a profit.

As no cable (see p. 202) had arrived by the 8th July 1953, the date on which Stuttard's letter of the 2nd July was received, I consulted the friends who had offered to sponsor the deep-freeze, and told them of the situation that had suddenly arisen. When they were originally approached about this, I had explained that the vessel would come via the Cape, as that was clearly understood from Stuttard at that time. After full consideration they now stated that they were not prepared to continue this offer under this sudden drastic change of conditions imposed by Stuttard, and the following cable was sent to Stuttard:

> Sponsors prepared refund cost deep-freeze only on arrival here and in time carry out full operations. Unless come via Cape impossible co-operate.

Three days later I received a letter from the Jersey Electrical Co. Ltd., dated 6th July 1953, the concluding paragraph of which was as follows:

We shall be pleased to receive your remittance at the very earliest opportunity and in order to prevent any further delays, we are assuming that this will be coming forward and consequently, we have asked Frigidaire Ltd., to get the equipment down immediately as we understand from Mr. Stuttard that he wishes to leave Jersey by the end of July.

Even by this date, 11th July 1953, there had been no cable from Stuttard, and now this letter had disturbing implications. It implied that by the 6th July, when this Jersey firm wrote, Stuttard knew that he could not leave before the 15th, and so, according to his own statement, if he came at all he intended to take the Mediterranean route.* Nielsen was due to join Stuttard on the 10th July, and I doubted whether he would be prepared to wait at Jersey until the end of that month, which this last letter indicated as sailing time.

Nothing further happened until the 16th, which brought a letter from Stuttard dated 9th July 1953 in which he informed me, among other matters, that Nielsen was due to arrive next day, and that he would be informed about the situation (see p. 206). He went on to say that to prevent an absolute fiasco he advised me to send everything to Mozambique and he would be there in ample time.

On the 16th July 1953 I sent a cable to Stuttard as follows: 'Letter ninth received confirm ending co-operation writing Smith.'

On the 23rd July 1953 a letter dated the 18th arrived from Stuttard, in which he informed me that he was in touch with the French and Danish Embassies with a view to participating in an international expedition, and that he had made a statement to the President of the South African Council of Scientific and Industrial Research and to the Acting Secretary of External Affairs.

On the 21st July 1953 a letter dated 16th July arrived from Nielsen, in which he stated:

> The 10th July I arrived at Jersey expecting to leave from there one of the next days with *La Contenta*. Mr. Stuttard informed me, however, that the actual departure would not take place before the end of July, and that moreover the boat would go via Suez instead.

* See letter from Nielsen, p. 206.

The failure of the *La Contenta* expedition must have been very discouraging to Smith and might have been the final straw as far as his decision to abandon further research on the coelacanth was concerned. From 1954 onwards he focused his field research on the other fishes of the Western Indian Ocean, with no emphasis on the coelacanth.

The Jago submersible off Grande Comore in 2008

Karen Hissmann

While diving in the *Jago* submersible at depths of 170 m off Grande Comore in 2008, Hans Fricke and Jürgen Schauer felt the jolt of dynamite depth charges that were being discharged by fishermen in the area. This harmful and indiscriminate fishing method is likely to devastate coral and rocky reef communities off the Comoros.

If the typical pattern of 'development' of rural fisheries is followed in the Comoros, i.e. the replacement of traditional methods with dynamiting and deep-set gillnets, this will represent a serious threat to coelacanth populations.

Many traditional fishing methods, if they are used in moderation, harvest fishes and other aquatic animals sustainably.

Constriction traps – a traditional African fishing method for catching snake catfish

Mike Bruton

of the original route via Cape Town. I was somewhat troubled both because of the altered route, firstly on account of the rainy season starting in Madagascar some time in November, and secondly because I had been looking forward to seeing you in Grahamstown.

I therefore altered my plans by booking a passage to Port Elizabeth in the *Edinburgh Castle*, Union Castle Line. I leave Southampton today and expect to arrive at Port Elizabeth on the 1st August.

In this way I get a whole month more time for the work in Madagascar, and, moreover, I get an opportunity to see you in Grahamstown one of the first days in August.

Nielsen arrived at Port Elizabeth on the 1st August 1953. He had arrived at Jersey on the 10th July 1953 and went to *La Contenta*, where he met a few men and women on board, introduced as the crew. Later he went to a maternity or nursing-home with Stuttard, who introduced him to his (Stuttard's) wife, still abed, having a few days before given birth to a child. According to Nielsen, he was informed of the delay in sailing and of the change in route, but nothing was mentioned about the possible severance of connection between Stuttard and myself. After weighing the situation Nielsen eventually decided to leave, as outlined in his letter. It was only when he got a cable from me at sea that he realised that something had gone wrong in the arrangements between Stuttard and myself.

So, after all those months of strenuous work, there I was in late July right back at the starting-post again, and once more rapid action and decisions were called for. A vast amount of material, equipment, and stores was waiting, including a ton of explosives, and while the idea of just abandoning everything was hateful, to go on meant finding a ship in a hurry. We must be at the Comores for October and November at the latest, which meant leaving before the end of August, barely a month.

As will be outlined in the following chapter, I had other troubles, for negotiations with the French had not been settled.

It was suddenly announced over the radio that I had cancelled the Coelacanth expedition. This I immediately contradicted. My first move was to telephone Minister Paul Sauer, to whom I outlined the position and asked him if there was any vessel in the Harbour Administration that would be suitable for my project, and if so how could I set about getting it. He said his staff would find out, and within twelve hours the answer came back that there was no vessel they could recommend or release.

Meanwhile I had telephoned prominent people in the shipping world at all the ports and others in Johannesburg and Pretoria, asking them to try to locate some vessel in South Africa that could be diverted to our project. A few doubtful prospects were notified, and on the 27th July 1953 I set out by air in the forlorn hope of finding at that late date some vessel that might still enable us to do what we had planned. I wanted to hunt Coelacanths, but had no intention that we should fill a Tiger Shark's belly by going in an unsuitable vessel. I ransacked Port Elizabeth, but found nothing suitable; then left by air and got to Durban that same evening, where within an hour of arrival I was inspecting vessels. One or two were possible, but were ruled out either by cost or by the time needed for essential repairs. On the 29th July 1953 I went to Cape Town and was whisked off at once on arrival to a conference. It was bitterly cold at that time, but I could visualise East Africa and the Comores. One vessel was hopeful but very costly, and when I telegraphed to explain the situation and to ask the South African Council for Scientific and Industrial Research if they could manage another thousand pounds, the answer came back: 'Yes, but we strongly recommend postponement until next year.' I could not ignore that, and with all the other difficulties not yet solved it spelled 'Finis', there was too little time.

Nielsen's ship arrived early in the morning and I went to meet him. We had only a few brief moments, in which I gave him the main outlines of the story and told him my wife would meet him at Port Elizabeth.

I had asked to see Dr. Malan—once again I was in that office where I had sat in such discomfort before. I told his Secretary that my need was past, and if the Dokter was very busy I should go; but he said no, he knew he would like to see me. When I went in Dr. Malan rose at my entrance and smiled as he held my hand. He asked what I wanted, but I said I would not burden him with my troubles. I wished just to greet him again, though I told him briefly that my Coelacanth hunt was off.

I slept most of the time in the plane on the way back to my wife and Nielsen at Port Elizabeth, from where we took him to Grahamstown, to the laboratory, and of course first to the

We were luckier in our quest for suitable research vessels for our 1991 and 1996 *Jago* expeditions in South Africa. After a brief search in Port Elizabeth and Cape Town we found *Deep Salvage I,* which was skippered by Peter Wilmot, an experienced scuba and hardhat diver who had first been inspired by Jacques-Yves Cousteau and Hans Hass. This fine ship was equipped with a one-man diving bell, mixed-gas tethered helmet dive apparatus to 120 m and scuba to 60 m, and proved to be a reliable mothership for our dives.

Hans Fricke

Deep Salvage I, *the mothership used during the 1991* Jago *dives in South Africa*

The 1996 *Jago* dives off Cape Town were carried out from the equally well-equipped *Zealous,* sponsored by the marine diamond mining company De Beers Marine.

Coelacanth model in the Iziko South African Museum in Cape Town

The coelacanth, though it becomes dull and brown when it dies, is a beautiful blue fish when alive, and has a mesmerizing effect on ichthyologists, palaeontologists and informed lay people. It has such a significant evolutionary story to tell that it elicits a potpourri of emotions ranging from awe and enchantment to respect and reverence.

Coelacanth. Nielsen had worked on Coelacanth fossils all his adult life. He had baked in heat and frozen in terrible cold hunting them from the equator to the Arctic circle, so that they were not mere fossils and academic abstractions to him, but an intimate part of his life. On his way to South Africa, Nielsen had been staggered by the comparative lack of interest in Coelacanths he found in an important institution in London. One scientist, indeed, had said he was 'Sick of Coelacanths'. It was beyond Nielsen. When we got into the room where old *Malania* lay, Nielsen walked all round and looked for some time in silence. Then he said, almost breathless with emotion, and with deep sincerity, 'How beautiful.' And so he is, old man Coelacanth, to a scientist, a wonderful and beautiful thing.

Nielsen is, of course, a Scandinavian, and we heard later that before his arrival our staff speculated on his appearance. It was decided he would be huge, blond, and genial, but instead Nielsen is small and dark, active in mind and body, and has a rapier-like humour. His English is excellent, as we learnt when he gave a public lecture we had requested on his work. He is no soft laboratory worker, but as tough as a scientist in his field has to be to get anywhere. He would have been at home on the shores of a Triassic swamp.

When he got to work on *Malania* it was interesting to see that he went first to examine that extraordinary puzzling cavity and channels in the nose, next to look for traces of the air-bladder, exactly as I had done myself. Smoking is not permitted in our highly inflammable Department, so that Nielsen got to know the outside of our building quite well.

After a week of intensive study, Nielsen was due to leave. Early that morning, at his special request, we went for the last time to the laboratory. He took the Coelacanth's pectoral fin in his hand and shaking it, said solemnly, 'Good-bye, Malania.' It was very sincere and quite touching.

It is interesting to record that by courtesy of the French authorities in the Union, we managed to send Nielsen's heavy baggage to Madagascar on a French warship that happened to be passing. Nielsen went over by air, and later wrote to tell me that he had got more than two thousand splendid Coelacanth fossils in Madagascar. One could hardly describe them as anything but valuable scientific material, but he did not indicate that there had been any difficulty in taking them from French territory.

The following is quoted from the *Evening Post* of Jersey, issue of 2nd February 1955.

END OF LA CONTENTA'S *THREE-YEAR STAY* *COELACANTH SEARCH RECALLED.*

The 160-ton motor yacht, *La Contenta*, which has been in St. Helier Harbour for nearly three years, left this morning for St. Malo.

This former Admiralty Fairmile Class craft came into the news in May 1953, when she was being fitted out to undertake an expedition in search of that extraordinary fish the Coelacanth. A specimen of this supposedly prehistoric fish had been found in the Indian Ocean, leading to arrangements being completed between the noted Professor J. L. B. Smith and the owner-captain of the *La Contenta*, Mr. W. J. Stuttard, to equip the craft to search for another specimen. However, after a period of time this expedition fell through and it was understood that *La Contenta* would then be used to search for treasure in the China Seas. This venture also fell through.

ONLY ONCE PUT TO SEA

In a stay of two years nine months in St. Helier Harbour, *La Contenta* only once put to sea, and that was to Gorey on the occasion of the regatta there in 1953. Recently *La Contenta* was thoroughly overhauled and successfully underwent tests in St. Aubin's Bay.

Mr. W. J. Stuttard stated shortly before leaving, 'We are bound for Vigo in Spain and from there we shall cruise in the Mediterranean where we shall stay. We are tired of Jersey.'

O.F.—14

Considering that the owner-master of the yacht *La Contenta*, Stuttard, set sail only once during his two years and nine months in Jersey, and that his next venture, a search for treasure in the China Sea, also fell through, it is fortunate that JLB did not risk his reputation, and his life, with this skipper.

A third coelacanth was caught off Mutsamudu in Anjouan on 24th September 1953 and reached Jacques Millot in Paris via the Scientific Research Institute in Antananarivo, Madagascar.

Chapter Eighteen

PORTCULLIS AND DRAWBRIDGE

THE global reactions to the extraordinary culmination of my long search for the Coelacanth had profound effects in France, where widespread and somewhat hysterical propaganda in the press aroused public feeling. I was denounced as a robber who had pounced on their national treasure. As a result, there was widespread agitation, and it was urged that the French Government should demand the surrender of the Coelacanth to France.

This could scarcely have arisen had the matter been presented to the French nation in its true perspective, for apart from my initial intense anxiety to see that it really was a Coelacanth and that it was safe, it had never been my intention to keep it to myself in any way. I intended the fish to be available to expert scientists of all nations. It was indeed my hope to be able to find more specimens for that purpose myself, and the expedition had of course been planned to that end. All specimens were to be for the benefit of science generally, not to belong to me or to be confined to the scientists of any one country, but for all.

Nothing of this outburst was communicated to me officially, but it was natural to expect 'that its effects would not be without repercussions in higher circles. As I wished to do scientific work in French waters, it would be necessary to request formal permission from the French Government, and those who had to consider my request could scarcely be unaware of this degree of national resentment that had been aroused against me, while they would almost certainly be unaware of the true circumstances.

As I had no direct contact with any personality in French Government circles, on the 15th January 1953, as a preliminary, I wrote as follows to Dr. Millot, who was then in Paris:

> I wish to record my great appreciation of all the assistance and co-operation we received from the authorities of your country, not only through their representatives in South Africa, but also

those in charge of the Comores Archipelago. The Governor, M. P. Coudert, did everything in his power to make our all too short stay on Pamanzi Island as pleasant as possible, and both he and his officials, realising the significance of this great discovery, gave us every assistance.

As a scientist I know that you will rejoice that this has occurred, and it will naturally be of added interest to you that the discovery has been made in waters under the charge of your Government.

I may inform you officially that although the knowledge of this has come via our leaflets and constant propaganda, I have requested Captain E. E. Hunt that the next Coelacanth he might get in French territorial waters should be handed over to the French Authorities. Even though the present specimen is considerably more damaged than I had hoped, and a complete fresh specimen is essential for the full study which is contemplated, I shall be happy to know that you get into your possession the next, and I trust complete, specimen.

Will you kindly accept, my dear Sir, once again my sincere thanks for the co-operation of the French Authorities and a renewed expression of my belief that all of us who are scientists in Africa must work together for science.

I received from Millot in Paris a cordial reply, dated the 19th February 1953, in which he expressed the hope that we might work together, and that he would later outline steps to that end. He also stated that the resentment which had been aroused in France had resulted in a decree prohibiting the export of scientific material of value, including Coelacanths, save with the authority of competent local scientific organisations.

To this letter the following reply was sent to Millot on the 23rd February 1953:

> . . . the reactions and objections in France . . . were surely made in the absence of knowledge of the true facts of the case. Only today have I had a letter from another source quoting some of the events you mention. For that reason I am enclosing a statement which I shall be pleased if you will kindly hand to the proper authorities, and to the press if you judge it correct to do that.*
>
> At the moment I am engaged in the preliminaries of arranging an expedition to the Comores and Madagascar. This involves very much work and a fairly large vessel (150 tons) which has to be

* I have not yet received comment from any source on this memorandum.

French scientists acquired most of the next 25 coelacanths caught in the Comoros, although there is no record of the location of specimens 9, 16 and 25. Jacques Millot, James Anthony and later Daniel Robineau made extremely detailed studies of coelacanth anatomy and published their results in a series of comprehensive monographs between 1954 and 1978.

Eugene Balon

Coelacanth skeleton on display in the Grande Galerie de l'Évolution in the Muséum National d'Histoire Naturelle de Paris

The extraordinary process of 'coelacanth détente' was initiated in the early 1960s when the French donated a specimen caught in 1956 to their arch rivals, the British Museum (Natural History) in London. Since then the French have officially donated specimens to Japan, Algeria, China, Kuwait, South Korea, South Africa and even the United Nations in New York.

By late 2011, 45 coelacanth specimens had been lodged in museums in France (the largest number in any country) with a further 10 in Madagascar. Altogether, 215 of the 299 coelacanth specimens whose capture locality is known have been caught off the former French colony of the Comoros.

Rotenone is a widely used fish poison that is derived from the roots of the tuba root, *Derris elliptica*, and the beach poison vine, *Derris trifoliata*, which are both common in East Africa. It interferes with the ability of cells in the gills to absorb oxygen, which kills or stuns fish or causes them to rise to the surface to gulp air, where they can easily be caught.

Mike Bruton

Beach poison vine, Derris trifoliata

Rotenone is still widely used by ichthyologists for catching fishes in habitats where traps, nets or lines are ineffective, such as in well-vegetated swamps or rocky or coral reefs. As it decomposes rapidly in sunlight and degrades in warm water, it leaves no harmful residues.

carefully arranged many months ahead. Will you kindly let me know specifically as soon as possible if we shall be permitted to work at the Comores and about Madagascar; we plan to come from about August–October, first at the Comores and then to voyage round Madagascar.

You may not know that I am engaged on a large-scale work on the fishes of the western Indian Ocean. During the past seven years we have made a number of expeditions to East Africa, covering the whole coast of Mozambique, Zanzibar, Pemba, and Kenya.

In order to use our time as efficiently as possible, we need to employ explosives and Rotenone poison, and have everywhere received permission for that purpose. Will this be permitted in your territory also?

We have got together very large and important collections from the east coast for our final study, and it will certainly be a great pity if the French Territories are not included in this work, which is expected to run to a number of volumes.

An early reply will be greatly appreciated.

The memorandum which I enclosed was as follows:

I have heard with great surprise that there has been strong resentment in France against my fetching the Coelacanth fish from the Comoro Islands.

May I present the facts. From the time that the first Coelacanth was found in 1938 I have constantly sought to find more and if possible where they live. My deductions led me to think that this would be on the east coast in the neighbourhood of Madagascar.

Immediately after the war I started an intensive campaign, one of the methods being a special leaflet in English, French and Portuguese, showing a picture of the Coelacanth and offering a reward of £100 for each.

By various means these were distributed along the East Coast. In 1948 I was in Lourenço Marques arranging this distribution with the Portuguese authorities, and with a Portuguese official called on the French Consul in Lourenço Marques, and explained the importance of the matter to him. He readily agreed to send a batch of the leaflets by air to the authorities of Madagascar, and assured me that they would be distributed there as requested. Neither then nor later was there any suggestion that if a Coelacanth were to be found, its removal would not be permitted by the French authorities.

Since 1945 I have made a number of expeditions to the east coast of Africa, always seeking the Coelacanth. A great deal of time and many thousands of pounds have been spent in this search.

The Coelacanth at the Comoro Islands was saved chiefly because of my search and propaganda leaflets. I regarded and regard that fish as ethically mine. Had it been found by purely French effort, I should have made no effort whatever to get it.

It may be recorded that I asked Captain Hunt if he got another Coelacanth in French waters, as a result of my search, to hand it to the French authorities.

I do not regard this fish as belonging to me personally, or even to one country, but to the world of science. I have already told the South African Council for Scientific and Industrial Research that it should be examined by a panel of International Experts, and we are at present considering how this may be arranged. Scientists in France will have an equal opportunity with those of other countries. A statement about this will appear in the journal *Nature* (London) on February 28 next.

Early in March 1953 I submitted a memorandum requesting from the French Government permission to hunt fishes in their territories,* exactly as we had been granted permission by all other foreign governments in Africa. Time was of course a vital factor, and it was disturbing that only in May 1953 did I receive in reply a request for further information relating to that application.

On the 8th May 1953, to acquaint him with what was happening and hoping that he might be able to assist, I wrote to Millot as follows:

. . . I now enclose a copy of a statement about the poison and explosives we wish to use. To you personally I stress again that we are exceedingly careful in our use of these from every point of view. Not only do we always aim at causing the least possible damage to natural life, but we are fully aware of the effect of such methods on the minds of primitive people. We have our own ways of impressing on the Natives that both these methods are not for the ordinary man, and that, especially in the case of explosives, disaster will almost certainly follow their use.

It is not only in your territories that these methods are forbidden, but in all, and it is because of our reputation and the great care that we always exercise that all the other Governments have given me this

* Up to September 1955 no specific reply to this request has been received.

JLB raises the ethical issue of the impact 'on the minds of primitive people' of the use of explosives to catch fishes. As explosives can be made from readily available ingredients (diesel fuel, soil fertilizer and recipes from the Internet), their use is increasing despite the fact that they are banned.

While JLB did exercise caution using explosives he had a close shave off Mozambique when a wave washed their boat over a site where a 'bomb' had been dropped and JLB and Margaret and their boat were blown out of the water; fortunately, they were unhurt. Also, in a caption to a photograph taken during their 1951 expedition to Mozambique, JLB writes, 'Silhouette Island, where bombing fish caused an island to disappear'!

Nothing came of JLB's correspondence with Jacques Millot except a meeting in Nairobi in October 1953 and later meetings in Paris. He never returned to the Comoros, or Madagascar, and did little further work in the French territories.

In 1956 JLB wrote a caustic letter to *The Times of London* suggesting that the French had sufficient specimens and that their present policy is 'debasing a once important scientific quest to the level of senseless slaughter of one of our most precious heritages in biology'.

He published his last scientific paper on the coelacanth in 1955. *Old Fourlegs*, published in 1956, was his swansong. Margaret published some reminiscences and reviews on her involvement in coelacanth research in 1979 and 1980.

free permission. Not a single one of them has ever had cause to regret this in any way whatsoever, and I am confident that at any time in all the Territories where we have worked we shall continue to be able to enjoy these special concessions.

In the matter of the Coelacanth I wish to emphasise to you that we are not coming on this venture with any sense of competing with you and your nation. Not at all, we are coming because I feel that no effort must be spared to find further specimens. We shall be only too honoured to be able to collaborate with you to the fullest extent. I have probably more experience of getting fishes under East African Tropical conditions than any other living scientist, and it would be foolish not to use that for this great hunt. We can indeed offer to share any methods or materials that we may have with you on the fullest basis of co-operation.

At the moment it does not appear likely that we shall be able to commence operations until the end of August at the earliest. That should enable you to work for some time before we come, and I shall all the time hope to hear that you have caught a Coelacanth. As you possibly know, we have decided that the specimen here is not to be dissected further until other specimens become available. It would, in any case, be desirable to keep this intact unless we get another specimen of the same type.

I do hope your Government will see its way clear to getting Captain Cousteau to assist in the Coelacanth hunt, as that deep diving will almost certainly be a most useful method of hunting. It would be a very great achievement if one of your countrymen could report on live Coelacanths in the sea, and the whole world would certainly take the greatest interest in anything he could tell of their habits. From what I have read of the method, in those clear waters it would appear to hold out one of the greatest hopes in our search, and I shall be greatly obliged if you would kindly send me your comments and report on this.* We ourselves use goggles and diving, but have of course nothing comparable with the magnificent equipment developed by your countrymen.

I appreciate your difficulties fully, but know that together we shall be able to overcome them all. The important thing is that we must secure more specimens for science. It does not matter to me who gets them, except that it would give me the greatest satisfaction for your country to have a complete and perfect specimen, whether by your own expedition or any other. I am hoping daily to hear that you have already found one.

* No reply was received on this point. But see p. 223. (Nairobi meeting).

And so my dear Sir, I do trust that you will be able to remove from the minds of any of your countrymen the idea that I am intending to go to your area to compete with you in these matters. I do certainly feel that it would be a very great pity if the fishes of the French Territories were not to be included in my large planned monograph, and feel that I can rely upon the same freedom and facilities for collecting as all the other countries where we have worked.

There was other correspondence with Millot, but nothing further from the French authorities until August 1953, when I received notification that they had decided not to grant authorisation to search for Coelacanths in their waters during 1953. It was stated that various French and foreign scientific organisations had made proposals similar to mine, and it was feared that if all were permitted to operate it might have undesirable results. Instead of separate expeditions, the French Government intended to invite the 'Scientific Council for Africa' to consider the amalgamation of proposed expeditions into an international expedition, under French leadership. During the second week of August 1953 a statement more or less to this effect was issued officially by the French authorities, and appeared in the press in various parts of the world.

The ending of negotiations with Stuttard and this decision of the French put an end to my hopes of working in the Comoro-Aldabra region in 1953. The exclusion from French waters at that time was an action I found regrettable in science, even with the prospect of an international expedition ahead. It had earlier been reported in the press that two expeditions (Italian and Swedish) were in East African waters to hunt Coelacanths, the leader of one had indeed written to me from East Africa. After the decision of the French, I wondered if these expeditions would do as I had planned under this ban, i.e. work at other likely places. Shortly after this (late August 1953) there was a report that at least one of these expeditions had gone over to Aldabra, and then it was stated that one had gone to the Comores and was actually working there. As this was apparently contrary to the decision of the French Government, both as published and in the form communicated to me, I doubted the accuracy of the report. It was, however, later confirmed that not only had the Italian expedition been working

Angered by the French response, JLB and Margaret Smith turned their attention in 1954 to the Seychelles, an archipelago of Western Indian Ocean islands. As about 92% of the marine fishes of the Seychelles occur over the whole Western Indian Ocean, including the Comoros and Madagascar, it was the ideal place for them to continue their work. In 1963 they co-authored a comprehensive book on the fishes of the Seychelles.

Fishes of the Seychelles *(1963)*

No coelacanths have been recorded from the Seychelles, although they might occur there. The Aldabra coral atoll (one of the Outer Islands of the Seychelles) was designated a UNESCO World Heritage Site in 1982 and there is strict control over fishing activities there.

In 1954 a group of Italian divers, called Spedizione Zoologica Italiana and led by Franco Prosperi, secretly dived off Mayotte Island (where no coelacanths have reliably been recorded) and claimed to have photographed a live coelacanth 'in its dim and aerie habitat' at a depth of 15 m (which we now know is far too shallow).

The image that they published in the popular media in Italy, France and England turned out to be a fake – they had photographed a crude, inflatable coelacanth! This farcical exercise might have hardened French attitudes towards foreign research on the coelacanth.

Franco Prosperi

Fake coelacanth photographed by the 1954 Italian expedition

The first live coelacanth to be seen by Western scientists was an immature female 142 cm long, caught off Mutsamudu on 12th November 1954 and towed back to the harbour, where it remained alive for about 24 hours. It was kept in a flooded boat and observed until it died at 15h30 the next day.

at the Comores, but when they returned to East Africa early in November 1953 it was with the startling news that they had succeeded in photographing a live Coelacanth under the water, and at the Comores themselves.

On the 9th November 1953 the following statement, emanating from Dar-es-Salaam, appeared in the press of the world:

> Only French scientists will be allowed to search for the Coelacanth off the French Comoro Islands, in the Indian Ocean between Mozambique and Madagascar, for the rest of this year.
>
> French authorities there have declared a complete ban until December 31 on expeditions by foreign scientists—two days after an Italian zoological expedition secured the first photographs ever taken of a living Coelacanth.

Another report of the 10th November 1953 emanating from Dar-es-Salaam, stated further:

> An Italian expedition which has been working at the Comoro Islands is convinced that there are many Coelacanths there. . . .
>
> This expedition was compelled to suspend its operations* because the French authorities placed a ban on further search for Coelacanths until the end of this year.

On my 1954 expedition to the Seychelles I made a long voyage in the same vessel that the Italians had employed. The owner confirmed that they had indeed worked in Comoran waters after the French ban of August 1953, and he told me that the Italians had handed over to the French authorities part of the collections made, which had been a condition of their being permitted to work there.

---

* In November 1953, at the Comores.

Once again JLB is whipping himself into a frenzy of self-doubt and recrimination, shouldering the 'terrible responsibility' of possessing the only relatively complete coelacanth specimen, and feeling 'like a bird in a cage, not broken but badly battered'.

Chapter Nineteen

MARCHAND DE BONHEUR

By the end of June 1953 the course of negotiations left little doubt in my mind that the French would not grant me permission to work on my own in their waters. Despite all I had learnt at the Comores, it was disturbing that no further Coelacanths had been caught although I was pleased to hear that the French were apparently offering the same reward for Coelacanths as I had paid, namely £100 each, to induce the native fishermen to bring them in. I felt that more than this was necessary, and in the absence of information about the precise measures being taken by the French could not shed the responsibility of feeling that I must go after Coelacanths myself, to make sure whether they really lived in those parts or not. Had the arrangement with Stuttard not broken down or if a suitable vessel had been available in time, I should certainly have gone to work in the western Indian Ocean. There were many places outside French territorial waters where profitable work could be done. I should never be able to rest until the home of the Coelacanth was surely found.

My chief recollection of that tense and difficult middle half of 1953 is of the weight of the terrible responsibility that pressed down on me. Not only was there this matter of the home of the Coelacanth, but the possession of this Coelacanth *was* a terrible responsibility. It was the only relatively complete specimen in the world, it was the first virtually complete specimen ever to be found, and for those reasons it should if possible be kept intact as an historical relic. I had to consider the needs of the world of science, but if others were found then this one could be kept intact. On the other hand, if no other specimens were found I could scarcely justify merely hoarding this one in that way. I was torn between these two considerations, and the only clear solution was to find another, more, as many as possible, and soon. Day and night this worry clouded my brain and my life, and when all hope

No coelacanth has as yet been recorded from Bazaruto but one was caught by a trawler off Malindi in Kenya on 26th April 2001 at a depth of 85 m. The only coelacanth caught to date off Mozambique was trawled off Pebane in relatively shallow water (40–44 m), which is nevertheless far too deep for a free diver.

On their September–November 1953 expedition to Bazaruto, the Smiths collected diverse fish specimens off Inhassaro, Vilanculos, Ponte de Barra Falsa, Inhambane, Vila Joao Belo, Inhaca and the Bay of Lourenço Marques. They discovered more than 80 fish species not previously known from the area, as well as several species new to science.

of a proper Coelacanth expedition of my own for 1953 was finally abandoned, I felt like a wild bird in a cage, not broken but badly battered.

My mind always framed that story the Mozambique native had told about the fish he got at Bazaruto. There was no valid reason why it should not have been a Coelacanth. While I was prevented from going to the Comores at that time, the Portuguese are my very good friends and I could certainly go to Bazaruto. I had already been there, but many years before, and had indeed never worked over the area as thoroughly as its richness deserved. It would repay closer investigation. So I sought and received from the South African Council for Scientific and Industrial Research permission to use some of the Coelacanth expedition funds for a small venture to the Bazaruto area. The Portuguese authorities, as always co-operative and prompt, at short notice arranged the requested facilities for work in September and October 1953.

Before we left for Mozambique I received a letter, dated the 3rd August 1953, from G. F. Cartwright of Salisbury. He wrote to say that he had been goggle-fishing at Malindi in October and November 1952, and that he had seen us at work there at the close of our stay. Later when out over deep water on a reef one day, equipped and armed with spear-gun, he suddenly saw not far below him a large fish whose appearance gave him a shock. It had a huge mouth and a 'baleful and ancient appearance'. To quote further:

> It was a large fish, heavily built, and from 100 to 150 pounds in weight. It was totally unlike any fish I had seen or ever saw afterwards. It looked wholly evil and a thousand years old. It had a large eye and the most outstanding feature was the armour-plate effect of its heavy scales, scales so heavy that it was set quite apart from other fish I saw.

Although its appearance was rather terrifying, he decided to try a shot, but the harpoon just glanced off the scales and the creature disappeared.

On the return to shore, Cartwright told of this fish and questioned other anglers and spear-gunners. The nearest they got was to suggest a large Rock Cod, but Cartwright had had wide experience of diverse Rock Cods and was quite positive it was not. The scales alone ruled that out, and the identity of the fish remained a mystery.

Soon after returning to Rhodesia, in some periodical Cartwright saw for the first time a picture of a Coelacanth, and was at once struck by its resemblance to the fish he had seen. Shortly afterwards he visited the Centenary Exhibition in Bulawayo, and to his satisfaction found a full-size model of a Coelacanth (from the East London Museum) on view there. Its close resemblance to the fish that had startled him was even more striking, and by contriving to put himself into the same relative position as he had been to his fish, as far as he could judge this appeared to confirm in every respect that the fish he had seen had indeed been a real live Coelacanth. What did I think? Well, it was clear that if the Comores was the home of the Coelacanth, Malindi was much nearer and much more easily accessible in every way than East London, which one Coelacanth at least had actually reached. Bazaruto fell in between these places, and Cartwright's experience at least lent colour to the Bazaruto idea. Furthermore, from my wide knowledge of the fishes of the western Indian Ocean I could think of no species that fitted Cartwright's description as well as a Coelacanth. Not one.*

Meanwhile, arising from the refusal of the French to permit any foreigner to work in their waters, this alternative of an International Coelacanth expedition had been taken up by the Scientific Council for Africa, one of whose functions is to co-ordinate scientific effort in Africa south of the Sahara. This body appointed a Committee consisting of Dr. Millot (France), Dr. Worthington (British), and myself to consider the matter. The meeting was arranged to be held in Nairobi during the last week of October 1953.

Early in September 1953 my wife and I and a scientist friend, H. J. Koch, set off for Mozambique. Our work about Bazaruto and the other islands, and at Ponte de Barra Falsa, proved extremely interesting and rich in results, as far as fishes generally were concerned, but not of Coelacanths. That was naturally only

---

* It may be noted that it has been pointed out that it was unlikely that the fish that Cartwright saw was a Coelacanth, because the French have reported that the first live Coelacanth to be caught showed such fear of light. (But see p. 242.)

An angler from Zimbabwe once sent us a photograph of a coelacanth that he claimed to have caught on rod-and-line near Beira in Mozambique. Unfortunately for him, we were able to recognize the fish with the help of Karen Hissmann (a member of the *Jago* submersible team), from the pattern of white dots on its body. It had apparently been flown to Zimbabwe from Grande Comore in a deepfreeze aeroplane and the angler had posed with it to give a false impression.

French diver Jean-Louis Geraud confirmed that coelacanths that had been caught off Grande Comore and held near the surface on a tether, constantly tried to swim downwards and appeared to be afraid of bright light. They might also have been seeking cooler, more oxygen-rich water.

Hans Fricke

Jean-Louis Geraud with his own coelacanth

The psychological strain that JLB imposed on himself at the time was almost overwhelming. His anxiety may have been tinged with guilt as he knew that he had poached the coelacanth from a foreign country and now, more than anything, he wanted that country to have a specimen of their own.

Even today Vilanculos is isolated, although it does have a small international airport. When I visited there recently, sea horses and sea cucumbers were being exported illegally by the bagful by Chinese merchants and we found the carcass of a slaughtered dugong. In the bay adjacent to the Sao Sebastiao Peninsula we saw a colony of enormous sawfishes, and purple sea pens were common in the sandy shallows.

Mike Bruton

Fishing dhow at Vilanculos, Mozambique

a forlorn hope, but each Coelacanth-blank day only served to add to the severe mental depression that clouded my life. Would this two-fold responsibility of the Coelacanth, the decision about the fate of this animal, and the suspense of waiting to hear if more had been caught at the Comores, never end? I did not look forward to the meeting in Nairobi, for I had a premonition that my frantic desire to get on with finding Coelacanths was going to be side-tracked, wrapped up in official and technical cotton-wool. It had happened before, but this was much worse, for the Comores were foreign waters and closed against me.

Living on a small vessel and on the islands at that time, we were completely isolated except for an occasional visit to the lofty lighthouse at the northern end of Bazaruto, and when we went to work at sea we had no contact with the rest of the world. Towards the close of our work in that area, we landed one day on the mainland opposite Bazaruto, and found the people there very excited, for at low tide a man had gone probing into the dirty water at the base of an eroded mushroom-like coral lump and had found hidden there an enormous 'Garrupa', or Rock Cod, which he eventually managed to kill. It was well over 200 lb. in weight, a terrific fish to have remained in what was only a puddle. As we stood by looking on, a Chinese who spoke Portuguese came up to me and said that there had been a radio report the night before about a curious large fish that had been caught by the French. He had heard my name mentioned, and that had made him take some notice of it, as he knew I was working among the islands. We certainly jumped at this. Had they mentioned where it had been found? He thought it had been Madagascar, but was not sure. Was it a Coelacanth? Yes, that was the name.

It was like the burst of a bomb to me, but although we questioned him closely he could tell us no more. We rushed round to everyone who had a wireless-set, but found out nothing much besides. Although it was not quite certain, for they might all have been mistaken, I felt many years younger. We could get no other information of any kind, for when newspapers reached that part they were at least six days old. It was not until we got back to civilisation a week later that we learnt that it was indeed a Coelacanth, and that it had been taken at the Comores and at Anjouan, the same island.

I shall always remember the sensation of terrific relief this gave me, as if a crushing burden had been lifted from my mind. So it was the right spot, they were there! I could see the end of this great strain, I could keep my *Malania*, and it would be only a matter of time now before all those high specialists were each and all wresting the secrets of the life of the long ages past from the tissues and structures of fresh Coelacanths. I could see in the near future long queues of eager sightseers filing past a tank where a living Coelacanth stared scornfully at his equally 'degenerate'* near descendants.

Two in the same place. It must be their home, so that my enduring aim had been achieved, and most of my burden would now fall on the French. Their exclusion of myself from their waters was based on mistaken ideas, for in doing so they had tried to keep me away from something I had never wanted for myself but for science.

As soon as we could get to a Post Office, I sent Millot a cable conveying my warmest congratulations.

Now that one home of the Coelacanth had been found, the Nairobi meeting took on a new aspect, and I looked forward to it more with interest and anticipation than with concern.

Cartwright's story had been constantly in my mind. From my wide knowledge of the fishes of South and East Africa, I could still think of no species of fish known from there that fitted his original spontaneous description as well as a Coelacanth. I wanted to question him further, so arranged my flight north from Lourenço Marques so as to be able to spend a night in Salisbury, where I met Cartwright and we talked at length. It is significant that nothing emerged from our discussion that made it in any way less likely that his fish had been a Coelacanth; rather the reverse.

When I got to Nairobi, it was to find that Millot and Worthington had invited Drs. Menaché and Wheeler to attend our meeting as well, and it certainly was interesting.

Despite my now relative detachment, my technical instinct was naturally to further the catching of Coelacanths, soon and many. As the proceedings developed, it was almost uncanny how they followed, only in much more condensed fashion, almost exactly

---

* This term is applied by some scientists to those forms which do not or are not likely to give rise by evolution to other forms. Thus the Coelacanth and *homo sapiens* both fall into this category.

After JLB died an international controversy arose as to whether the capture of live coelacanths would threaten their survival. The failure of extremely well-equipped expeditions from Japan, the USA, Belgium and elsewhere to overcome the difficulties of catching one, and the outcry from conservationists (myself included), resolved the problem temporarily but the issue is being debated again today.

The choice of Bart Worthington, a British limnologist with experience in African freshwaters, and Alwyn Wheeler, an expert on British fishes and curator at the Natural History Museum in London, was rather odd; it reflected the dearth of British expertise on Indian Ocean marine fishes.

Fifty years later South Africa did launch a successful multi-disciplinary oceanographic expedition that included a search for coelacanths. The ACEP project investigated coelacanths living in canyons in the iSimangaliso Wetland Park and carried out valuable oceanographic research.

Coastline in the iSimangaliso Wetland Park south of Sodwana

The broad-based oceanographic survey proposed by Jacques Millot never got off the ground, although independent research projects on the coelacanth were later carried out by scientists from France, Britain, Belgium, the USA and Japan.

the pattern of the A.C.M.E. expedition talks of six years before. Once again I was apparently alone in my Coelacanth single-mindedness, and to me they all appeared less interested in catching Coelacanths than in seeing how much else they could hang on to it. I listened to a commendably elaborate scheme for carrying out oceanographical investigations covering an enormous range of scientific effort. Most of the first day I listened, and it certainly was an impressive project, but to me pure phantasy and of little value as a Coelacanth-hunting venture. Next day I had my turn and talked a good deal, casting doubts on the feasibility of so extensive and costly a scheme and on any possibility of raising the relatively large sum such a project must cost. We had a battle about estimates and got to a compromise figure, and even that was far outside my ideas of practical finance. Though South Africa might be prepared to assist towards hunting Coelacanths I could not see our country financing extensive oceanographical work in seas so far away. I told them it seemed impractical to visualise such a scheme when we now knew at least one place where Coelacanths lived. I wanted to concentrate first on the Comores alone. I wanted to see Coelacanths caught for science, and outlined the simple, direct, and relatively inexpensive scheme I had in mind.

However, I was assured that the Scientific Council had expressly directed that this expedition should have the oceanographical character that was now outlined, and that the Council did not want any other. They did not share my doubts as to whether the necessary funds would be forthcoming.

Certainly of all the accessible parts of the oceans near populous countries, the region of the Mozambique channel is probably the least known scientifically. Oceanographical investigations there are highly desirable and would probably pay handsome dividends. But my common sense told me that if the French caught more and more Coelacanths in their closely preserved waters, no foreign Government was going to spend large sums of money on an expedition to find Coelacanths. I had no faith in this international project purely as far as hunting Coelacanths was concerned.* If

* It has recently (June 1955) been announced that the International Expedition to hunt Coelacanths has been postponed, but that Dr. Millot is going on an expedition of his own, equipped with special cages to keep Coelacanths in deep water, and with a type of bathysphere designed by French experts so that he may go down and study live Coelacanths in their natural home.

they wanted to do general oceanographical work, that was quite another matter; so I let it go, but told them that I at least could see no part in it for me, and did not intend to participate. I had plenty to do besides. I wanted to get on with my work on the fishes of the western Indian Ocean. I made it plain that as soon as the home of the Coelacanth had been established, my own work in that field was done. Until then my knowledge and experience would always be at their service, and I visualised the possibility of assisting them while engaged in my own work. I told them my own plans to hunt Coelacanths, and outlined how I had intended to try to keep them alive by using a fairly large decked boat, partly filled with sea-water, as a temporary aquarium.* They were startled when I said the chief problem was that Coelacanths transported in this way or in tanks might get seasick and die, for strange as it may sound fish in aquaria at sea do get seasick. I noticed that when possible vessels were discussed, Millot made no mention of Stuttard's *La Contenta* (see p. 205), and when I raised the issue of Cousteau and his research vessel (about which I had written earlier to Millot, see p. 214) as part of the expedition, the others indicated that this would be useless as the Coelacanth lived in water too deep for such methods to be of any value!

This meeting had its lighter moments. We were discussing personnel and equipment for the vessels, and it was established there would be quite a number of scientists and assistants on each. One Frenchman said, 'It weel be necessairy to provide a wench on each ship.' This certainly shook the Britishers. As they looked up sharply I could see in their faces the unspoken comment, 'It may be the tropics, but really. . . .' I said mildly, 'He means winch'; and there was some laughter, which the French at first did not understand.

The Committee composed a statement about the international project, which was later given due prominence in the press.

Millot showed me photographs of the third Coelacanth. It was wonderful to see them. It was complete, and the first dorsal and that little extra tail made it look as if my old *Malania* was indeed a freak, a gigantic practical joke on the part of nature. Still, I

---

* It has been interesting to note that the French used this method in attempting to keep alive the first live Coelacanth to be brought in (see p. 239).

JLB's plans to keep a coelacanth alive on board a boat partly filled with sea water never came to fruition.

The erroneous conclusion, first mooted by British scientist Errol White, that the coelacanth lives in very deep water, continued to cause confusion among European scientists some 15 years later at the meeting in Nairobi in 1953, despite all Smith's arguments to the contrary.

JLB was right – fish do get seasick in sloshing aquaria aboard ships! Like humans, they lose their sense of balance and their orientation.

The absent first dorsal fin on the second coelacanth was probably the result of predation but the missing extra lobe in the tail fin is more difficult to explain except as a variation on the normal coelacanth form.

If *Malania anjouanae* had been the first living coelacanth found, and had therefore become the type specimen on whose description the new species would be based, then later specimens would, at least initially, have been regarded as variants. Eventually, after a large number of specimens had been studied, scientists would have realized that *Malania* was the variant.

JLB Smith

The second coelacanth, initially named Malania anjouanae *by JLB Smith*

The French *Calypso* expedition of 1953 was indeed clothed in secrecy, perhaps because they did not find coelacanths, and very little is known of their findings, except in French circles, even today.

wanted to see a good many more before that was certain. I suddenly wondered what would have happened if old *Malania* had wandered away to East London instead of *Latimeria*. That would have been a poser!

I left that meeting feeling, as a scientist, a solid satisfaction at the prospect of extensive oceanographical investigations in so interesting and virgin an area of the seas. But against my Coelacanth-hunting obsession, as a method of finding soon where Coelacanths lived and of catching numbers quickly, it had very much the flavour of an English fox-hunt, in which the formalities are of more account than the quarry. Left to myself I felt I could do it more expeditiously and at far less cost. (Although it was their own idea to stop individual expeditions by suggesting the international project, it would almost appear as if the French soon came to be as sceptical about its early practicability as myself, for only a month or two after our meeting in Nairobi the French research vessel *Calypso*, Cousteau's famous ship, set out on a six-months' cruise and covered a wide area of the tropical western Indian Ocean where Coelacanths might be expected to live, including the Seychelles and the islands and banks over the thousand-mile-long arc from there to Aldabra, where they spent some time, and then to the Comores. There, under the direction of Professor Millot, they carried out extensive submarine electronic flash photography in areas where Coelacanths were likely to be found, using for this purpose a special camera designed by an American scientist. In that part of the Indian Ocean, outside French waters, they covered almost the exact field which I had informed the C.S.I.R., and later the press, was to be the area of operations for my 1954 Seychelles expedition. This French expedition had left Seychelles not very long before I arrived, and we heard a good deal about their operations. They were apparently greatly taken by Aldabra; which is not surprising, for not only did it prove to be the richest virgin area for fishes I have ever seen, but I confidently expect that Coelacanths will be found there or at other islands not far off, such as Astove. This French project was apparently accorded unusually little publicity from any source, since few people have heard of it.)

Although every due moment was given to Coelacanth matters, Nairobi, at that time in the throes of racial conflict and crisis, was very interesting. I spent every spare moment talking to those who were informed and in walking many miles through the streets, noticeably the only European to do so outside the central shopping area. It was a city without lightness or laughter, life was a grim business, you could feel the tension in the air. The windows of my room in the hotel were barred and there was a warning to keep the door closed and locked when inside. I heard many stories, and, after their physical characteristics had been described to me, could easily pick out the Kikuyu from the rest. I reflected somewhat grimly that even if this happened in South Africa, we should not be prepared to endure its dragging on like this. It would come to a quick end, one way or the other; yes, one way or the other. Such things are not new in our country; we know all about it, for what was happening in Kenya was in effect almost the exact modern counterpart of what our forefathers had endured in South Africa a century ago, remote control, wild country, disregard for ownership and for human life. In South Africa those hard times had bred tough citizens, both men and women, as the conditions in Kenya are clearly doing now. Everything has its points.

The bestiality of the Kikuyu in their slashing murders is world news, but I learnt many other things about them. One is typical. At night they will go into a field of mature potatoes, and working through the soil with cunning fingers will remove most of the tubers without killing the plants or leaving any trace.

The last of that visit to Nairobi, as we drove to the aerodrome through the streets, was a kaleidoscopic compression of uniformed figures, grim faces, armed sentries, barbed-wire fences, and sandbag defences.

I do a good deal of flying and like it, but never climb into a plane without having a good look round at earth and sky, with the conscious thought that this may be my last view. So many crashes occur within a few moments of taking off. As I strapped myself in I realised that my heart was filled with deep content, happiness if you like. It was really quite remarkable, as if I was living in the heart of a 'Happy ever after' ending in a novel. After all the uncertainty and agony that had followed the trail of the Coelacanth, it had now all been ironed out. Almost miraculously, everyone was happy.

o.f.—15

JLB Smith is referring to the Mau Mau Uprising, a military conflict in British Kenya between 1952 and 1960. It involved Kikuyu-dominated groups (called 'Mau Mau'), white settlers and elements of the British Army, including the local Kenya Regiment. The capture of the rebel leader, Dedan Kimathi, on 21st October 1956 signalled the defeat of the Mau Mau.

King's African Rifles regiment during the Mau Mau Uprising in the 1950s in Kenya

JLB's roller coaster of emotions had now reached its conclusion – he was happy with the outcome and ready to move on to new challenges, with or without international collaboration.

The coelacanth had become a peacemaker between nations and a source of unending fascination among scientists and lay people alike. It had more than fulfilled its promise.

One wonders whether 'old fourlegs' would have become such an icon if it had been discovered by an obscure scientist working in, say, Antananarivo, who covertly beavered away, describing its anatomy in technical journals. Would it then have been doomed to the same obscurity as the lungfish?

The coelacanth needed someone with the *chutzpah* of JLB Smith to bring it to the world's attention and provide everyone with insight into the fascinating process of science.

The 'Dokter' had been pleased. Being a Prime Minister is as near as any human comes to being a god, but even he must feel pleasure after making a decision to cover an unusual situation, to harvest almost universal approval, even if it is grudged by his opponents. The French were happy. They had the first really complete Coelacanth, got by themselves; they had the prospect of an undisturbed monopoly of more, and, for some time at least, they would get world headlines for every one they found. They would be able to point out any mistakes I had made, and all hurt feelings would be soothed. The authorities of the Comores were happy. Coelacanths had put their islands very much 'on the map'. They would probably issue a set of 'Coelacanth' stamps, that collectors and ichthyologists all over the world would eagerly seek. At least three Comorans were in such a position that they need not work for a good while, and the lives of all the islanders must be brightened, for almost every one of them could at least fish for that wonderful possibility. Everyone round there would be happy in the increased supplies of fish. Even Stuttard had apparently not been unhappy.

I just could not help smiling as I thought of my ugly old *Malania*, snug in his asbestos-cement coffin, in the role of a 'Marchand de Bonheur', a dispenser of human happiness.

The plane rumbled to the end of the runway, and as we shook to the roaring test of the engines my heart was filled with fierce deep content, for I had shed the worry and responsibility of the Coelacanth; one of the greatest ambitions of my life, to find the home of the Coelacanth, had been fulfilled; I was leaving the horror of the Mau Mau and going back to my own beloved country, to my beloved fishes, to my old *Malania*, who could now for ever sleep in peace.

APPENDICES

APPENDIX *A*

SOME CHARACTERISTICS OF COELACANTHS AND HOW THEY DIFFER FROM MOST MODERN FISHES

IT has been emphasised that in its bony jaws and overlapping scales even the earliest Coelacanths showed characteristics that have endured to the present day. While the Coelacanths have changed little over vast ages, there have been modifications in structure in other forms of fishes, so that the primitive Coelacanth differs markedly in many ways from most modern bony fishes.

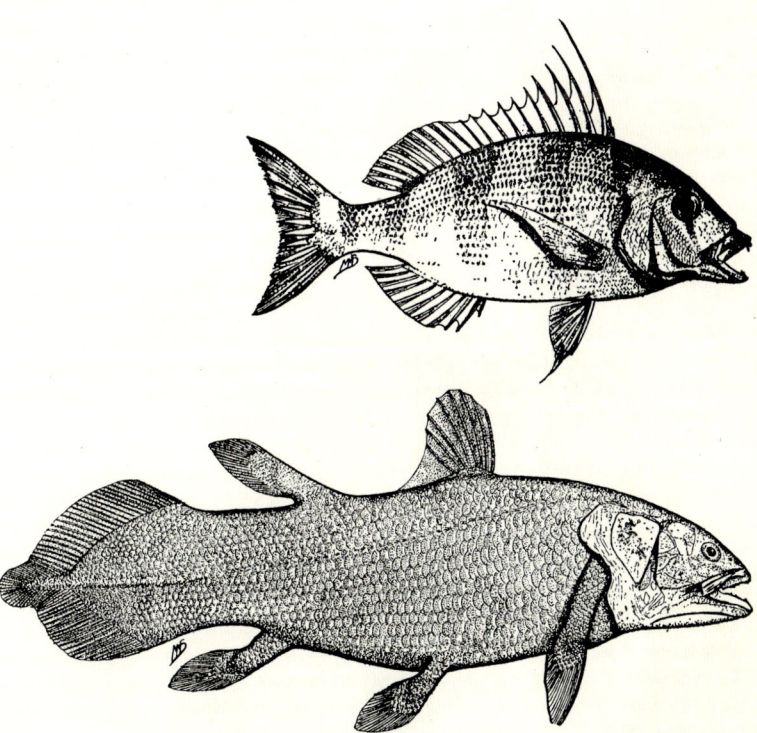

FIG. 7.—Above, a typical modern fish, the 'Soldier' of Natal ; and below, a Coelacanth.

The extent to which JLB Smith, a self-taught ichthyologist, became intimately familiar with the anatomy of fishes during his relatively short, 36-year ichthyology career is extraordinary. This was achieved through dogged determination, a very logical and analytical mind, and the hands-on study of thousands of fishes.

His experience as an angler would also have added to his intimate knowledge of fishes.

The coelacanth's skull is very unusual. No other living animal (and only a few extinct frogs) have the ability both to lower the lower jaw and lift the upper jaw in order to create a wide gape.

Although the coelacanth has a relatively large brain cavity, most of it (98.5%) is filled with fat, and the brain is very small.

In the head of the Coelacanth quite massive bones lie right in the surface. The skull itself is not a single unit as in most fishes, but is in two almost separate parts across behind the eyes. The brain cavity is quite a reasonable size. There are no nostrils like those of modern fishes, but in the cartilage in the front of the head before the eyes is a peculiar central and quite large cavity, from which six tubes lead to the surface, two to the front and two to each side of the head in front of the eye. At first sight, externally, one might think the two openings on each side to be nostrils, like those of modern fishes, but they are not, and up to the present no one has been able to say exactly what this structure is. When my account of the first Coelacanth was published, the description of this particular part puzzled all scientists, and one went as far as to say that I must have been mistaken. The very first thing I looked for in *Malania* was this structure, and, sure enough, there it was, exactly the same. It was also the first thing Nielsen looked for in *Malania*. He told me, apologetically, that they all had doubted my description and figure. We scratched our heads together.

Most bony fishes have a well-developed soft 'air bladder', which lies above the intestines in the body cavity. (It was mentioned on p. 92 that when a trawl-net comes up, the fall in pressure causes the air-bladders of the fishes to expand.) The modern Coelacanth has no true air-bladder. In some fossils there is a peculiar structure in the belly cavity in the position in which the air-bladder is generally found, but in those earlier Coelacanths this was 'calcified', i.e. it had hardened walls. It has been assumed, therefore, that Coelacanths had air-bladders, but that these had become hardened. An air-bladder is apparently used to adjust the average density of a fish to suit its environment, or else as a breathing organ. What use an air-bladder with hard walls would be it is difficult to imagine, but in some existing fishes there has been found a degree of calcification of their air-bladders. At any rate, there is no true air-bladder in the modern Coelacanth, only a shred of skin that may be the remnant of an air-bladder that might once have been present. It may be mentioned that sharks and rays have no air-bladder.

In the Coelacanth the gills are not soft cartilage like modern fishes, but bony and hard. They bear teeth and not ordinary soft gill-rakers.

The scales of the Coelacanth are both peculiar and characteristic. The main basal part is almost horny and with less bone than a comparable modern fish. The 'tubercles' on the scales are separate units, each set on its own little plate, and each is hollow. The scales overlap so much that the whole body has a covering three scales thick, a powerful protective armour.

While the Coelacanth has bony jaws and powerful jaw muscles, the structure of the upper jaw is different from that of the *Rhipidistia* and of modern bony fishes. The maxillary bones at the side of the upper jaw are missing, there is only a thick fold of skin. The teeth in the upper jaw are set differently from those of modern bony fishes. They are in clusters in separate but adjacent plates.

The scales bear tubercles, each of which is a strong, separate, enamelled structure stuck on the main base of the scale. Some of these tubercles are smooth all round, but some have a sharp point behind. These tubercles are hollow inside. It has long been held that teeth in jaws are developed from scales that 'migrated' inwards. In the development of a shark it is possible to see something of this, and in the Coelacanth this shows very clearly, for the teeth are in groups on adjacent but separate bases, and examination soon reveals that each of these is merely a modified scale, in which the base has become thicker and stronger, and the hind pointed part of the tubercles has become longer, more bony and strong, sharply pointed, and, in fact, a hollow canine tooth. Even the teeth on the gills are of this form. In addition, some of the surface bones of the head show themselves as no more than modified scales.

On the floor of the mouth of the Coelacanth there is a hard, bony, toothed structure different from anything in any modern fish. Below the lower jaw are two bony, reptile-like plates, 'Gular Plates', that are found in a few only of the more primitive types of living fishes.

The fins are certainly curious. The pelvics and pectorals give clear indication of being used as limbs, and it is plain that the fish can crawl about, in the water at least. Ancestral Coelacanths probably crawled out on land. (Old Fourlegs!)

The tail is characteristic. That small extra tail is not found in modern fishes. It is indeed a remnant of the true tail that is supposed to have been present in ancestral forms. (See also the figure of a Rhipidistian fish, p. 17.) The 'tail' of modern fishes is really evolved from two fins, one above and one below, into which the hinder true 'tail' eventually shrank away.

The 'skeleton' is remarkable. The 'backbone' or 'axial' column is not of bone, but a hollow tube of cartilage, that fits in front into a hollow in the skull, and behind tapers to a thin rod in the tail. On the backbone of an ordinary fish you see hard, bony spines above and below. The Coelacanth has something like these, but they are hollow tubes and not very hard. (This is how the Coelacanth got its name. 'Coelacanth' means 'Hollow spine'.) At the base of each fin, however, there is a rather large and heavy bony plate, the 'basal' plate, well known from fossils.

The intestines of a Coelacanth are short and have in them a 'spiral

The lobed fins of early coelacanths might have enabled them to crawl around in shallow water, but they never developed the other accoutrements needed for life on dry land, such as lungs and strong skeletons. It is possible that they waddled onto the fringes of the land, and ate insects and other small prey, in the manner of modern mudskippers.

Bjørn Christian Tørrissen, Wikipedia CC BY-SA 4.0

Mudskipper, Periophthalmus *species*

Some animals, having 'retooled' themselves to live on land, then returned to the water, such as the crocodiles, turtles, penguins, platypuses, dugongs, seals and whales. Freshwater terrapins have apparently done a double-act, returning from water to land a second time and then becoming aquatic yet again.

Many aspects of the coelacanth's anatomy are 'truly remarkable', including the bony skull, tube-like notochord, elongate scales, teeth in clusters, the unique array of fins, extra lobe in the tail, spiral valve in the intestine and oil cells underneath the skin.

Like the peripatus, sea dragon, yeti crab, dumbo octopus, lamprey, goblin shark, blobfish, axolotl, platypus, kangaroo, okapi and narwhal, the coelacanth has 'torn up the script' and redefined the norm.

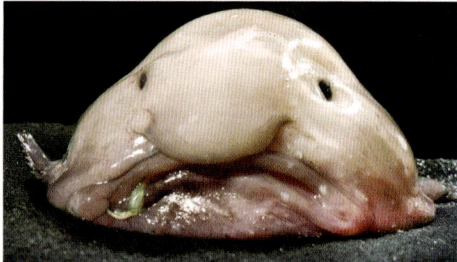

Blobfish, Psychrolutes marcidus

valve' very like that of sharks and rays. The intestines generally remind one of those of a shark rather than of a bony fish.

While some modern fishes are oily, probably none are as oily as the Coelacanth. Below the skin there is a layer of cells full of oil, and even after months of capture this oil continues to ooze from the body. In *Malania's* 'coffin' there are always big blobs of oil floating on the surface of the preservative solution in which he lies.

APPENDIX *B*

WHY THE DISCOVERY OF THE COELACANTH AROUSED SUCH WIDE INTEREST

ONE of the main questions that we have heard many thousands of times is, 'Why is the Coelacanth so interesting and important?' There are many things involved.

As was explained in Chapter Two, Coelacanths as such lived over a longer period than any other known type of creature, certainly of vertebrates. They apparently spread over most of the earth, for fossils have been found in very many places. They left apparently one of the most constant and unbroken series of fossils, often almost perfect, that one could desire, covering 250 million years, during which they lived almost unchanged in general form. This fossil record was so good that it apparently gave an index of numbers as well as distribution. According to that record, from about 100 million years ago, Coelacanths steadily became fewer in numbers, and the very last and comparatively rare fossils occur in rock strata earlier dated at about 50 million years old, but which more recent estimates put at about 70 million. Compared with the age of the earth, 70 million years is not very much, but it is a terrific stretch of time in comparison with life as we know it. A good many profound changes have taken place on the earth during that 70 million years. Almost all the creatures that lived 70 million years ago, both of the land and the sea, have vanished, and most of them would look strange or startling if they appeared now. Most people have heard of the Dinosaurs and other giant reptiles, the enormous fish-eating lizards and flying reptiles, and other similar creatures of past ages. It requires little imagination to picture the sensation that would be caused if one of those gigantic Dinosaurs ambled into civilisation today. Indeed, the appearance of any piece of that long-buried past is an event. While the Coelacanth is not the size of a Dinosaur, its appearance, still alive, is in many ways much more startling.

As far as scientists are concerned, this appearance of a living Coelacanth was a terrific shock. All those who worked on fossils had been quite confident, had indeed repeatedly said, that Coelacanths were all extinct, and had been so from at least 50 million years earlier. This proved them wrong, and it is good for dogmatic statements to be proved wrong, at least in science, since it induces a caution that is

---

We now know that the coelacanth fossil record dates back over 420 million years, longer than any other backboned animal. By this measure, the legendary dinosaurs are relative rookies, having evolved only about 250 million years ago.

JLB is right to condemn dogmatic statements in science. Science is not about final explanations and certainty but about doubt, scepticism, testing alternative ideas and constantly questioning the status quo. It is, and always will be, work-in-progress, attempting to find the best explanations for natural phenomena based on available evidence.

In fact, the most significant advances in science have often been made when the 'current wisdom' has been unceremoniously discarded and replaced with a new (but still imperfect) understanding.

Although my main research has been on freshwater catfishes and tilapias, and on the life-history styles of fishes, all people want to talk about when they meet me is the coelacanth!

Furthermore, virtually everyone has an anecdote to relate on their own engagement with 'old fourlegs', no matter how obscure. I doubt whether any other aquatic animal, except perhaps the great whales, has had such a dramatic impact on the public consciousness.

The coelacanth saga continues to be enriched by the spectacular films made by the *Jagonauts*, the discovery of populations in eastern and southern Africa and of another living species in Indonesia, and the secrets revealed by the creature's DNA.

proper to science. Any scientific statement or theory should be preceded by 'As far as we know at present. . . .' It was a salutary lesson.

All intelligent human beings have an almost instinctive deep interest in living things, and there are few who are not fascinated by the life of past ages.

During the past century more and more scientists have turned their attention to unravelling the threads of the course and development of life on the earth, and a marvellous story it is. Not only have many books about it been produced, but some countries have life-size exhibits portraying extinct creatures and plants of past times based on reconstructions made by scientists.

Portraying or modelling a creature from fossil remains, often incomplete or put together from dispersed fragments, is, of course, bound to be partly guess-work. How can one know that this work is sound? There is always a tendency to go too far. For example, some workers have managed to produce models of the brains of some of the long-extinct forms. It is marvellous work. One method is to take a fossil skull that is now all stone and to grind it down a fraction of an inch at a time, making a wax film at each stage, and when these wax films are all put together in their proper order, there is a model of the brain. Those that have been prepared in this way look quite convincing. However, one sceptical worker has cast doubt on the value of this, for he has shown that casts of the inside of the skulls of modern fishes, at least, are by no means accurate models of the actual brains that come from those skulls.

Nevertheless, the appearance of the Coelacanth did give much increased confidence in the ability of palaeontologists to reconstruct with accuracy, for their models and pictures of Coelacanths were on the whole fairly close to the real live fish. Reconstructions exhibited in museums make the Coelacanths of past ages look rather stodgy and wooden, and indeed they are labelled that way. This certainly does not apply to the 1938–55 models. By human standards the appearance of the Coelacanth can best be described as 'tough and terrifying'. This is, however, not as important as the fact that the reconstructions were fairly accurate, so that we may reasonably expect models of other creatures to be equally close to what they really were. It has given everyone increased confidence in this type of reconstructive work.

The discovery of the Coelacanth has focused attention on how very little we really know of life in the sea. That 'Man's dominion ends with the shore' is indeed very true. While we control virtually all life on land, we have a far from complete knowledge of life in the water and practically no control there at all. Within the confines of London or Paris, for example, there is very little wild life of any kind on the land, hardly any that is not under full human control, excepting only minute forms. Yet right in the heart of those old, civilised, and densely populated centres, in the rivers Thames and Seine, life goes on exactly as it did a million, 50 million, even longer ages before, primitive and savage, hide or be eaten, fly or be eaten, eat or be eaten. In no water is life entirely subject to man-made laws.

After all the work that has been done on the sea, the emergence of the unknown Coelacanth, a large robust creature, does show clearly how little we know, and it brings at least a hope that there may be other primitive forms still surviving somewhere there. It tells us even more. It has shown that comparatively large creatures can live for long ages in the sea and leave no easily accessible fossil traces. It may, and almost certainly does, mean that there have been other creatures who lived always in the sea of which no traces have been found and of which we have no knowledge at all. It leaves a hope that there may still be such unknown forms alive in the sea, and that they may be discovered when man achieves greater mastery of that region beyond the tidal fringe. There is, indeed, probably a greater possibility of this than of the existence of another known type like the Coelacanth.

It is astonishing how much an intensive study of fossils can tell us about the creatures that left them. Often even habits and other characteristics can be deduced. Nevertheless, there are big gaps in our knowledge of past life. We know next to nothing of the soft parts of the early forms of life, nothing of that very early big change-over from invertebrate to vertebrate.

Flesh (protein) is composed chiefly of remarkable aggregates of substances known as amino-acids, and different amino-acids in varying proportions are found in the flesh of different animals, while the protein of plants is again very different from that of animals. In the course of evolution, profound changes in amino-acid components of flesh protein may have accompanied other structural modifications. I learnt at the Comores that the flesh of the Coelacanth when cooked becomes jelly-like. It will probably be found to have a composition different from the flesh of ordinary fishes.

In the course of evolution, there will almost certainly have been continual change in the form of the intestines, in the composition of the digestive fluids and enzymes, and in many of the soft parts. It is unlikely that we shall ever know very much of this, but the Coelacanth holds out a hope of gaining some of that knowledge.

One of the most outstanding characteristics of the Coelacanths is that they have changed remarkably little during the vast ages they have lived. The bony structures of our modern Coelacanth are almost exactly the same as those left by Coelacanths of several hundred

JLB mentions 'other creatures who lived always in the sea'. In fact, most coelacanth fossils are from fishes that lived in freshwater swamps or estuaries. Few marine coelacanth fossils are known.

The discovery of living coelacanths has certainly focused more attention on the oceans, but we still know relatively little about them. Four times more men have walked on the moon (12) than have ventured into the deepest part of the ocean (Marianas Trench).

The discovery of the coelacanth also reminds us that there might be other undetected, large marine creatures. The huge, 5.5 m-long, megamouth shark, *Megachasma pelagios*, was only discovered in 1976. It is a rare, deep-water plankton-eater; only 61 specimens had been caught or sighted by April 2015, despite intensive searches.

FMNLH Ichthyology, Wikipedia CC BY-SA 4.0

Megamouth shark, Megachasma pelagios

JLB did not know at the time that the coelacanth is not an egg layer, as most fishes are, but that it gives birth to live young, the most advanced breeding strategy.

At the age of about 28 days human embryos have simple gill openings and a tail, and look like fishes. This led the German scientist Ernst Haeckel to propose that 'ontogeny recapitulates phylogeny', i.e., when an animal develops from an embryo to an adult, it passes through phases that are representative of the successive stages in its evolution.

million years ago. There is, then, at least a hope that the soft parts of the modern Coelacanth may also be little changed from those of earlier times, and from study we may be able to deduce something of the finer details of the earliest vertebrate types.

There has been a good deal of controversy about the origin of the important oil deposits in the earth. Some scientists consider that they had a purely inorganic origin, being formed by the action of pressure and moisture on carbon. Others consider it at least as likely that the oil resulted from the action of heat and pressure on great numbers of oily fishes, possibly killed in some upheaval of nature. It is certainly interesting that the Coelacanth is oily, very oily indeed, and a study of that oil may throw some light on this whole question.

The development of embryos is a most fascinating study, for it has been observed that many show characters of the earliest forms of life from which the creatures have evolved. For example, at certain stages the human embryo has gill-slits in the throat, and a tail, indicating our fishy origin.

One rather wonderful discovery was the fossil of a fairly large Coelacanth with remains of two others, both very much smaller, situated near the hind part of the belly region. That could mean merely the fossils of three Coelacanths, one big and two small, but it could also have meant that these two smaller ones were the well-developed but still unborn young, indicating that the Coelacanth brought forth the young alive. If this was the case, it would mean that the study of the embryo in the living Coelacanth would likely have enabled scientists to gain some knowledge of life still earlier than the origin of Coelacanths, a wonderful possibility that we all have cherished. As Coelacanths have been caught one by one, so we have gone on hoping that each would be a female, especially one with unborn young. By some curious trick all the first were males; then came an immature female; and now, finally, a gravid female, but she has eggs, not embryos. This takes away some of our hopes, but not all, for the embryo develops in the egg, probably in a case like that of the sharks and rays, and there will be much to be learnt when we can find these.

APPENDIX *C*

THE LATEST POSITION ABOUT COELACANTHS

THE position regarding recent Coelacanths up to July 1955 is as follows: The French have got in all seven more Coelacanths, all apparently at the Comores. One report that stated that a specimen had been caught on the coast of Madagascar has not been confirmed.

Almost all information about these Coelacanths has come from the French, who report them as all having been taken by native fishermen at depths ranging from 80 to 150 fathoms, in each case the exact depth of capture has been stated (see note, p. 242).

All have been large fishes, the largest weighing about 150 lb., all plainly the same species, with fins like *Latimeria* but body shape like *Malania*, so that probably only the former genus is valid.

The first were apparently all males, and this led to the suggestion that the females live in much deeper water, implying, in fact, that they all do, and that it is only the males who occasionally rise to lesser depths. This theory was invalidated by the capture of a female at about the usual depth. Not long after this the first egg-bearing female was caught, and by the most curious coincidence the fisherman in this case was one of Hunt's own crew. Hunt wrote to me from Majunga, Madagascar, about this fish and said that it was caught only about a couple of hundred yards from the schooner and quite early in the evening. The man got his reward from the Government. The eggs were rather like those in a chicken, being in a cluster of varying sizes, three of them large and well-formed. A man who broke one open and sucked it said the flavour was the same as that of a chicken. Next morning the fish left for Paris.

This is the first report about the flavour of a Coelacanth's egg. Two things emerge from this: one is that our hopes of delving into the still more remote past within the embryo of a Coelacanth are less than if the creature had brought forth its young alive, and secondly the Coelacanth doubtless sheds its eggs inside a special case, quite possibly like those produced by some sharks and rays. Who will be the first to find one? When they are found, it is well within the bounds of possibility that Coelacanth egg-cases of bygone ages will be identified among fossil remains.

Judging by what has been published, only the French, mainly Dr.

After the first specimen had been caught off the coast of South Africa, the next 160 coelacanths (for which the place of capture is known) were caught off the Comoros. The 162nd coelacanth was caught off Mozambique, and the first specimen reliably known from Madagascar (number 173) was landed in August 1995.

Most of JLB's predictions about the coelacanth proved to be correct but he was wrong when he predicted that their egg cases would resemble those of sharks and rays. As the coelacanth's eggs hatch inside the mother, they do not have egg cases at all.

Shark egg case

Chloe Langton/Shutterstock.com

In fact, the proportions of male and female coelacanths caught between 1952 and 1991 was about the same (53% females and 47% males). There is no evidence that males venture into shallower water than females. In fact, the opposite might be true as the large, pregnant female caught off Mozambique in August 1991 was trawled in very shallow water (40–44 m).

Of the first 161 coelacanths caught (all in the Comoros), 160 were landed by traditional fishermen using hand lines. The 162nd specimen caught off Mozambique in 1991 was landed by a trawler, as was the first off East London in 1938. Of the total of 242 coelacanth specimens for which the method of capture is known, 190 (78%) were caught on hand lines, 44 in deep-set gillnets, one in a ring net and three in trawl nets; four were found floating on the water surface and were caught by hand.

Millot, have worked on the Coelacanths so far. The reports already issued are in a sense preliminary and have not revealed anything of especial importance. That can emerge only after long specialised study. There has been criticism of the French for keeping this marvellous scientific material of such wide interest to themselves, but as long as the essential work is properly carried out, it is not important who does it.

The eighth Coelacanth was, like all the others, taken on a line, but in this case the fisherman managed to tow it alive to the shore. There it was put into a sunken boat which was covered with nets. It died next day.

Here is the account of the event given by Dr. Millot, published in *Nature*, London (February 1955):

'The organization for the fishing and conservation of the Coelacanths of the Comoro Islands created by the Institut de Recherche Scientifique de Madagascar, with the invaluable support of the Administration supérieure and of the Commandement de l'Air, reports a new success: on November 12 last a further *Latimeria* was captured at Anjouan. This brings the total since 1938 to eight and is the finest yet, as regards both size and state of preservation, and by far the most interesting because it is the first near adult female specimen which has come into our hands as well as the first of these precious fishes which anyone has been able to observe alive; for although an Italian expedition claimed to have photographed one last year, at 15 metres depth, the circumstances were quite incredible.*

'As a matter of fact the principal objective, once an adequate number of specimens for anatomical investigation had been acquired, was to capture a living one and keep it alive sufficiently long to make the biological observations desired. This was a difficult proposition. Hitherto, almost as soon as the fish had been brought to the surface, the fishermen had promptly battered it to death with oars or dispatched it with harpoons or knives in order to prevent it from struggling, and to be able to hoist it into their narrow pirogue without too much trouble. We had to put a stop to this deplorable behaviour and, on the contrary, persuade them to do their utmost to bring the fish, alive and uninjured, to the nearest harbour. This they were never willing to attempt, fearing, not without good reason, that on the way a shark or a shoal of barracuda might wrest their prize from them and lose them the promised reward. It took a great deal of persuasion to obtain their compliance, with an express promise that should they be successful the reward would be doubled.

* (It would be interesting to know why.)

'We, for our part, had to ensure a prison containing a sufficient quantity of sea-water for the captive fish. At first the provision of a fishpond on the seashore was considered; but on these rugged shores the work of constructing one would certainly be difficult. That solution had, besides, the inconvenience of creating a predetermined and immovable rendezvous, although the place of the next capture could not be foreseen. Used as a kind of aquarium, a sunken small boat seemed to have many advantages to commend it—cheapness, simplicity, rapid installation, mobility—and it was decided to adopt this course.

'The eighth *Latimeria* was pulled in from 140 fathoms depth (255 metres) at 20.00 hr. on November 12 by a fisherman, Zema ben Said Mohamed, assisted by Madi Bacari, both of the Maijihari quarter of Mutsamudu. Their pirogue was then about 1,000 metres offshore, opposite Mutsamudu jetty. The sea was very calm and the tide ebbing; it was two days after full moon and the moon had just risen.

'By the way in which the fish had taken the bait, the usual hunk of "roudi" (*Promethichthys prometheus* (Cuvier)), Zema, an excellent fisherman, immediately guessed that it must have been a Coelacanth. Nevertheless, he took half an hour and every precaution over hauling it in and, having made sure that it really was a *combessa* (the local name for the fish), he decided to try for the double reward by keeping it alive. He succeeded, "en le tenant par la main", as he said, in passing a cord in through its mouth and out through the gill-opening, and by means of this cord and of the line (which remained attached to the centre of the anterior part of the floor of the mouth) he led the creature all the way back to Mutsamudu jetty; though sometimes it was the fish that towed the pirogue.

'Administrateur Lher, as soon as he had been notified (it was by then 20.50 hr), decided, as we had agreed in advance, to sink a whaler immediately in which to place the *Latimeria* and keep it under the least unfavourable conditions practicable. The receptacle was ready by 21.30 hr. and anchored at a few tens of metres off the end of the jetty. The basin of sea-water so contrived measured about 7 metres long by $1\frac{1}{2}$ metres wide and 80 cm. deep. The bung had been removed from the bottom of the boat, so as to provide a small but continuous current of water. Besides this, every half-hour the boat was violently rocked in order to renew the greater part of the water. A net covered the top of the whaler to prevent the Coelacanth from escaping, which it never seemed to want to do. The greenish-yellow luminescence of its eyes was very pronounced and could be seen at quite a distance. The colour of the fish was very dark greyish-blue, recalling that of the steel of a watchspring, with fins having clearer grey-bluish reflections.

The opportunity to examine a live coelacanth was a very significant step in coelacanth research and the French researchers were well prepared for this eventuality. While many anticipated that the coelacanth, a predator, would be aggressive, it turned out to be docile, even timid, and did not try to escape from its 'sunken boat' aquarium.

The subdued behaviour of captured coelacanths, as well as those observed later in their natural habitat by the *Jagonauts* and mixed-gas divers, is best explained by their energy-conserving lifestyle that minimizes any movements that would waste energy.

Photograph by Barbara Brou of the Jago *submersible being launched in 2000 off Sodwana Bay during the ACEP programme, with Phil Heemstra on the submersible and Karen Hissmann on the left.*

NRF-SAIAB

The first description of the curious sculling movements of the coelacanth's paired fins was a revelation. No other fish swims in the same way.

NRF-SAIAB

Sculling movements of the coelacanth's paired fins

Coelacanths are very sensitive to low oxygen concentrations. As warm water contains less oxygen than cooler water, the stress that this coelacanth experienced at the water surface was probably caused by asphyxia, a shortage in the supply of oxygen to the blood. Asphyxia causes generalized hypoxia (oxygen deficiency), which eventually affects all the tissues and organs.

'Throughout the night—which the delighted population of Mut-samudu passed in singing and dancing to celebrate the capture—the Coelacanth was watched over with admirable care by the chef de circonscription, taking turns with his adjoint, M. Solére. It seemed, although quite bewildered at the sequel to its ascent to the surface, to be taking the situation very well, swimming slowly by curious rotating movements of its pectoral fins, while the second dorsal and anal, likewise very mobile, served together with the tail as a rudder.

'After daybreak it became apparent that the light, and above all the sun itself, was upsetting the animal very much, so several tent canvases were put over the boat to serve as some kind of protection. But despite this precaution and the more or less constant renewal of the water, the fish began to show more and more obvious signs of distress, seeking to conceal itself in the darkest corners of the whaler.

'At 14.45 hr. it was still swimming feebly; but at 15.30 hr. it had its belly in the air and only the fins and gill-covers were making agonized movements.

'It was then covered with a sheet and taken immediately to the hospital. There was not a scratch on it, apart from a tiny incision in the centre of the anterior part of the floor of the mouth made by the fisherman when recovering his hook. Altogether, it was in remarkably good condition, without any rupture of the viscera or suffusions of blood.

'It measured 1·42 m. in length and weighed 41 kgm.

'Chemical and histological investigations could be made under the best possible conditions on perfectly fresh tissues.

'Notified by telegraph of the capture and rushed from Tananarive by a special aircraft, I arrived just in time to witness the last moments of the fish.

'Two principal conclusions emerge from the corroborated statements made by local observers and by myself: (1) the extreme photophobia of *Latimeria*—the sunlight seemed literally to hurt it; (2) the exceptional mobility of the pedunculate fins, correlated with the wealth of muscula-ture which is revealed by anatomical studies. The pectorals, in partic-ular, can move in almost any direction and show themselves capable of assuming practically every conceivable position.

'There can be no doubt that death was brought about by decompres-sion combined with rise in temperature. The previous water-samples taken by Menaché (1953) and by Millot and Cousteau (1954) in the precise positions in which previous captures had been effected, showed important temperature differences between the stratum frequented by the Coelacanths and the surface water (26° C., more or less) during the daytime off Moroni or Mutsamudu.

'It must also be noted that the *Latimeria*, which appeared greatly distressed on its arrival at the surface, seemed to have recovered appreciably after an hour or so and passed the rest of the night quite comfortably without any too obvious inconvenience. It was daybreak, with the appearance of sunlight and the gradual warming of the water, which initiated the progressive discomfort that led rapidly to its death.

'The trial having thus been made under satisfactory conditions, it does not seem likely that substantially better results can be anticipated from the employment of the same technique in future. We must be prepared to make other arrangements. The only procedure offering any hope of keeping a live Coelacanth for a longer time would seem to be the construction of a great trellis-work case in which we could place the fish immediately after capture; there we should keep it normally submerged at a depth of 150–200 m., and only haul it up for limited times when someone wanted to observe or photograph the animal. Such a cage will be put in hand at Anjouan.'

As a result of this report, I sent the following to *Nature*, and it appeared in the issue of 3rd September 1955.

LIVE COELACANTHS

'From the discovery of the first Coelacanth at East London in 1938 it was my aim, not only to discover their true home, but also I hoped to live to see a living Coelacanth; and for mankind generally to be able to see this living link with the incredibly remote past. When *Malania* was found at the Comores, I planned to catch Coelacanths alive and to keep them alive. It is therefore gratifying that the French are plainly making every endeavour to achieve this. The article by Professor J. Millot in *Nature* of February 26 1955 on the experience of the first living Coelacanth at the Comores is of special interest.

'The failure of the French to keep their fish alive for more than a few hours is attributed by them to decompression combined with rise in temperature of the water, while a high degree of photophobia on the part of the fish is alleged.

'While there may be something in this, in my view the cause is probably quite different. Professor Millot and his collaborators are possibly not aware of the experience that large fishes taken alive after a struggle on a line, even with no visible laceration, rarely live long after, certainly not in aquaria, and even when liberated many die very shortly. Curiously enough, fishes taken by harpooning, even when extensively gashed, show a greater survival rate than those taken on hooks. Coelacanths caught by net or trap and kept in a closed vessel

O.F.—16

JLB's comment, 'The failure of the French to keep their fish alive for more than a few hours' was unfair. It was a remarkable achievement to keep it alive for nearly 20 hours, especially as it had been caught at a depth of over 250 m and then dragged over 1,000 m to shore. They then made valuable observations on the living animal, something that Smith had hoped to do himself.

JLB Smith published many meaningful articles in *Nature* but this one contains little of scientific merit. It was his last scientific paper on the coelacanth.

While Smith had a broader knowledge of fish behaviour than Millot, he should have known better than to extrapolate from other fishes to the coelacanth as its physiology and behaviour are so different. As the coelacanth's swim bladder is filled with fat, which is about the same density as fish flesh, it is not subject to the same pressure variations as the gas-filled swim bladders of many other fishes. The low oxygen concentration in warm, shallow water probably caused the death of this fish.

Many pregnant coelacanths have been caught and have yielded eggs and pups for study. They have contained between 2 and 197 eggs and 2 and 26 pups.

The first scientist to provide evidence that the coelacanth is a live-bearer was Professor DMS Watson, the London palaeontologist who found clear impressions of two embryos inside a fossil of the Triassic coelacanth, *Undina*, in 1926.

Fossil coelacanth (Undina) *embryo*

will almost certainly have a greater chance of survival even at normal pressure.

'It is doubtful whether the view about decompression or small variation in temperature is tenable, since after being hauled to the surface in a trawl-net near East London the first Coelacanth lived for more than three hours, out of the water, on the deck of a trawler on an unusually warm day.

'It is interesting to note that the French used a boat as an improvised aquarium. At a meeting in Nairobi in October 1953 I suggested using a decked boat, since it seemed likely that an important factor in survival would be to shield the fish from shocks until such time as it could become accustomed to a new environment. An open whaler, however, was used at Mutsamudu so that the fish had a clear view. We are told that "Throughout the night—which the delighted population of Mutsamudu passed in singing and dancing to celebrate the capture—the Coelacanth was watched over with admirable care", by officials, doubtless with constantly flashing torches, and only those who have experienced a night such as is indicated can have any idea of the noise and lights. That poor live Coelacanth at Mutsamudu must have passed the night in a state of high nervous tension.

'What the French considered "Photophobia" on the part of the Coelacanth is in my view no more than the natural uneasiness that any large and intelligent fish would experience as unfamiliar surroundings and objects become increasingly obvious from dawn.

'The "luminescence" of the eyes of the live Coelacanth is interesting. This phenomenon is, however, quite common in sharks and other large fishes of shallow waters, and on this night there was bright moonlight.

'It is a notable feature of the reports that the depth to a metre at which each Coelacanth was caught has been stated. As all of these were apparently taken by natives fishing from drifting canoes at night, and the slope of the bottom offshore at the Comores is stated to be at least 50°, it would be of general interest to know how this high order of accuracy is achieved.

'In the matter of the first egg-bearing female Coelacanth, it is a strange coincidence that this was captured by one of the crew of Captain Hunt's* vessel, only a short distance from where this was anchored. The fish was apparently cut open and seen to contain a cluster of eggs at all stages of development "such as is observed in a chicken" or in oviparous sharks. We may therefore expect Coelacanths to have egg-cases like those of Elasmobranchs.'

* It was Captain Hunt who took our Coelacanth leaflets to the Comores which resulted in the discovery of *Malania* there (see Chapters Eight and Thirteen).

The age of the Earth is now estimated to be 4.543 billion years, one-and-a-half times older than the opinion available to Smith in the mid-1950s. Furthermore, the latest evidence suggests that life originated about 3.8 billion years ago.

APPENDIX *D*

COPIED BY PERMISSION FROM *THE TIMES*, LONDON, 2ND JANUARY 1953

OLDEST OF FISHES

ORIGINS AND IMPORTANCE OF THE COELACANTH
By Professor J. L. B. Smith

GRAHAMSTOWN, *January 1st*

THE word 'Coelacanth'—pronounced 'seelakanth'—means 'hollow spine'. Only 14 years ago probably not more than 1,000 human beings had any notion of what the word meant, and probably not one in every 100,000 had ever heard the word at all. On the other hand, over the past 100 years to a small and select group of scientific intellects this word has stood for a remarkable race of fishlike creatures of almost incredible antiquity.

These fish were some of the first to appear in that dim and distant past when life on this planet began. This is no guesswork; brilliant men working from often only fragmentary fossilized remains have, step by step, built up a chronological picture of the main stages in the development of life on this planet. The age of the earth as a separate entity is estimated at about 3,000 million years. It was, in the beginning, no more than hot viscous matter and gas, rapidly cooling in its whirling course. By about one thousand million years ago the earth was settling down with a solid crust, mostly bare rock, and the ocean was completely enveloped in dense cloud, its surface lashed by storms and torrential rain much more terrible than anything we know today.

By about sixteen hundred million years ago something queer had happened. What we call 'life' had come to the earth. If you take a fragment of iron ore and treat it with water and heat, it changes to a soft, slimy substance which can flow and adapt its shape to the surface on which it rests. In that state it is much more 'reactive' than the hard rocky ore, and can absorb other substances, which, while profoundly changing its fundamental structure, leave it still able to go on absorbing still other substances. Some time, somewhere, there was possibly formed from other elements a jelly with power to move on its own—not only

O.F.—16*

JLB Smith did science and the general public a great service by publishing popular articles on his fish research and giving radio interviews; few other serious scientists in his day bothered to do so. Even today some scientists regard popular publications as a waste of time. How wrong they are!

One wonders, though, whether a modern newspaper editor would be prepared to publish such a scientific article? Although we live in the Information Age, and science is more important now than ever before, modern media tend to cover science in a relatively superficial way.

when it was taken by currents or pulled by gravity. It was able to choose its own path and to protect itself.

This can only be guesswork. We know hardly anything of this phase of life, but development must have proceeded at a great rate, for quite suddenly there appeared on earth clumsy, monstrous fishes with large armoured heads. It is generally assumed that this armour was for ordinary protection. This is hard to believe. With only the head encased a creature of this type would be likely to be vulnerable to attack and destruction, as indeed their ultimate disappearance proved. It is more likely that the heavy casing of the head was to protect the delicate brain from increasing osmotic action as the sea became more salty.

Hollow Spines

Be that as it may, among other fishes of the very early past—some 300 to 350 million years ago—appeared the Coelacanths. These were easy of recognition because of numerous features, among them the hollow spines, resembling tubes, from which they got their name. While the pattern of life showed a constant series of changing forms that came and passed, the sturdy Coelacanths went steadily on. Many left easily recognizable fossil remains over about 300 million years. There were not many species, but they showed relatively little fundamental change over that vast period.

All fossil records ceased about 60 million years ago, and the Coelacanth was said to have become extinct; but in 1938 a living Coelacanth was caught off East London, Cape Province. For 14 years since then I have looked for another. I realized that the Coelacanth, if another existed, must be sought in water of moderate depth, with uneven rocky bottom, probably with swift currents and wind-lashed seas. There it would be difficult to catch by any means. With its thick, heavy scales and its ability to hide, it would be safe from almost every kind of attack.

My deductions told me that the best place to concentrate our search would be the area about Madagascar, and my collaborators and I have for years flooded the whole East African region with a leaflet in English, French, and Portuguese, giving a picture of a Coelacanth and offering £100 reward for each of the first two found.

Now off the north-western tip of Madagascar, from the islands of Anjouan, in the Comoro group, another Coelacanth has been caught— a second species, new to science, which I have named *Malania anjouanae*. It is a great relief that the 1938 Coelacanth was not a last hoary survivor. The ancient line still goes on. On my brief visit to the Comoro Islands last Monday to collect the newly caught fish, I learnt that the natives had got odd Coelacanths over a very long time, and it will not be surprising if there are still more species in those waters.

Why is this discovery so important? It is a stern warning to scientists not to be too dogmatic. Not only is there a Coelacanth still in existence; there are at least two species still doggedly carrying on their ancient line. It is not unlikely that more will be found in other seas. We have in the past assumed that we have mastery not only of the land but of the sea as well. We have not. Life goes on there just as it did from the beginning. Man's influence is as yet but a passing shadow. This discovery means that we may find other fishlike creatures supposedly extinct still living in the sea. Some may be even more important than the Coelacanth itself.

Another important aspect of this discovery is that it has established the uncanny accuracy of the work of the palaeontologists, for their deductions about the Coelacanths from fragmentary fossils—a bit here and another bit there—have now been proved correct. It is therefore justifiable to assume that comparable work on the fossils of other forms of life is equally sound, and it gives us confidence in the views of scientists on the procession of life.

I am asked repeatedly what we may expect to prove with this fish. I am exhausted from strain after my hurried flight to the Comores, and wearied by the attempt to cope with a host of those who wish to tear something of my thoughts for the many millions to whom 'Coelacanth' is now almost a household word. It is difficult to co-ordinate the chaotic thoughts that flood my brain. I have scarcely had time to do more than satisfy myself that it is a Coelacanth, that it is a new genus and species, probably a new family; and that most of the flesh and intestines are intact.

Brain Destroyed

All the soft parts of the first Coelacanth were lost. The native who caught the second one beat it on the head. The man who got the fish from him left it to his native sailors to cut for salting, and they sliced the creature open from snout to tail; most of the brain and other soft parts of the head are gone. All this, however, does not perturb me. It cannot be stressed enough that one most important aim has been achieved. We have established where some Coelacanths live, and it is only a matter of time until we get other specimens.

One thing of which we know nothing at all is the nature of the soft parts of those very early creatures 300 million and more years ago. There is every reason to believe that the early Coelacanths may have had soft parts at least something like those of other creatures of that dim past; and since Coelacanths retained their hard parts almost

The opportunity to examine the anatomy of a 'living fossil' that had changed little over time confirmed the accuracy of predictions made by palaeontologists. These predictions do, however, need to take into account that the living coelacanth is specialized for living in marine reefs and canyons, and would differ from coelacanths that lived in estuaries or freshwaters, as was the case for many extinct species.

The discovery of living coelacanths allowed scientists to examine their soft anatomy, which is not preserved in the fossil record, and to observe them in the wild. This enabled us to describe their physiology, behaviour and ecology in detail, which, in turn, made it possible for us to reconstruct the lifestyles of extinct species.

The 'one remarkable Coelacanth fossil' that JLB refers to is probably that of *Rhabdoderma exiguum*, from Mazon Creek, Illinois, which contains the fossilized traces of two 5 cm-long 'yolksac juveniles'. This indicates that coelacanths may have been live-bearers as long as 200 million years ago.

Rob Gess

Rhabdoderma exiguum from Mazon Creek

JLB's lonely appeal for 'a good seaworthy vessel … to meet my wife and me at the Comoros to help us search' unfortunately came to nothing.

unchanged through the centuries, it is at least possible that their soft parts may also still be much the same.

We may therefore be able to learn something of what the internal organs of creatures of so long ago were like. This may go a long way to clearing the evolutionary picture. Most people know that a developing embryo shows features which are believed to be clues to ancestral forms.

One remarkable Coelacanth fossil suggests that they produce the young alive. This means that once we start getting female Coelacanths with unborn young, it may be possible to peer into the remote past of organic life. I can imagine the astonishment of a biologist if he finds an early Coelacanth embryo with no jaws and a shell-cased head.

Some Questions

Here are some of my jostling thoughts. What is the composition of the flesh of the Coelacanths? What are its component amino-acids? The natives report that when boiled it goes to jelly; that is interesting. The Coelacanth just drips oil; what is its nature, and will it help us to decide whether fish-oil was really the origin of our mineral oil-deposits? What was the nature of the cells in the earliest creatures? Did they have a liver? Did they have spiral valves in the intestines? What sort of digestive juices did they have? Have they perhaps not characteristic unchanged internal parasites? How did jaws develop? (The first fishlike creatures had only soft mouths.)

There is hardly a limit to what we may learn through the Coelacanth. It may indeed prove to be a sort of H. G. Wells's 'Time Machine', only always in reverse. I hope to get yet more information from the Coelacanth when an absolutely complete fresh specimen is caught.

It will need the services of a team of experts before all the secrets of this ancient fish lie exposed. After one partial and one not quite complete Coelacanth, I should like a real whole fish. Will some person with a good seaworthy vessel of fair size fitted with refrigeration, give up next August and September and meet my wife and me at the Comores to help us search—even if others get there before us?

APPENDIX *E*

COELACANTH BROADCAST FROM DURBAN
29th December 1952

THIS broadcast was first sent over the National Network of South Africa about 10 p.m. on the 29th December, 1952. As a result of nation-wide requests it was repeated the following day and was also sent out over the British Broadcasting System and in the U.S.A. Translated into many different languages, at least parts of it were broadcast in virtually every country in the world.

Reproduced below is the text of the broadcast; this followed a brief interview with Dr. Vernon Shearer, M.P., and introductory comments by the announcer.

'It is my astounding privilege to announce to the world the discovery of a second Coelacanth. This all started fourteen years ago—no, of course I am wrong, it really started 300 million years ago. For that is the time that scientists estimate as the first appearance of the Coelacanth fishes on earth—it would take too long to tell you how this estimate is made, but that figure has been arrived at after long study by some of the best brains of mankind.

These rather curious fishes were evidently a vigorous line, for they flourished and multiplied, their fossil remains being found over a great area, and they kept on almost unchanged for a far longer period than any other type of creature we know. After about 200 million years of existence they began to decline in numbers and there are no fossil remains in rocks less than 60 to 70 million years old. Scientists therefore assumed without question that this powerful and ancient line had become extinct about that time. It can therefore be well understood, and many of you will remember something about it, that the discovery of a living 5-foot undoubted Coelacanth near East London in South Africa in 1938, was the greatest shock to scientists everywhere, and their excitement was so great that the man in the street was infected, too, so that the South African Coelacanth became probably the best-known biological curiosity in the world. It was discovered on the 23rd of December 1938 and was kept for me to examine and identify, but the unfortunate dislocation of normal life by the Christmas holidays eventually resulted in the loss of all the flesh and skeleton of

This historic broadcast had a lasting impact on the public's perception of science. Many listeners experienced the agony and ecstasy of making a scientific discovery for the first time. Decades later, people told me how they, or their parents or grandparents, were moved to tears by JLB's emotional broadcast.

It is an extraordinary interview, given by an exhausted man without reference to any notes. The interview captures the spirit of his maniacal determination to secure a second coelacanth, and JLB also describes in detail the logistics of the complicated flight and the salient features of the fish. I doubt that a more dramatic interview has been given by a scientist in South Africa, either before or since.

The Colour.com

The coelacanth is an ambush predator.

Remarkably, all of JLB's predictions on the behaviour and habitat preferences of the coelacanth, based on his detailed examination of its anatomy, proved to be true. It does 'move quietly about reefs', 'catch its food by stealth and cunning', and live in 'moderately deep water with rough rocky bottom washed by a strong current and probably rough or dangerous seas'. Furthermore, 'it is not speedy' (except in short bursts), it is 'a pouncer', and 'would easily take a baited hook'.

this wonder fish, though the head and skin were almost intact. These important remains enabled me to ascertain a tremendous amount of information about Coelacanths, and, in fact, I have written virtually a book about that. Can you imagine a more tantalising situation? Here had been found this wonderful, almost incredible, relic of a past so remote as to be almost beyond the grasp of the ordinary mind. It was almost exactly like its early ancestors, and we were by such unfortunate circumstances prevented from being able to find out what most of its body and organs were like. It therefore became more than normally desirable, really imperative, to find more—even one would do. I naturally gave more thought to this matter than any other living man— not only that but I have probably a more informed and intimate know- ledge of the coast and seas of South and East Africa than anyone else. With my wife, I have tramped many hundreds of miles of that coast, possibly thousands, on foot—parts of it many times. The first thing that struck me was that this East London Coelacanth was most likely a stray. I know fish and can tell with reasonable certainty from the appearance and shape of a fish both its habits and a good deal about where it is likely to live.

That Coelacanth looked to me like a fish that moved quietly about reefs—the kind of fish that would catch its food by stealth and cunning. Its body shape shows clearly that it is not speedy. Its heavy scales would be perfect protection from casual bumps on rocks or coral, and its very powerful jaw muscles and the nature of its quite formidable teeth told me clearly that it was a pouncer which, with larger prey, would grab and hold on grimly until that prey was exhausted and overcome. It was probably rare at any time, but looked the type of fish that would easily take a baited hook. As it had apparently never been seen before, it was not likely to live anywhere where many people fished or where it could easily be caught on a line. This to me said clearly, moderately deep water with rough rocky bottom washed by a strong current and probably rough or dangerous seas. For with such conditions, bottom-line fishing is almost impossible. No Coelacanth had ever been caught before in any trawl-net, and the East London area has been swept by trawlers very thoroughly for very many years. The type of habitat I have outlined would also be consistent with its not being caught normally by trawlers, for they could never operate with such a bottom.

One European scientist who had not seen the fish, attempted to explain the mystery of the sudden appearance of the Coelacanth by saying that of course it lived in the great depths and that this one had come up casually. I did not share his views. No fish from the depths ever bore so powerful an external armour of bones and scales as the

Coelacanth, and I could find not one vestige of evidence in support of his view. He can hardly hold it now!

I have outlined the conditions under which I believe that Coelacanths are likely to live. It could hardly be somewhere too remote, like Alaska or Iceland. My mind at once turned to the vast reef system of East Africa, and there, over thousands of miles, are to be found the very conditions I have outlined. Further, one of the places where numbers of Coelacanth fossils have been found is Madagascar, and the whole coast of Madagascar and all the islands off East Africa, is just a series of reefs of the type I could not help feeling would be just right to hunt Coelacanths. So I turned my eyes and my mind towards Madagascar, and said many times that our hope lay there. In 1939 the war came, and most scientists had to mark time until that madness wore itself out. During all this time my mind was busy on the Coelacanth, and as soon as things began to clear I set about trying to organise an expedition to East Africa and Madagascar to hunt for more Coelacanths. So convinced was I that I came to look on attempts to find more of them about East London as sheer waste of time. In the end all our plans came to naught. I was determined to continue, but had no funds for any big venture. An expedition can do much, but the mind of a determined scientist can do more. After long thought I decided to prepare a leaflet giving a picture of the Coelacanth and a brief account of it in English, French, and Portuguese, thus covering the major languages of East Africa. Without great difficulty I persuaded my University and the South African Council for Scientific and Industrial Research each to offer £100 reward for the first two Coelacanths to be found, and this was stated in the leaflet. I then had printed many thousands of the leaflet, and it was distributed by various means over the whole long coast-line of East Africa. I sought and received the aid of the foreign governments of those territories, and numbers of the leaflets were sent to their wide-spread officials, who distributed and explained them to the natives. Since that time my wife and I have covered a vast area of that coast in our work, always doing Coelacanth propaganda. It has been a continual thrill to find that leaflet in the weirdest places, stuck on a pole in a remote native hut, posted up in lighthouses, shops, isolated posts. Now and again in some wild part a native fisherman learning who we are, with gestures and grimaces from some fold in his garment produces a dirty cloth or paper-wrapped incredibly tattered Coelacanth leaflet, and shows it to us with great pride. It was a great comfort to feel that thousands and thousands of eyes were always scanning their catches for this reptile-like fish, though most natives probably doubted whether this loony white man would really pay so vast a sum for just a fish. Our main aim has all along been to ensure that if a Coelacanth

JLB's 'coelacanth propaganda' campaign was ultimately successful – he found a second coelacanth – but one wonders what other benefits this campaign yielded. Perhaps it heightened awareness among fishermen of the diagnostic features that distinguish different fish species? Or maybe it made traditional fishermen aware, for the first time, that research was being done on their fishes, and that it would benefit them to support this research?

JLB's vivid recollection of the sequence of events that led to the recovery of the second coelacanth captures the spirit of discovery in science better than most science fiction stories. He even re-created, on the spur of the moment, the first-hand conversations between Eric Hunt and Margaret on the remote possibility of finding a coelacanth!

JLB was living his dream like few other scientists, though sometimes in agonising circumstances (which he didn't hesitate to describe), and he carried the public along with him.

did turn up anywhere, it would not be lost for lack of information.

This year my wife and I went on a scientific expedition to Zanzibar, Pemba, and Kenya to investigate the fishes. Here, as in all our East African work, I never went out to any reef to bomb without hoping that I might one day see a Coelacanth's belly breaking the foam after a blast. But none came. At Zanzibar we met Captain E. E. Hunt, a man who owns and runs a fine schooner trading between Zanzibar and the Comoro Islands. Now I can't tell you why, but for the past three years those remote and little-known islands have been nagging at my brain. I felt I had to go there, and often said so to my wife. This Eric Hunt is no fool, and when he saw the Coelacanth leaflet he was more than normally interested. When we returned from Kenya the ship touched at Zanzibar, and Eric Hunt sought out my wife and got from her a good deal of extra information about the Coelacanth 'so that if I ever come across one I shall be able to be certain'. I have a great respect for my wife's judgment, even when she slashes my work I think good. When she told me about her talk with Hunt she said, 'That man is all there, I think we can rely on his judgment if ever he gets a Coelacanth. He is sound.' As he said good-bye to her at the steps of the landing-stage of Zanzibar on the 13th December 1952, he said, 'Okay, Mrs. Smith, when I find a Coelacanth I'll send you a cable.' One need hardly guess that both of them smiled with amusement, and yet ten days later he sent just that very cable.

My wife and I have been baking and stewing in tropical heat for five long months, and we wanted to get home. We have a vast collection of fish, making an endless vista of work ahead. Our ship had reached Durban. A pressman had got me in a corner when an officer came with an urgent telegram, one of many at that time. . . .

I finished what I was saying and then opened the telegram. My heart turned right round or it felt like that, for two words leapt to my eyes— 'Coelacanth' and 'Hunt'. My wife was looking at me as I got up dizzy from reaction and trying to read the telegram of dancing letters. I said 'Coelacanth', and she jumped to her feet. The message came from one of the Comoro Islands and from Hunt—a 5-foot Coelacanth. Then my brain got to work. The Comores are remote and primitive— no refrigeration, did Hunt have formalin?—my wife said he should have some. We looked at one another with the same fear. The cable said caught on the 20th; this was the 24th—no cold store—little formalin, and the December heat of the Comores. For a while I went almost insane. Was the same thing to happen again? I sent an urgent cable to Eric Hunt asking for a statement of the condition of the specimen— it came 36 hours later—injected 5 litres formalin.

Christmas holidays—how I hate them! People eat too much and

drink too much and all work stops. They gave me a bad time now. Communications with the Comores at any time are virtually non-existent. From our ship in Durban Harbour I set the Post Office on to tracking two Cabinet Ministers I know, one after the other, and neither, on account of these confounded holidays, was in any position to do much. I tried to find another who might be able to help, but though I never left the telephone for seven long hours that Christmas Day he could not be found. Then I tried a certain high official in the Government, with the same result. Desperately I made contact with a high military officer, unknown to me personally, and it took me some time to overcome his utter disbelief that a fish could mean anything to the armed forces of South Africa, but even so he could do nothing without authority from higher up—I was almost frantic and then as a climax a cable arrived which made it plain that unless I appeared in person to claim this fish, it would be lost to my country. I sought the aid of Dr. Vernon Shearer, M.P., of Durban, and eventually, after overcoming one line of defence after another, late on the night of the 26th December 1952 he was able to speak to Mrs. Malan, wife of our Prime Minister. He was resting at a Cape seaside resort and, already in bed, was on no account to be disturbed. His good lady asked for a brief account of the matter, which she said would be given to the Prime Minister next morning. Once more time was passing, and somewhat despairingly I sat to drink some tea which our kind hostess had prepared—my mind far away in the Comoro Islands. The telephone went—Dr. Shearer went, and almost at once his excited voice called me urgently, and I went. Dr. Malan wished to speak to me, and at his request I gave this amazing man an outline, stressing that I could not be certain it was a Coelacanth but that of all the laymen I knew, Eric Hunt was in the best position to know a Coelacanth. As soon as I had finished Dr. Malan said he could see that it was important and urgent—it was impossible to do anything so late at night, but he would next morning make contact with his Minister of Defence and ask him to allocate a plane to take me to the fish. We waited next morning—I left the ship and at noon my wife went on down south. We then heard that the critical telephone system had broken down, but at 3.00 p.m. I was informed that all arrangements had been made and that a plane would come from Pretoria early next morning to take me north, to certain heat and to an uncertain Coelacanth.

We left Durban Airport at 7.00 a.m.—little me in a big Dakota with six huge South African Air Force Officers, all somewhat astonished at this quest. Once the initial reserve wore off and I was apparently human, they bombarded me with questions and eventually they got hold of it and became my firm allies in what had seemed an insane adven-

JLB embarked on an almost impossible quest during that seven-hour-long telephone marathon on Christmas Day and Boxing Day in December 1952 to find a government official who would authorize a flight to a foreign country to fetch a dead fish. But, amazingly, he succeeded where a less determined mortal would have failed.

No reader could fail to be moved by the drama that unfolded on that hot, summer day in December 1952 in the remote Comoros. Smith had, against all odds, got *his* fish, and no-one would stop him from taking it back to South Africa.

The second coelacanth was caught at a depth of 160 m, not 20 m (although JLB gives the depth more accurately as '20 fathoms' on page 156), and about 800 m from shore (not 200 yards). It is not surprising that JLB muddled some of the facts during this exhausting broadcast.

The fish was a male measuring 37.5 kg and 135 cm. It was alive at the water surface but was killed in order to haul it back to shore.

The brain was, in fact, lost.

ture. We refuelled at Lourenço Marques and Lumbo, where we slept. I asked for a 4.00 a.m. departure—they all groaned so we made it 4.30 and got away then. I sat and sweated in a 'Mae West' vest as we flew over a blue sea and terrifying cumulus clouds that our commander eyed with more apprehension than he wanted me to see. Seven o'clock and the island of Mayotte appeared. Our destination was the tiny islet of Pamanzi near by. Could we land there. Desperately the operator thumbed his keys. Suddenly the answer came over 'Yes, the strip was useable', and down we went. We landed. Eric Hunt ran up—I said, 'Where's the fish?' He said, 'On my boat; it's true, don't worry.' I went quickly to the car, but the Governor wished to meet me so I had to go there first. Unable to accept the luscious food and drink so freely offered, my mind was elsewhere. Eventually we got away, and there on the deck, swathed in cotton-wool, was the fish. I could not bring myself to touch it and I asked them to open it, and they did and I knelt down to look, and I'm not ashamed to say that after all that long strain, I wept . . . for it was true . . . it was a Coelacanth—and what was more wonderful, a species different from that of 1938—another Coelacanth. It was more than worth while all that long strain.

Eric Hunt told me the story—a line-fisherman, Ahmed Hussein at the village of Domoni on the Island of Anjouan, was fishing in 20 metres of water about 200 yards from shore and caught a large fish on the evening of 20th December 1952. He took it home—fortunately did not clean it—thank God for native indolence—and next morning took it to the market. As the fish was being sold a native came up, looked at it and said urgently, 'Don't sell that—this is the fish Bwana Hunt was telling us about', and showed the paper. 'There is much money.' So the fish was carried that long hot day 25 miles over difficult mountainous country right across the hills to a village called Matsamudu, for they had been told that Eric Hunt's vessel was there. He first saw it at 5.00 p.m. that day and recognised it immediately as a Coelacanth and it was going bad fast. He had no formalin, only salt. My wife had said to him, if you have no formalin for heaven's sake use salt. He ordered the natives to make cuts to put in salt and most unfortunately they sliced it open all along the body, but no part was lost, and then they covered it with a heap of salt. Hunt set out at once for Pamanzi and once there enlisted the aid of Dr. Le Coteur, Director of Medical Services, and got a syringe and 5 litres of formalin which he injected all over the fish.

And so there was the fish—smelly—but the soft parts all there and in good order. Hastily we completed our formalities with the French. I sent cables to my wife and the Prime Minister, and we left at 10.00

a.m., refuelled at Lumbo and set off down south with the Coelacanth and its smell and away from the heat. Heavy clouds made us fly low and so we did not touch down at Durban until about 9.00 p.m., but I was terribly thankful to be in my own country with that fish. In the island I thanked the French authorities for their co-operation, but I also told them that if that fish had been found on the steps of their Governor-General, I would have gone to claim it because ethically it was mine—it had come as a result of these fourteen years of hard endeavour.

G. MOORE: Professor Smith, one question, please. What is it you hope to prove now that you have your Coelacanth?

PROF. SMITH: Of all creatures the Coelacanths retain the structures in their body unchanged over vast periods of time more than any other creatures, and so we hope in the soft parts of this fish to learn something of the early types of life. Our great hope is that we shall find one of these Coelacanths with young inside them, because from the embryo we may learn more. I should have told you one important thing, and that is that the natives round there have told Eric Hunt that they catch two or three Coelacanths every year, not only that, there is a smaller kind, so we may even hope for another species as well. But I do believe that it is only a matter of time now before we get one of these wonder fishes in such a condition that it will give biologists information that none of us dared to hope for fifteen years ago.

G. MOORE: Thank you, sir. Have you named this fish at all?

PROF. SMITH: It is my present intention, subject to further study, to name it first in honour of the Prime Minister and secondly to commemorate the locality, and the name I have in my mind, but that is to be confirmed, is *Malania anjouanae*, the new Coelacanth.

G. MOORE: Thank you, sir—and may I say on behalf of South Africa how much we congratulate you and how proud we are of you.

PROF. SMITH: Thank you, and I say again publicly as I have said to him personally, we owe a very great debt to our Prime Minister for his foresight in providing a harebrained scientist with a plane to go and look for a dead fish.

G. MOORE: Thank you, sir, thank you. (*Clapping of hands, etc.*)

G. MOORE: And that is the actual description of the newest page in South Africa's history.

ANNOUNCER: You have just listened to a re-broadcast of an eye-witness account by George Moore of the arrival of Professor J. L. B. Smith at Durban Airport late last night, together with an interview with Professor Smith.'

Many letters about this broadcast were received, not only from

JLB's radio broadcast introduced many people, for the first time, to the arcane worlds of fish anatomy, physiology, evolution and palaeontology. The significance of the interview therefore transcended ichthyology – it gave meaning to the whole scientific endeavour.

Subsequent research revealed that the second coelacanth, initially named *Malania anjouanae* (after DF Malan and Anjouan island) by Smith, was the same species as the first specimen, *Latimeria chalumnae*. All coelacanths caught in the Western Indian Ocean belong to this species. The coelacanth discovered in September 1997 in Indonesia is a different species, *L. menadoensis*.

Latimeria menadoensis *from Indonesia*

Hentje Lumentut, Manado Post

Fortunately JLB and Margaret were ardent archivists and kept almost all their correspondence from these dramatic times. This material is now carefully curated by Senior Librarian Sally Schramm in the archives of the Margaret Smith Library in the South African Institute for Aquatic Biodiversity in Grahamstown.

South Africa, but from all over the world. Some were anonymous, not all amiable, but one is reproduced here:

Professor J. L. B. Smith,   *December 30th 1952*
GRAHAMSTOWN.

Dear Dr. Smith,
    Thank you for one of the most moving broadcasts it has ever been my privilege to hear, and for not pleading exhaustion as an excuse at a time when you might well have done so.
    Thank you indeed for sharing with us, the listeners, your hour of triumph. With you we were, each one of us for a few moments, hare-brained scientists in quest of a dead fish; with you we wept on the deck of a boat at the islands which we shall probably never see.
    Thank you, and God bless you.

                                        From one of the many.

INDEX

A.C.M.E., 71, 97, 222
African Coelacanth Marine Expedition, *see* A.C.M.E.
Afrikanerdom, 118
Airbladder, 92, 208, 230
Air Force officers, 131
Albany Museum, 7
Aldabra, 215, 224
American journalists, 97
Amino-acids, 235, 246
Amphibians, 14
Ancestral Coelacanths, Appendix B, 233
Angling, 6, 27
Anjouan, 156, 220, 244, 252
anjouanae, see Malania anjouanae
Anonymous letter, 190, 254
Ape-man, 14
Aquarium for Coelacanth, 223, 239, 242
Archeozoic, 14
Aristea, 148
Art school, opening of, 119
Astove, 224
Author, biography of, 4
'Axial' column, 231

Bacari, Madi, 239
Backbone, 37, 231
Backman, F., 183
Backman, L., 183
Ban on Coelacanth expedition, 215
Barnard, Dr. K. H., 8, 33 *et seq.*
Barros, Comandante P. Correia de, 134
Barros, Senhora Donna Maria Emilia, 134
Basal (fin), 231
Bazaruto, 74, 78, 134, 163, 218
Beek, J., 109, 113
Beira, 134
Bergh, Lieut. W. J., 133 *et seq.*, 164
Birds, 14
'Bishops', 4
Blaauw, Commandant J. P. D., 131, 133 *et seq.*, 164
'Blondie', 92
Boer War, 4
Bowker, T. B., M.P., 94, 97
Boxing Day, 107
Brain, 183, 245
Brain cavity, 230

Brains of fossils, 234
Brink, Cpl. F., 133 *et seq.*, 164
British Museum, 23, 48, 55, 190, 198
Broadcast from Durban, 160, 190, Appendix E, 247
Broom, Dr. R., 98
Bruce-Bayes, Dr., 25, 42, 55

Cables re Second Coelacanth, 85, 89, 101, 109, 130
Cainozoic, 14
Calypso, 224
Cambrian, 14
Cambridge, 5
Campbell, Dr. George, 86, 90, 128
Campbell, Mrs., 131
Cape Delgado, 77
Cape Times, 98
Cape Town, 94
Caranx, 159
Carboniferous, 14
Cartwright, G. F., 218 *et seq.*
Cells, 246
Center, Mr. (taxidermist), 37
Chalumna River, 40, 43
chalumnae, see Latimeria chalumnae
Channel Islands, 196 *et seq.*
Chemistry, 4, 6, 67
Chemistry Department, 69
Chief of Staff, 107
Christmas Day, 105
Cod-end, 61, 92
Coelacanth(s), characteristics of, Appendix A, 229
— colour, 61, 239
— described as 'degenerate side-line' 19
— diet of, 20
— habitat, 59 *et seq.*
— head of, 230
— home of, 51, 59, 193, 226; confirmed, 221; idea of hunt for, 51
— latest position, Appendix C, 237
— leaflet about, 72, 78, 80 *et seq.*, 110, 149, 156, 244, 249
— live, 238 *et seq.*; keeping alive, 223; photographing, 216; reason for death, 240 *et seq.*; seen by goggler, 218
— meaning of name, 231

Coelacanth(s)—*contd.*
— origins of, Appendix D, 243
— picture of, Ch. 2, p. 11 *et seq.*
— prohibition of export, 211
— pronunciation of name, 3, 243
— reconstruction of, 19
— skull, structure of, 51, 230
— stamps, 226
— tail, 231
— teeth, 20, 231
Coelacanth, Eighth: 238
— — Length of, 240
— — Weight of, 240
Coelacanth, First:
— — Axial skeleton, 37
— — Backbone, 37
— — Capture of, 24
— — Depth trawled, 40
— — Dissection of, 51
— — Flesh of, 37
— — Gills, 30, 31
— — Length of, 24
— — Letter re, 27, 30
— — Monograph on, *see* Monograph
— — Name of, 45, 46
— — Oil of, 37
— — Photograph of, 38, 39, 45
— — Police guard for, 48
— — Positive identification, 39
— — Press and, 46
— — Provisional name, 36
— — Publicity re, 45
— — Return to East London, 54
— — Skeleton of, 37
— — Sketch of, 25, 26, 28, 30, 35
— — Stomach of, 37
— — Viscera, 30, 31
— — Weight of, 24, 37
Coelacanth: First identification, 33
— First sight of, 41
— How differs from modern fishes,
      Appendix A, 229
— Internal nostrils, 52
— In time, 16
Coelacanth, Second:
— — Capture of, 156
— — Depth, 156
— — Identification of, 146
— — Investigation of, 187, 188
— — Photograph of, 169
— — Preservation of, 158
— — Telegram re, 85
Coelacanth, Third:
— — News of capture, 220
— — Photographs of, 223
Collecting box, 126
'Combessa', 239
Commander-in-Chief, 165

Commander-in-Chief, Cape, 171
Comorans, 226, 244
Comores, 73, 78, 84, 85, 215, 250
Comores, Authorities of, 226
— Governor, *see* Governor, Comores
— Ichthyology of, 88
— Structure of, 155
Comoro Islands, *see* Comores
— natives, 155
Comoros, *see* Comores
Coudert, Governor, M.P., 211. *See also*
      Governor, Comores
Council for Scientific and Industrial
      Research, *see* South African Council
      for Scientific and Industrial Re-
      search
Courtenay-Latimer, Miss M., 23 *et seq.*,
      41, 54, 174
Cousteau, Captain, 214, 223, 240
Creation of life, 14
Cretaceous, 14
Crossopterygian fish, 35
Crossopterygii, 15, 16, 30, 35
C.S.I.R., *see* South African Council
      for Scientific and Industrial Re-
      search
Customs, 165
Cyclone, 141, 148, 185

Dagger, Stanley, 85
Daily Dispatch, see East London Daily
      Dispatch
Daily News, 104, 169
Dakota, 130, 251
Daniel, Brigadier, 129
Danish deep-sea expedition, 161
Dar-es-Salaam, 216
Deep-freeze, 202 *et seq.*
Deep-line fishing, East Africa, 155
Deep-sea expedition, 60
Deep-sea fishes, 61
Deep-sea refugee, 53, 59, 161
Defence, Minister of, 105, 116, 122,
      251
'Degenerate', 221
Dennis, N., 183
Department of Ichthyology, 70
Depths at which caught, 60, 242. *See
      also* Coelacanth, First; Coelacanth,
      Second; Coelacanth, Eighth
Devonian, 14, 15
'Dez Contos Peixe', 76
Diego Suarez, 133, 136
Digestive fluids, 235, 246
Dinosaurs, 233
'Discovery' Committee, 71, 75
Dogmatic statements, 232, 245
Domoni, 156, 252

Donges, The Hon. T. E., 101 *et seq.*
Dragon, 190
Dragons, flying, 189
Dunnottar Castle, 83, 100, 106, 120, 122, 128 *et seq.*, 173
Durban, 84, 165
Durban Radio Station, 109
du Toit, Colonel Louis, 171
du Toit, Dr. P. J., 100, 135
Dzaoudzi, 85, 136, 140 *et seq.*

Earth, history of, 12, 14, 243
East Africa, 65, 249, 250
East Africa, fishes of, 73
East Africa, reefs, conditions, fishes, 74
East African Airways, 138
East London, 247
East London Daily Dispatch, 46
East London Museum, 23, 41, 48, 54, 182
Economic Affairs, Minister of, 102
Education, Minister of, 119
Eggs, Coelacanth, 236, 237, 242
Eighth Coelacanth, *see* Coelacanth, Eighth
Election, General, 99, 191
Embryo, 236, 253
Enzymes, 235
Eocene, 14
Eozoic, 14
Erasmus, The Hon. F. C., 105, 116
Eusthenopteron, 17
Evans, Frank, 86, 114 *et seq.*
Evening Post, Jersey, 209
Evolution: Trend of, 34
Expedition, 64, 71, 249
— Deep-sea, 60
— Italian, 215
— Swedish, 215
— to seek Coelacanths, 56
Expeditions, ban on Coelacanth, 215
— to Comores, 215
Explosives, 202, 203, 212, 213
External Affairs, Secretary of, 205
Eyes, luminescence, *see* Luminescence of eyes

Female, 185, 236, 237, 246
Fernão Velhoso, 160
Fins, 231
— movement of, 240
First Coelacanth, *see* Coelacanth, First
Fish, modern, Appendix A, 229
Fish mortality, 93
Fishes, 14
— South African, book on, 68, 70, 74, 81, 124, 169
Fishing, deep-line, East Africa, 154

Flesh, 37, 235, 246
Flowering plants, 14
Flying dragons, 189
Fossils, 9, 11, 52, 80, 233, 244, 247, 249
— Coelacanth, 66, 208
— formation of, 18
— size of extinct Coelacanth, 20
France, 210 *et seq.*
— Consul for, 72, 104, 105
French, 111, 226
French Authorities, 88, 109 *et seq.*, 197, 211 *et seq.*, 253
French Consul, *see* France, Consul for
French Government, 210, 213
Freshwater habitat, 63, 64

Ganoid fish, 25
'Garrupa', 74, 220
General Election, 99, 191
Geological Time-scale, 14
Geology, 13
Gilchrist, Dr. J. D. F., 8
Gill, Dr. E. L., 38
Gills, 37, 184, 230
Godetia, 91
Goetsch, Miss M., 181
Goosen, Captain N., 24, 43, 55
Government Fisheries vessel, S.A., 63, 196
Governor, Comores, 81, 83, 145 *et seq.*, 195, 211, 252
Graaff Reinet, 4
Grahamstown, 27, 46, 119, 174
Grand Comoro, 78
Great War, 4
Greenland, 20
Groote Schuur, 95, 117
Gular plates, 231

Haughton, Dr. S. H., 71
Helium, 12
Holocene, 14
Hundred Pound Fish, 76
Hunt, Capt. E. E., 80 *et seq.*, 104, 111, 143 *et seq.*, 185, 237, 242, 250
hunti, see Malania hunti
Hussein, Ahmed, 156, 252
Huxley, Julian, 181

Ichthyologist: to become, 4
Importance of the Coelacanth, Appendix D, 243
Inhambane, 134
Institut de Recherche Scientifique de Madagascar, 159
Interest, Reason for, Appendix B, 233
Internal Affairs, Minister of, 102
Internal nostrils, 52

International Coelacanth Expedition, 219 *et seq.*
Intestines, 184, 231
Invertebrates, 14, 15
Investigation, Second Coelacanth, 187, 188
Irvin and Johnson, 23, 24, 91
Island of Mozambique, *see* Mozambique, Island of
Isotopes, 12
Italian expedition, 215, 238

'Jakob', 50
Jaws, 229, 246
Jersey, 196 *et seq.*
Jersey Electricity Co. Ltd., 204
Jersey *Evening Post*, 209
Jurassic, 14

'Kambesi', 159
Kenya, 80, 82, 225
Key, Bransby A., 68, 99
Kikuyu, 225
Knysna, 27 *et seq.*, 41, 170
Koch, H. J., 219
Koen, Miss R. M., 181
Kombessa, 159

La Contenta, 196, 223
Latimer, Miss M. Courtenay-, 23 *et seq.*, 41, 54, 174
Latimeria, 53, 54, 112, 194, 224, 237
Latimeria chalumnae, 36, 45, 46, 70. *See also* Coelacanth, First
Le Coteur, Dr., 252
Leguan, 190
Letley, Captain P., 133 *et seq.*, 164
Lher, Administrateur, 239
Life, beginning of, 14, 243
Line-fishing, East Africa, 154
Lions, 126, 127
Living fossil, 104
Locke, Mrs. Hester, 170
London Illustrated News, 53
Lourenço Marques, 69, 72, 133, 163, 251
Louw, The Hon. E. H., 101 *et seq.*
Lumbo, 133, 138, 160, 251, 252
Luminescence of eyes, 239, 242
Lung-fish, 24, 25
Lung-fishes, 52, 112
Lurio mouth, 126

McGahey, Mayor Patrick, 174
McMaster, Mrs., 101
Madagascar, 20, 66, 71, 73, 77, 81, 88, 110, 136, 249
Majunga, 237

Malan, Dr. D. F., 88, 90, 114 *et seq.*, 121, 166, 168 *et seq.*, 172, 207, 251
Malan, Mrs., 120 *et seq.*, 171 *et seq.*, 251
Malania, 146, 158, 194, 221, 223, 224, 226, 237
Malania anjouanae, 147, 167, 244, 253
Malania hunti, 147
Male Coelacanths, 236
Malindi, 218
Mammal-like Reptiles, 14
Man, 19
Man: Ape, 14
— Modern, 14, 19
— Piltdown, 12
— Stone-age, 14
Marietjie (Malan), 172
Mau Mau, 22, 225, 226
Maxillary bones, 231
Mayotte, 136, 142, 144, 159, 252
Melville, Brigadier, 129
Memorandum for France re Second Coelacanth, 212
— to C.S.I.R., 70
Menaché, Dr., 221, 240
Mesozoic, 14, 35
Military Authorities, 130
Millot, Dr. J., 88, 211 *et seq.*, 238 *et seq.*, 240
Miocene, 14
'Missing Link', term applied to Coelacanth, 50, 87, 181
Mofamede Island, 137
Mohamed, Zema ben Said, 239
Mohilla, 155
Mokambo Bay, 160
Mombasa, 5, 82, 83
Mombasa Times, 83
Monograph, First Coelacanth, 52, 56, 67
Moore, George, 165, 253
Mozambique, 73 *et seq.*, 126, 205
Mozambique, Island of, 136, 138, 160
Mozambique Channel, 70
Mozambique Current, 63, 65, 77
Mutsamudu, 158, 239, 240, 242

Nacala Bay, 160
Nairobi, 219 *et seq.*
Naso rigoletto, 195
Nationalism, 4
Nature, 42, 49, 53, 185, 188
Nielsen, Dr. Eigil, 199 *et seq.*, 206 *et seq.*, 230
Nostrils, 230
Nostrils, internal, 52
'*Noticias*', 134
Nuffield Trust, 201

Oceanographic investigations, 71, 222 *et seq.*
Odour, 24, 48, 51
Oil, 37, 48, 158, 232, 236, 246
Oil-fish, 155, 194
Old Fourlegs, 64, 231
Oligocene, 14
Ordovician, 14
Organisms new to science, procedure re naming, 45

Palaeontologists' reconstructions, 234, 245
Palaeontology, 11, 13
Paleozoic, 14
Pamanzi, 85, 141, 144, 252
Pamanzi landing strip, 133
Parasites, 246
Parrot-fish, 82
Parrot-fish, giant, 106
Pebane, 137
Pemba, 80, 82
Permian, 14
Phillip, South African Vice-Consul, 133, 163
Photophobia, 240, 242
Piltdown Man, 12
Pinda, 75, 160
Plants, First lowly forms, 14
— Flowering, 14
— Primitive, on land, 14
Pleistocene, 14
Poison, Rotenone, 212, 213
Politics, 4
Ponte de Barra Falsa, 219
Port Amelia, 76
Portuguese, 69, 134, 139
Portuguese Authorities, 72, 73, 79, 218
Portuguese East Africa, 89. *See also* Mozambique
Portuguese officials, 163
Post Office, 100 *et seq.*, 251
Post Office, Durban, 100 *et seq.*
Pre-Cambrian, 14
President, C.S.I.R., 70, 99, 100, 201, 205
Press, 44, 182
Prime Minister, 3, 67, 71, 88, 90, 95, 114, 226, 251
Princeton, U.S.A., 18
Prior, Desmond, 114, 128
Protein, 235
Proterozoic, 14
Protozoa, 14
Pungutiachi Island, 83

Quaternary, 14
Querimba Islands, 80

Radio Station, Lumbo, 138
R.A.F. officer, 189
Ralston, Lieut. D. M., 133 *et seq.*, 164, 165
Reconstructions by Palaeontologists, 234, 245
Reptiles, 14
— giant, 233
Research Fellowship, 69
Reward, 56, 64, 72, 147, 150, 151, 249
Reward, First Coelacanth, 55
— First Offer, 39
— offered by French, 217
— Presentation of, 185
Rhipidistia, 14, 17, 52, 231
Rhodes University College, 6, 72, 119
Rhodesian Centenary Exhibition, 182, 219
Roberts, (Mrs.) Natalie, 87, 169
Rock, age of, 11
Rock Cod, 49, 62, 74, 218, 220
Rotenone, 212, 213
Rovuma, 76
Royal Society, 40
Ruvettus, 155, 194

S.A.B.C., 165
Saccalaves, 77
S.A.C.S.I.R., *see* South African Council for Scientific and Industrial Research
S.A. Government Fisheries Research vessel, 63, 196
Salisbury, 218, 221
Salted sharks, 82
Salting pits, 82
Sauer, The Hon. P., 101, 103, 206
Scale sent to U.S.A., 49, 191
Scales, 61, 158, 159, 229, 231, 244
— in Australia, 53
— in Johannesburg, 53
— of First Coelacanth, 39
Schonland, Dr. B. F. J., 99
Scientific Council for Africa, 219, 222
Scientific Institute of Madagascar, 159
Scientific naming of new species, 45
Scientific work in South Africa, 22
Sea Fishes of Southern Africa, see Fishes, South African, Book on
'Sea Missing Link', 87
Sea-serpent, 29
Seals, 92
Second Coelacanth, *see* Coelacanth, Second
Seychelles Expedition, 216, 224
Shark(s), 82, 84, 92
Sharks, salted, 82

Shearer, Dr. Vernon, M.P., 114 *et seq.*, 128, 247, 251
Shearer, Mrs., 117, 120, 123
Shimoni, 82
Silurian, 14, 15
Size, importance of, 104
Skeleton, 37, 231
Smell, *see* Odour
Smith, G. G., 174
Smith, Mrs. M. M., 80 *et seq.*, 169 *et seq.*
Smith, William, 169 *et seq.*
Smuts, General J. C., 5, 87, 90, 94 *et seq.*, 191
Smythe, Captain Patrick, 84, 90, 125 *et seq.*, 173, 174
Solére, M., 240
South Africa, Early conditions, 22
South Africa, Scientific work in, 22
South African Broadcasting Corporation, *see* S.A.B.C.
South African Council for Scientific and Industrial Research, 66, 68 *et seq.*, 97, 100, 197, 113, 124, 130, 135, 187, 207, 218, 249
South African Museum, 8, 33, 38, 53
South-west African coast, 93
Spiral valve, 184, 231, 246
Stamford Hill, 130
Stellenbosch, 4
Stone-Age man, 14
Stonefish, 160
Stuttard, W. J., 196 *et seq.*, 226
Sunderland flying boat, 89, 107, 129
Sutton, Guy Drummond, 86, 128
Swakopmund, 98
Swamp conditions, 20
Swamps, 17
Swedish expedition, 215
Symons, George (photographer), 116, 170

Tananarive, 159
Tanganyika, 80
Telephone conversation with Dr. Malan, 121
— — with military man in Pretoria, 107

Television, 183
Temperature of water, 240
Tertiary, 14
Tetragonurus, 194
Third Coelacanth, *see* Coelacanth, Third
Time Scale, 11
— — Geological, 14
Times, The, 180, 195
Times, The, article, Appendix D, 243
Torres, Carlos, 72
Transport, Minister of, 103
Trawl, 61
Trawler(s), 9, 43, 91
Triassic, 14, 20
Trustees, Fish Book, 98

Unicorn fishes, 195
Union-Castle Mail Steamship Co. Ltd., 83
University of South Africa, 28
University (Rhodes), 249
Uranium, 11

van Niekerk, Cpl. J. W. J., 133 *et seq.*, 164
Vertebrate fishes, 14
Vertebrates, 15
Vessel, search for, 206
Vice-President, C.S.I.R., 201
Victoria College, 4
Viscera, 183

Walvis Bay, 93
War, 56, 67. *See also* Great War
Water, temperature of, 240
Western Indian Ocean, 154, 224
Western Indian Ocean, fishes of, 223
White, Dr. E. I., 53, 59, 181
White River, 105
William Scoresby, 75
Worthington, Dr. E. B., 219 *et seq.*
Wynberg, 4

Ysterplaats Airport, 171 *et seq.*

Zanzibar, 80, 82, 83, 249, 250

COELACANTH DISCOVERIES

After JLB Smith died in January 1968 the tempo of coelacanth research and conservation escalated sensationally. Dramatic discoveries and far-reaching conservation initiatives followed one another at a furious pace and the coelacanth remained at the forefront of public consciousness and scientific enquiry. The turn of the century yielded some especially exciting developments.

From 1953 to 1971 the French placed an embargo on the foreign acquisition of coelacanths from the Comoros, which lasted until the Comoros gained their independence in 1975. During this period the French scientists Jacques Millot, Jean Anthony and, later, Daniel Robineau published a series of descriptions of the coelacanth, arguably the most detailed anatomical studies of a fish ever carried out.

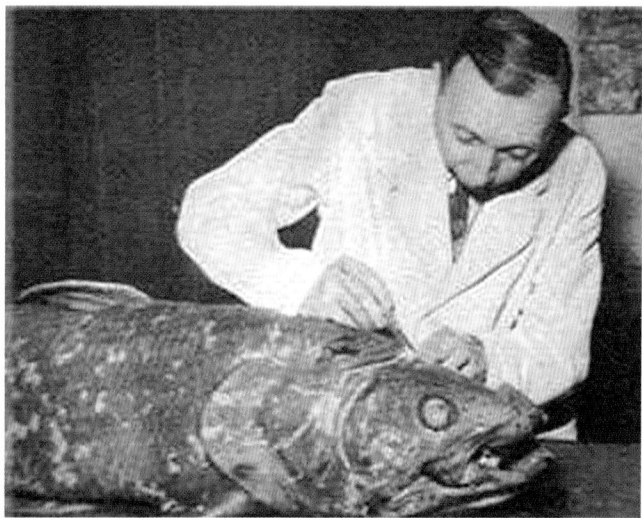

Jacques Millot in Paris, dissecting a coelacanth

In the 1960s and 1970s coelacanth research expeditions to the Comoros were mounted from Britain, France and the USA, some of which attempted to catch live specimens. In 1969 and again in 1975 a youthful Hans Fricke unsuccessfully searched for coelacanths using SCUBA off Madagascar and the Comoros. At the conclusion of his 1975 trip he told his wife, 'Next time I come here, I'm coming with a submarine'. Nobody believed him at the time but his perseverance paid off and he returned to the Comoros in 1986 with the *Geo* research submersible and carried out a series of dives to depths of 200 m off Grande Comore.

A youthful Hans Fricke on an expedition to the Sahara Desert, 1963

Between 1970 and 1990 coelacanths from the Comoros were openly traded and purchased by museums and research institutions worldwide. According to Fricke, for several years the international museum trade offered between US$400 and US$2,000 per specimen, but specimens sold on the black market attracted far more. Scientists and museums also bought coelacanths for pitifully low prices (about $6.00 or less) directly from traditional fishermen in the Comoros.

In 1981 and 1983 Thys van den Audenaerde from the Royal Museum for Central Africa in Tervuren, Belgium, carried out coelacanth research in the Comoros; and from 1981 to 1983 coelacanth research

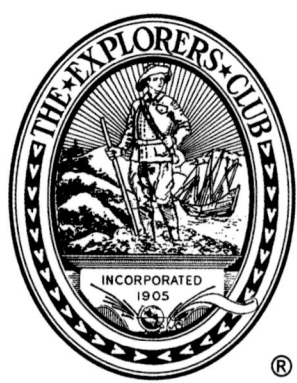

there intensified, with several expeditions by the 'Japanese Scientific Expedition of the Coelacanth' (JASEC), which unsuccessfully tried to catch a live specimen. In 1986 the Explorer's Club from New York and the New York Aquarium mounted further coelacanth research expeditions to the Comoros.

In 1986 Phil Heemstra from the Ichthyology Institute in Grahamstown and Malcolm Smale from the Port Elizabeth Museum mounted an expedition to Grande Comore during which they found that the overexploitation of large reef fishes, and of the giant triton snail, had resulted in the proliferation of long-spined and burrowing sea urchins, which had caused the coral reefs to deteriorate. They recommended the establishment of a marine reserve around the north shore of Grande Comore to allow sea urchin predators to recover and the reefs to recuperate. In December 1986 the Minister of the Interior of the Federal Islamic Republic of the Comoros declared the coelacanth a 'Heritage of Humankind' with Comorians as its custodian.

In January 1987 Hans Fricke and his team located the first live coelacanths off Grande Comore using the *Geo* submersible and, in September 1987, a coelacanth was pictured on the cover of the journal *Nature*, in which their discovery was announced.

Between 1986 and 2008 Hans Fricke, Karen Hissmann, Jürgen Schauer, Rafael Plante and other members of the *Geo* and *Jago* dive teams conducted detailed research on the coelacanth in its natural habitat in the Comoros and later in Madagascar, Tanzania, Indonesia and South Africa, and published a series of ground-breaking scientific articles on coelacanth biology and conservation. During over 200 dives to depths from 200 to 400 m in the *Geo* and *Jago*, their research covered fields as diverse as habitat preferences, home range, population size, conservation status, morphology, physiology, growth rate, movement patterns, social behaviour, breeding biology, locomotion, biogeography and genetics.

They also compiled an extraordinary database of 145 individually recognizable adult coelacanths, based on the pattern of white dots on their bodies.

This database allowed them to track the movements of individual fishes over two decades. Thanks to this team's tenacity, resourcefulness and bravery, and their passion for the coelacanth, it is now one of the best-known fishes in the sea. In June 1987 Hans Fricke showed his footage of live coelacanths at a conference in Grahamstown – to great popular acclaim.

Fricke's research revealed that the population size of adult coelacanths off Grande Comore was low (300–500 individuals) but stable over an 18-year period and that suitable caves in which coelacanths can shelter during the day are rare below depths of 220 m. He concluded that the availability of suitable caves may limit their depth distribution and population size.

Hans Fricke

Hans Fricke's famous photograph of the coelacanth, taken off Grande Comore

Further studies in the late 1980s, 1990s and 2000s by the author, Mike Armstrong, Eugene Balon, John Wourms, James Atz, Hans Fricke, Karen Hissmann, Jürgen Schauer, Olaf Reinicke, Lutz Kasang, Raphael Plante and others on the population characteristics of the coelacanth revealed that it grows slowly (females take at least 20 years to reach 170 cm, possibly much longer) and reaches sexual maturity at about 13 years. Furthermore, it was shown to be a live-bearer that

Hans Fricke

The Jago *submersible in its natural habitat*

produces very few young (2–26) per breeding event, although annual survival rates are high (about 85% for males and 82% for females).

The consensus among researchers was that the catches made by traditional fishermen in the Comoros do not represent a significant threat to coelacanth populations, as natural mortality, mainly from predation, is a far more important factor limiting population size. Overfishing of coelacanth prey fishes, which is probably occurring off the west coast of Grande Comore, where many coelacanths live, could, however, affect food availability and growth and survivorship rates. Reducing the number of large sharks, which prey on coelacanths, may contribute to their conservation. The use of large-mesh, deep-set gillnets, and especially of explosives, in Tanzania, Madagascar and the Comoros, is a major threat to coelacanths. If these irrational fishing methods are not controlled they could have disastrous consequences.

In April 1987 the Coelacanth Conservation Council/ Conseil pour la Conservation du Coelacanthe (CCC) was founded in Moroni, Grande Comore, during an expedition led by the author and including coelacanth scientists Eugene and Christine Balon and Richard Cloutier. The CCC was modelled on the Desert

Fishes Council that helped to save desert pupfishes from extinction. The CCC played a major role in promoting and co-ordinating coelacanth research and conservation efforts worldwide and introduced the 'CCC numbers' that are still used internationally to sequentially number coelacanth specimens that are caught. CCC newsletters were published regularly in the journal *Environmental Biology of Fishes*, edited by Eugene Balon, from early 1988 until 1999. These newsletters included reports on coelacanth research and conservation as well as (from 1991) updates of the coelacanth inventory and bibliography.

One of the tasks of the CCC was to compile a list of every coelacanth specimen caught anywhere in the world. This seemingly impossible task was achieved through a network of collaborators in museums and universities around the world, on the Western Indian Ocean islands, on the East African mainland, and later in Indonesia, and was co-ordinated by a succession of dedicated scientists and archivists.

The coelacanth inventory was started by Jacques Millot, Jean Anthony and Daniel Robineau in 1972 (for catches from 1938–1971), continued by John McCosker (1972–1977), Suzuki & Tanauma (1959–1983) and then by the author, Sheila Coutouvides and Jean Pote (1984–1996, until specimen CCC 175). Since then, the inventory has been updated and extensively

Eugene Balon

The author with a gyotaku *(traditional Japanese fish print) of the coelacanth in Mutsamudu, Anjouan, April 1987*

300

developed by Rik Nulens in Maaiseik, Belgium, in collaboration with Lucy Scott from the UNDP GEF ASCLME project in Grahamstown and Marc Herbin from the Muséum National d'Histoire Naturelle de Paris. They published the updated inventory through SAIAB in 2011 to specimen number CCC 285 (plus 12 additional uncatalogued specimens for which the date of capture is unknown) and the list has since been extended by Nulens to CCC 314. Rik Nulens is a mine of information about the coelacanth for coelacanthophiles worldwide.

In 1989 a decree by the President of the Comoros, Said Johar, introduced control over trade in coelacanths and stipulated that all specimens caught must be sold to the government. In May 1989 the author led an international expedition to the Comoros to investigate the conservation status of the coelacanth. The Comorian President presented the Ichthyology Institute with a coelacanth specimen in recognition of its contribution to environmental conservation in the Comoros.

In September 1989 the *Pacific Oak* left Cape Town for the Comoros where it acted as mothership during a further Japanese attempt to catch three live coelacanths for the Toba Aquarium in Japan. During this expedition

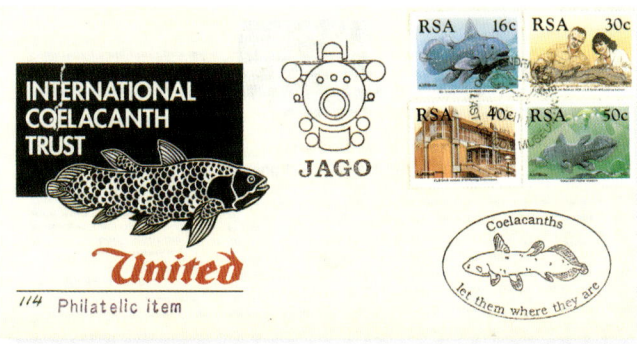

Commemorative cover for the International Coelacanth Trust with the four 1989 South African postage stamps featuring the coelacanth and related images

Cover of the coelacanth inventory produced by Rik Nulens, Lucy Scott and Marc Herbin

Hans Fricke and his team secretly placed a message in the Japanese coelacanth trap using the *Jago* submersible. The message read, 'Let them be where they are'. The shocked crew upped anchor and left the Comoros the next day. Their attempts to catch a live coelacanth were therefore unsuccessful but they did obtain valuable footage of live coelacanths at 184 m using remote-controlled cameras.

In the 1990s Hans Fricke and the author, as well as Jerry Hamlin from

the Explorer's Club (and many other scientists and conservationists) mounted a worldwide campaign to conserve and de-commercialize the coelacanth. Eventually, *Latimeria chalumnae* was elevated to Appendix I of CITES (which means that it cannot be traded for financial gain) and to 'Endangered' on the IUCN Red Data List.

On 16th May 1990, during a coelacanth research expedition to the Comoros led by the author and including Eugene Balon, Richard Cloutier, Robin Stobbs and the artist Roy Reynolds, a report entitled 'Récommandations sur la Conservation marine dans

Karen Hissman painting anti-Pacific Oak graffiti on a wall in Moroni, Grande Comore, in 1989

Members of the May 1990 Coelacanth Expedition (from left): Robin Stobbs, Mike Bruton, Richard Cloutier and Eugene Balon

Illustration by Roy Reynolds

la République Fédérale Islamique des Comoros' was handed to the Comorian Minister of the Interior, M. Ali Moroudjae, and the Director of the Centre National de Documentation et de Recherche Scientifique des Comoros (CNDRS) in Moroni, Grande Comore. Many of its recommendations were subsequently implemented.

In late 1990 the International Coelacanth Trust (ICT) was established in South Africa to raise funds for the first phase of dives by the *Jago* submersible off South Africa. The first expedition by the *Jago* to South Africa took place in May 1991, organized by the author. Fish communities and bottom topography were examined in the Tsitsikama Coastal National Park and off Port Elizabeth and East London. These dives revealed that this part of the South African coast does not appear to offer suitable habitat for coelacanths but also demonstrated that a manned submersible could successfully be deployed off our treacherous coast. In 1991 scientists from the Virginia Institute of Marine Science reported that coelacanths off the Comoros carry significant burdens of toxic insecticides, such as the organic chlorine compounds PCB and DDT, in their tissues.

On 11th August 1991 a large female coelacanth weighing 98 kg was caught off Pebane on the north coast of Mozambique by a side trawler, *Vega 13*. It contained 26 late-term pups. The specimen and its pups were studied and reported in the scientific literature by the author (Mike Bruton), Augusto Cabral, Director of the Museu de Historia Natural in Maputo, Mozambique, and Hans Fricke.

A coelacanth from the Comoros dissected at the Ichthyology Institute in Grahamstown on 12th December 1991 was found to contain 67 eggs. The chromosomes of this coelacanth were described in 1992 by James Bogart, Eugene Balon and the author in 1994. In 1991 the most comprehensive book on the coelacanth, *The Biology of* Latimeria *and Evolution of Coelacanths*, edited by Jack Musick, Mike Bruton and Eugene Balon, was published by Kluwer Academic Publishers through the journal *Environmental Biology of Fishes*. It included 27 chapters by 43 authors.

In 1992 Hans Fricke established a Coelacanth Tissue Bank at the Max Planck Institute in Munich to preserve coelacanth tissues for detailed analysis. Over 30 research institutes have used this facility. In September 1994 the Comorian President, M Djohar, signed the CITES Convention that forbids trade in coelacanths.

In November 1995 an Association for the Protection of Gombessa (APG) was established in Dzahadjou, Grande Comore, and a Coelacanth Centre was set up in a small house. In 1996 the ability of the *Jago* submersible to operate off South Africa's exposed coast was further tested with

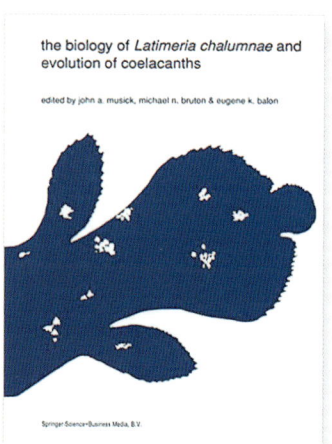

The Biology of Latimeria chalumnae and Evolution of Coelacanths, published in 1991

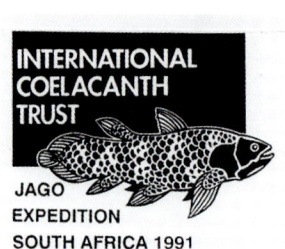

The Indonesian coelacanth

On 5th August 1995 the first coelacanth known to Western science from Madagascar was caught in a baited net off Anakao. Then, to everyone's surprise, the first specimen of a different species of living coelacanth was caught off Manado in North Sulawesi Island, Indonesia, on 18th September 1997. It was recognized in a fish market by American marine biologist, Mark V Erdmann, and his wife, Arnaz Mehta.

On 30th July 1998 a second specimen of the Indonesian coelacanth was caught, also in a bottom-set gillnet, and Erdmann and Mehta were able to observe its swimming behaviour. On 6th October 2009 Japanese scientists filmed a live coelacanth juvenile at a depth of 160 m off North Sulawesi Island in Indonesia.

The Indonesian coelacanth was found to be a different species and, in 1999, was named *Latimeria menadoensis* by a group of Indonesian scientists (L Pouyaud, S Wirjoatmodjo, I Rachmatika, A Tjakrawidjaja, RK Hadiaty and W Hadie) that did not include Mark Erdmann, even though he had been the first to identify the Indonesian coelacanth.

Coelacanth stamp issued by Madagascar in 1989

The Indonesian coelacanth Latimeria menadoensis

The Indonesian coelacanth, Latimeria menadoensis

a series of dives off the Cape Point Nature Reserve and Table Bay, sponsored by De Beers Marine. Its ability to operate under harsh South African conditions was now further proven.

Jerry Hamlin created a website for the coelacanth (www.dinofish.com) in 1996, which serves as a message board and discussion forum on coelacanth conservation and research (the website www. coelacanth is owned by an engineering company in the USA). The site received 6,494 hits in one day when the Indonesian coelacanth was discovered. Through the website Jerry also launched a 'Deep Release' project that raises money for, and distributes, kits that Comorian fishermen can use to release coelacanths accidentally caught.

In 1996 the pros and cons of keeping a live coelacanth in captivity were discussed at an international zoo and aquarium conference, and in July 1997 the author gave a talk at the Two Oceans Aquarium in Cape Town entitled 'Should we keep a coelacanth in captivity?'

On 28th October 2000, mixed-gas divers Pieter Venter, Peter Timm and Etienne le Roux discovered coelacanths living at a depth of 104 m in Jesser Canyon at Sodwana Bay in the newly proclaimed iSimangaliso Wetland Park (formerly the Greater St Lucia Marine Reserve) in Maputaland, South Africa – the shallowest colony of coelacanths that had been discovered so far. On 27th November 2000 they filmed three coelacanths at a depth of 108 m.

Logo of the 1996 Coelacanth Expedition to South Africa

Jerry Hamlin

Coelacanth 'Quick Release Kit' developed by Jerry Hamlin

On 26th April 2001 the first coelacanth known from Kenya, a female with 17 eggs, was caught off Malindi. This was the third coelacanth caught by a trawler (after the first specimens from South Africa and Mozambique).

On 8th September 2003 the first specimen from Tanzania was caught in a deep-set gillnet (at a depth of about 1,000 m). Between 2003 and 2008 over 80 coelacanths were caught off Tanzania (possibly more) at a rate of 13 per year, far higher than the catch rate anywhere else. On 17th August 2008, one group of fishermen caught six coelacanths off Mtwara in Tanzania in one day! In 2003 the Tanga Coelacanth Marine Park was proclaimed off the mainland coast of Tanzania to conserve coelacanths but deep-set gillnetting for sharks (which kill coelacanths as a bycatch) continues within the Park.

In response to the discovery of live coelacanths off South Africa and with the objective of initiating a new phase of research on the coelacanth and its habitats, the African Coelacanth Ecosystem Programme (ACEP) was established, with SAIAB as the lead organization. It was officially launched by the then Minister of Science & Technology, Dr Ben Ngubane, in April 2002, at Sodwana Bay. ACEP 1 was co-ordinated by Tony Ribbink (2001–2006), ACEP 2 by Tommy Bornman (2007–2011) and ACEP 3 (2012–2015) was led by the CEO of SAIAB, Angus Paterson.

From 2002 to 2004 the *Jago* team, operating from the mothership *R/V Algoa*, carried out 47 survey dives with a total bottom time of 166 hours at depths ranging from 46 to 359 m. Twenty-four different coelacanths

NRF-SAIAB

In 2011 the Sea-Eye Falcon remote-operated vehicle owned by SAIAB took spectacular photographs of coelacanths in the iSimangaliso Wetland Park. The data shows the depth, date and time of the dive.

were identified in three submarine canyons at depths from 96 to 133 m along a 48-km stretch of coast in the iSimangaliso Wetland Park. It was estimated that coelacanth populations in the Park are 'relatively small'.

A pregnant coelacanth caught off the coast of Tanzania in December 2007, containing 23 fully-developed juveniles

From the outset ACEP 1 was a multi-disciplinary programme. Acoustic bathymetry was used to determine the location of caves in which coelacanths might hide, and physical oceanographers monitored water temperatures, oxygen concentrations and ocean-current strengths to identify the most likely places to find coelacanths. Marine biologists made collections of the animals that shared the habitat of coelacanths.

Tissue samples were collected for stable isotope analyses to determine feeding relationships and energy flows between predators and prey. Tissues for genetic and genome studies, and for growing coelacanth cells in the laboratory, were collected from live coelacanth scales extracted underwater by scientists in the *Jago*. The data collected by taxonomists, oceanographers, zoogeographers, ecologists, geneticists and resource managers were brought together for analysis in a Geographic Information System (GIS).

Over time, ACEP 1 extended northwards into other countries on the east coast of Africa and the Western Indian Ocean islands. Further data on coelacanths, their habitats and the communities in which they live was collected, and national management committees were set up in Mozambique, Tanzania, Kenya, Mauritius, Madagascar, the Comoros and the Seychelles. These committees managed the research within their borders, and made arrangements for the *R/V Algoa* to visit their waters and for their nationals to participate in some cruise legs. Other countries that participated in ACEP 1 included Indonesia, Japan, Canada, France and Belgium.

In 2003 an international conference on the coelacanth, organized by ACEP, was held in the East London Museum. Fourteen papers from the conference were published in the *South African Journal of Science* in 2006 and a further 17 papers were published in the conference proceedings.

Poster for the 2002 Coelacanth Expedition in the iSimangaliso Wetland Park

In 2005, after research cruises involving many partner countries, ACEP 1 was recognized as a NEPAD 'Coastal and Marine Programme'. The programme was also extended to include the environmental education of schoolchildren who

Coelacanths were found in 2000 near Jesser Point at Sodwana Bay in the iSimangaliso Wetland Park.

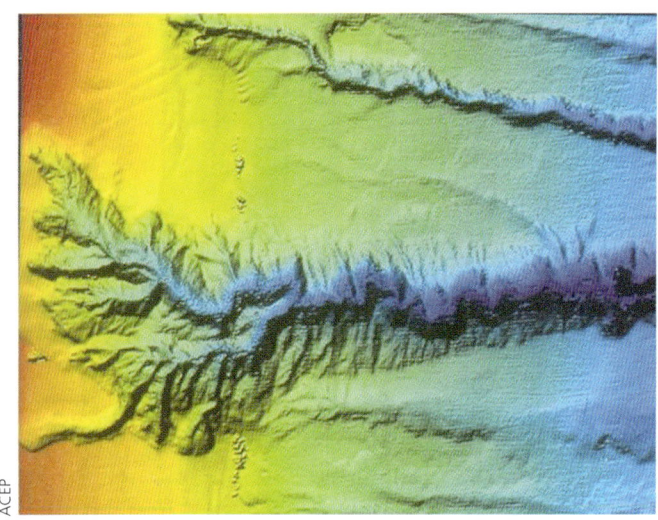

ACEP

Colour-draped bathymetric map of Leven Canyon, iSimangaliso Wetland Park (SAIAB ACEP project)

explored the ship, participated in educational activities onshore and occasionally joined the research cruises. ACEP 1 also laid the foundation for the launch of the GEF-funded United Nations Development Programme (UNDP) 'Agulhas and Somali Current Large Marine Ecosystems' (ASCLME) project, which was also hosted by SAIAB.

In 2009 Hans Fricke continued his coelacanth research in the Comoros using the motor yacht *My Octopus* belonging to the co-founder of Microsoft, Paul Allen, and sophisticated remote-operated vehicles that could penetrate to depths of more than 1,000 m. They found that ecological conditions were less suitable for coelacanths in deep water but their research was cut short by threats of piracy.

ACEP 2 was officially launched in Port Elizabeth on 25th March 2010. Research was carried out in collaboration with the UNDP-facilitated ASCLME project and included 30 research cruises to the participating African countries in ACEP 1, as well as to Somalia and various seamounts using the *R/V Algoa* and the Norwegian vessel *R/V Dr Fridtjof Nansen*. Through ACEP 2, SAIAB also acquired a coastal research boat *R/V uKwabelana* ('to share' in isiXhosa), as well

as a Sea-Eye Falcon Underwater Remote-Operated Vehicle (ROV). The ROV was launched in 2009 and was an immediate success, with seven coelacanths being filmed on two days of diving off Sodwana.

Fifteen different research institutions participated in ACEP 2 activities, resulting in an unprecedented level of interdisciplinary ecosystem research in the Southwestern Indian Ocean. The research disciplines covered included genetics, conservation, education, oceanography, palaeo-climates, biodiversity and data management, and 37 students completed their postgraduate degree research through ACEP 2.

In March 2011 a new 'Coelacanth Centre', built by Comorians and funded by Jerry Hamlin in New York, was opened in Itzounsou on Grande Comore. In September 2013 the author gave a second talk at the Two Oceans Aquarium, during which he proposed that the time had come for the international ichthyology community to seriously consider the benefits of displaying and studying a coelacanth in captivity in a public aquarium. In April 2015 the author published his autobiography in which he weighed up the pros and cons of studying a live coelacanth in captivity. He proposed that a coelacanth caught accidentally in deep-set gillnets off Tanzania should be kept alive for captive study in a public aquarium. In support of this changed stance, he argued that coelacanths are now known to be more widespread than previously thought and that coelacanth conservation methods have been successfully implemented worldwide. Furthermore, the catch rate in their main habitat, the Comoros, has dropped sharply.

Proceedings of the 2003 ACEP Coelacanth Conference, held in East London

Wikipedia CC BY-SA 3.0

Paul Allen's superyacht, My Octopus

As our knowledge of the physiology, behaviour and habitat preferences of coelacanths is excellent, we should be able to develop successful captive-breeding techniques for them. The author further argued that important aspects of coelacanth biology and behaviour that are vital for their conservation, such as their genetics, growth and breeding rates, and whether or not they provide parental care to their young, can only be fully understood through captive study – a proposal that had been opposed in the summary report on ACEP published in 2006.

In early 2012 ACEP 3 established the *Phuhlisa* programme, a major initiative that aims to transform the marine science landscape in South Africa by building capacity at historically Black universities. *Phuhlisa* was hugely successful, with 24 Honours students, eight MSc students and two PhD students receiving academic, professional and financial support through workshops, lectures and research supervision. ACEP also established a highly effective Environmental Education Programme that produced an outstanding range of educational resources, engaged with over 450,000 learners and 72,278 teachers, and raised awareness of marine conservation issues among the general public.

In 2012 Rosanne Thornycroft of Rhodes University used an image recognition method to identify individual coelacanths based on their patterns of white dots and other features. Kerry Sink from the South African National Biodiversity Institute manages the catalogue of individual coelacanths sighted off Maputaland (which now numbers 32 individuals) and has also pioneered the use of ROVs for researching coelacanths and their deep-water habitats. The total number of coelacanth encounters in the undersea canyons of the iSimangaliso Wetland Park now stands at 108, of which 46 were made by mixed-gas diver, the late Peter Timm, who had been diving with coelacanths for 13 years when he passed away while diving in 2014.

At the inauguration of the iSimangaliso Wetland Park, President Nelson Mandela said, 'iSimangaliso must be the only place on the globe where the oldest land animal (the rhinoceros) and the world's biggest terrestrial mammal (the elephant) share an ecosystem with the world's oldest fish (the coelacanth) and the world's biggest marine mammal (the whale)'.

Bathymetric research conducted along the East African coast through ACEP revealed that the highest density of submarine canyons that are likely to provide ideal habitat for coelacanths is found along a 222-km stretch of coast off northern Mozambique between Olumbe and Porto Amelia. In fact, Peter Ramsay and Warwick Miller have suggested that the founder population of Indian Ocean coelacanths may be located there. Other submarine canyons that might be suitable coelacanth habitat have been located off Tanzania and Madagascar.

The results of an ambitious programme initiated by Rose Dorrington and Greg Blatch of Rhodes University to sequence the genome of *Latimeria chalumnae* was published in a paper, co-authored by 99 scientists, in *Nature* in 2013. The DNA and RNA samples were obtained from a coelacanth caught off Hahaya on Grande Comore by a traditional fisherman on 18th September 2003 and were preserved by Sahid Ahamada, who had been trained by Dorrington of the Association pour la Protection du Gombessa. The size and complexity of the coelacanth genome meant

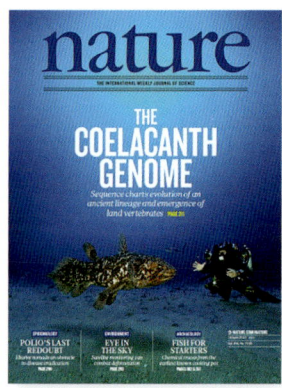

The analysis of the coelacanth's genome was reported in Nature *in April 2013.*

that a team of scientists from 40 institutions around the world toiled for 10 years to crack the code.

When it was finally deciphered, analysis of the coelacanth genome showed that its genes were evolving more slowly than those of other animals, presumably because it has been subject to less selection pressure, which explains its prehistoric appearance. The analysis also confirmed that the lungfish is closer to the ancestry of tetrapods than the coelacanth. However, the genome of the lungfish is so large that the sequencing of it is not possible with present technology, so the coelacanth genome provides the most complete record available for further studies of tetrapod evolution. These studies also revealed that the African and Indonesian coelacanths diverged from one another between 40 and 30 million years ago.

In 2013, as part of ACEP 3, an international coelacanth research expedition was launched to the iSimangaliso Wetland Park off Maputaland. This Park, an expansion of the Greater St Lucia Marine Reserve originally established in 1895, was declared South Africa's first World Heritage Site in December 1999. The discovery of coelacanths in the Park led to its rapid development and the promulgation of stricter fishing and diving regulations.

Mixed-gas and ROV diving techniques were used on the 2013 expedition and the team included the award-winning underwater photographer Laurent Ballesta and his divers from *Andromède Océanologie*. The aim of the expedition was to address key research

NRF-SAIAB

Trainees during the ACEP Phuhlisa *programme*

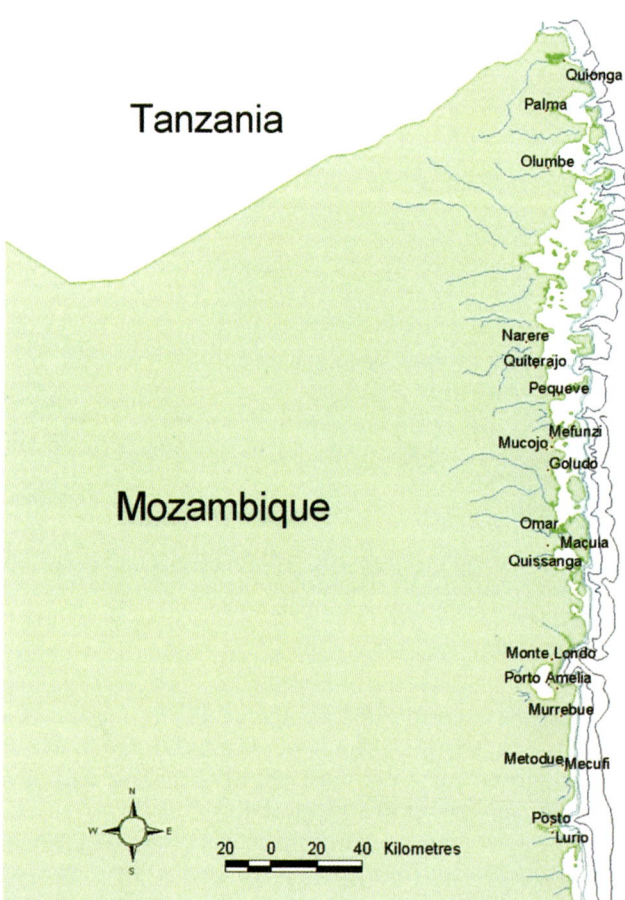

Bathymetric map of the canyons off the coast of northern Mozambique that might be suitable habitat for coelacanths

NRF-SAIAB

Dr Angus Paterson, Managing Director of SAIAB, with a coelacanth on the 75th anniversary of the first capture

priorities that had been identified in the Coelacanth Management Plan published by ACEP in 2004, including an improved understanding of coelacanth distribution, habitat and population dynamics.

During this expedition several coelacanths that had been identified before, and given names, were located again, including 'Jessie' (the first coelacanth seen in 2000, now sighted in eight different years), 'Taggi' (a fish that had swum at least 5 km between Jesser and Wright Canyons, and was last seen in 2003), and 'Sydney' (named after the pioneering South African molecular biologist Sydney Brenner). A new coelacanth, 'Tot', was also sighted and is the smallest observed so far (about 105 cm). A new non-invasive technique was also developed to collect swabs from the skin of coelacanths. These tissue samples will be used to study, for the first time, kinship relationships between individual coelacanths.

New species, ecosystems, technologies, research results and scientists have emerged from the discovery of coelacanths off Sodwana and ACEP. The coelacanth has furthermore helped South Africa to assume a position of leadership in Southwestern Indian Ocean research and Marine Protected Area development, and an exciting range of further multi-disciplinary expeditions is now being planned, including one to the site of capture of the first coelacanth off East London.

The coelacanth story has come full circle and continues to enthral scientists and non-scientists alike.

Debbie Zaloumis

Kosi Bay Estuary in the iSimangaliso Wetland Park

Rob Gess

Reconstruction from fossils of the extinct South African estuarine coelacanth, Serenichthys kowiensis, *discovered by Rob Gess of the Albany Museum, Grahamstown*

THE SIGNIFICANCE OF THE COELACANTH

Assigning intrinsic values to rare and significant animals has become an important ingredient in the modern conservation message. JLB Smith wrote *Old Fourlegs* because he felt strongly that the coelacanth is 'a priceless heritage from the past'. His view has been reinforced by subsequent research and conservation developments and is worth re-appraising here.

I doubt that there is any fish with a higher public profile than the coelacanth, even the popular commercial species such as menhaden, elf, kob, cod, mackerel, bass, salmon or trout. Other fishes may claim to be more dangerous (stonefish), beautiful (chaetodons), uglier (blobfish), faster (sailfish), and even warm-blooded (marlin), but none has the suite of characters of the coelacanth.

In comparison with 'old fourlegs', even the charismatic megaherbivores (elephant, rhinoceros, hippopotamus, giraffe, buffalo and large antelope), and the large cats and great whales, are relatively humdrum in terms of their evolutionary importance, anatomy and behaviour. Some extinct species, such as the dinosaur *Tyrannosaurus rex,* may have a stronger 'brand'; and the duck-billed platypus and tuatara are extraordinary animals, as are the velvet worm, horseshoe crab, nautilus, lamprey and aardvark; but the coelacanth enjoys a mystique of its own.

Why is the coelacanth so significant? Below are some of the many reasons.

Natural history and importance to science

Enormous longevity: The coelacanth fossil record can be traced back at least 420 million years, the longest of any vertebrate, and this date is constantly being pushed further back as new discoveries are made. This means that they evolved at least 170 million years before the dinosaurs appeared and then outlived them by surviving the deadly Cretaceous extinction event 80–65 million years ago. Strangely, no coelacanth

Fossil of the coelacanth Rhabdoderma exiguum. *The coelacanth fossil record can be traced back 420 million years*

fossils have been found dating from the years since then, which is why it was such a surprise to find living species. Even if living coelacanths had not been discovered they would have been regarded as amazing creatures that are near the roots of tetrapod evolution.

The ultimate survivor: We therefore know that coelacanths survived four major extinction events (about 373, 252, 208 and 65 million years ago) that wiped out about 70%, 90%, 75% and 75% respectively of all animal species – a remarkable feat of perseverance that is equalled by few animal groups. During this huge span of time coelacanths coexisted with an array of ferocious predators in marine and freshwaters; they also endured epochal changes in marine and freshwater habitats caused by continental drift, major changes in sea level, and the formation, breaking up and reshaping of vast land masses, ocean basins and river catchments.

They outlived 15 m-long dolphin-like reptiles in the Triassic Period, giant, voracious pliosaurs during the Jurassic and 6 m-long bony fishes (*Xiphactinus*) in the Cretaceous. Freshwater coelacanths had to contend with enormous semi-aquatic dinosaurs (such as *Spinosaurus*) and 12 m-long prehistoric crocodiles (*Sarcosuchus*). They also, of course, outlived over 800 species of land-based dinosaurs, including the legendary *T. rex,* and a formidable array of brachiosaurs, triceratops, stegosaurs and velociraptors.

Conservative anatomy: The coelacanth has evolved extremely slowly in some respects (although in other respects, such as their breeding strategy, they have evolved quickly and are quite advanced), and the modern coelacanth shares many anatomical features with its extinct relatives that lived over 375 million years ago, such as *Diplocercides*. These features include the rostral organ, intracranial joint, paired fins, hollow notochord and teeth. The earliest fishes, which evolved over 500 million years ago, had no jaws, teeth or fins and sucked up their prey, like their modern descendants, the lampreys and hagfishes.

Evolution of advanced features: Coelacanths were the first animals to evolve many advanced features, such as bony skulls, jaws and teeth, armoured scales, lobed fins and live-bearing of the young.

Relationship to four-legged animals: In terms of cladistics, whereby animals in the same taxonomic group must have a single common ancestor, coelacanths (and their close relatives, the lungfishes) are more closely related to the four-legged animals on land (and in the sea) than to fishes. If we take the 'clade' further back in time, then tetrapods and all bony fishes (lobe-finned and ray-finned fishes) form a natural group that does not include the cartilaginous fishes (sharks, skates and rays).

Elucidating the transition from water to land: Few scientists have proposed that coelacanths are the direct ancestors of tetrapods but most believe that they are our ancestors' 'first cousins'. However, none of the primitive lungfish or amphibian relatives that are closest to our roots have survived, so evolutionists who try to reconstruct the water-to-land transition are forced to compare modern and extinct coelacanths (and lungfishes) with the fossil record. The conservative anatomy of the coelacanth makes it a particularly useful model as it still looks similar to coelacanths (such as *Diplocercides* and *Nesides*) that were alive when the epic water-to-land transition first took place. If the coelacanth lineage had diversified as much as, say, the ray-finned fishes, they would be far less useful in this regard.

As a 'living fossil': Charles Darwin coined the term 'living fossil' in *The Origin of Species* (1859) to describe living species that have changed so little over time that they resemble extinct species closely enough to appear to be identical. Although the term is an oxymoron, and coelacanths are not relics from the past that have been brought back to life, it does have some explanatory value, given that a group of animals can be known from fossils as well as from living representatives.

The coelacanth is probably the most famous 'living fossil' but there are many other examples, including lampshells and horseshoe crabs (whose lineage dates back over 445 million years), various beetles, wasps, lobsters and mantis shrimps, nautiluses, velvet worms, some sharks, hagfishes, lungfishes, bichirs, paddlefishes, sturgeons, gars and bowfins, as well as crocodiles, platypuses, echidnas and aardvarks.

As a 'missing link': Peter Forey (1990) stated, 'So many of the features of *Latimeria* are unlike those expected in a tetrapod ancestor and, in this sense, the coelacanth has not lived up to its reputation as a "missing link".' However, research on the coelacanth has unquestionably taught us more about that crucial evolutionary step from water onto land, so it has served the purpose of a 'missing link', even though it is not in the direct line of four-legged animal evolution.

As a 'Lazarus species': Another epithet for the coelacanth is a 'Lazarus species', an evolutionary line that disappeared from the fossil record for a long period only to re-appear either as a fossil or a living species. Uniquely, coelacanths were known from their fossil record for a long period (99 years) before a living representative was discovered (the living lungfish was described nine years after its first fossils were found). Another Lazarus species among fishes is the Kokanee, a Japanese salmonid fish that was thought to have gone extinct in 1930 but was rediscovered in 2010. Other examples include various ants, snails, toads, island birds such as the Chacoan peccary and takahe, rats and possums.

IrinaK/Shutterstock.com

Kokanee salmon (Oncorhynchus nerka*) in its spawning colours*

Superb window into the past: Because of its conservative external anatomy, which has changed little over hundreds of millions of years, the living coelacanth provides a spectacular window into the past. Palaeontologists have mused, if the hard anatomy revealed by the fossil record is so unchanging, would the soft anatomy, physiology and behaviour also be as uniform over time? The answer is, 'No' as the coelacanth's soft anatomy, physiology and behaviour have evolved significantly over time and, in some instances, rival those of modern fishes in terms of their modernity.

Coelacanth anatomy

We have learned, though, that interpretations from modern to ancient coelacanths must be made with caution. *Latimeria chalumnae*, for example, is a specialized fish designed to live over marine reefs and canyons and differs from many of its ancient predecessors that lived in shallow swamps or estuaries. For instance, a recently discovered 400 million-year-old coelacanth fossil, *Shoshinia arctopteryx*, found in Wyoming by scientists from the University of Chicago, has an arrangement of bones in its fins that matches the asymmetrical fin patterns of primitive, but living ray-finned fishes, such as sturgeons, paddlefishes and sharks, but differs from that of the living coelacanth. The limbs of *S. arctopteryx* have more in common with those of four-legged animals than with the fins of living coelacanths. In another group of extinct coelacanths in the genus *Laugia* from Greenland the pelvic fins have moved forwards and the pectoral fins upwards into a similar position to the fins of highly-adapted modern ray-finned fishes, suggesting that these 220 million year-old coelacanths were capable of complex manoeuvres.

What the living coelacanth has taught us is that an ancient lineage of animals can retain a relatively conservative external anatomy yet develop advanced physiological and behavioural responses to the challenges faced by modern animals.

Unique internal anatomy: Coelacanths have a combination of the anatomical features of bony fishes and cartilaginous fishes, as follows:

Characters shared with bony fishes	Characters shared with cartilaginous fishes
Bones	Rostral organ
Bony fin supports	Hollow, cartilaginous notochord
Scales	Spiral valve in intestine
Symmetrical tail	Brain structure
Lateral line	Haemoglobin insensitive to urea
Presence of intracranial joint	Presence of a rectal gland
Many details of the bony head	Low-density fat in the swim bladder
Structure of the nostrils	Structure of pituitary gland and pancreas
Low amount of DNA content in cells	Structure of the thymus and thyroid
Gill structure	Large eggs and live-bearing
	Fatty liver
	Osmoregulation by urea retention
	Ability to synthesize urea in the liver
	Structure of the eye
	High salt levels in the blood
	Nitrogen-rich nature of the blood

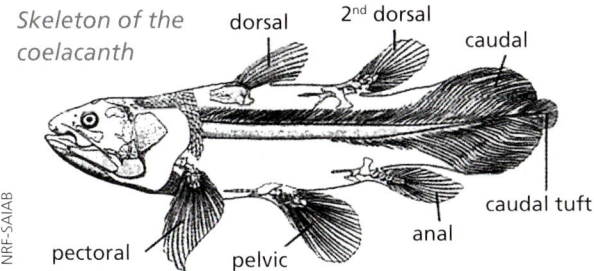

Skeleton of the coelacanth

dorsal · 2nd dorsal · caudal · caudal tuft · anal · pelvic · pectoral

NRF-SAIAB

Coelacanths also share some physiological and anatomical features with tetrapods, such as the structure of the inner ear (with birds) and a slow metabolic rate (with some amphibians).

They also have several anatomical features of their own that appear to be 'out of sync' with the rest of their anatomy, such as ventral gill muscles, absence of the principal upper jawbone (maxilla), teeth only on the palate and lower jaw bones, and a primitive brain. Furthermore, in contrast to other fishes, their heart chambers are arranged in a straight line rather than curled. According to Peter Forey, some extinct coelacanths even had lungs, like the modern lungfish. They are also the only living animals that are able to lower their bottom jaw *and* lift their upper jaw to create a wide gape.

Their scales are elongate and armoured and uniquely have nano-sized optical structures on their tubercles that may improve the visibility of their white patches, a novel way of harvesting light in the ocean's twilight zone! Another unique feature is that their skeletons comprise both bony (skull, lobe-fin supports) and cartilaginous (notochord) elements. Their lobed fins are shared only with living lungfishes, extinct osteolepiform fishes and extinct forms that are transitional between fishes and tetrapods, dubbed 'fishapods', like the remarkable fish-like amphibian, tiktaalik. Coelacanths, far from being degenerate, are therefore advanced, though ancient, fishes.

Unusual external anatomy: Coelacanths have large scales that overlap three times in any one place on the body to create a strong but flexible body armour.

Eugene Balon

The tapetum lucidum *in a coelacanth's eye*

Like crocodiles, bats, kangaroos, kiwis and some cats, coelacanths have a reflective layer, the *tapetum lucidum*, behind the retina in their eyes. It reflects visible light back through the retina, thus increasing the light available to the photoreceptors, and makes it easier for them to see objects at low light intensities. This reflective layer may explain why coelacanths are so sensitive to bright light. They also have many rod cells in their retinas for seeing in low light.

www.leonexus.com

Extinct giant coelacanth, Mawsonia gigas, *on display in the National Geographic Museum's* Spinosaurus *exhibit*

Size: Coelacanths are large fishes, with females reaching 179 cm and 98 kg. An extinct species of coelacanth in the genus *Mawsonia*, which lived about 100 million years ago, reached a length of 6 m.

Unique fins: They have a unique array of eight fins comprising two pairs of paired fins (pelvic and pectoral) and four single fins in the midline (two dorsals, tail and anal). The first dorsal fin (which can be folded down) and tail fin are fan-like and rayed, whereas all the other fins, including the second dorsal and the anal fins in the

The rayed first dorsal fin of the coelacanth

midline, are lobed and have well-developed, muscular, limb-like bases. The lobed fins are able to bend and rotate like paddles. The tail fin has a unique middle lobe and upper and lower lobes. No other fish has this peculiar arrangement of fins.

Distinctive swimming action: During normal swimming the paired fins move in an alternating pattern, similar to the leg movements of a four-legged animal such as a trotting horse. The coelacanth also sculls its second dorsal and anal fins together, slowly to one side then the other, similar to the swimming action of the ocean sunfish, *Mola mola*, which paddles its median dorsal and anal fins back-and-forth in unison.

Colour: The African coelacanth, *Latimeria chalumnae*, is iridescent blue in colour with distinctive pink-white patches, whereas the Indonesian species, *L. menadoensis*, is brown with white patches and has brilliant gold flecks over its upper body and fins.

The remarkable ocean sunfish, Mola mola, *has a swimming action similar to that of the coelacanth.*

Coelacanth biology and behaviour

Idiosyncratic respiration: Coelacanths have the lowest known haemoglobin count of any fish and a very small gill surface area, which means that they have a restricted ability to absorb oxygen and must live in well-oxygenated waters.

Metabolic rate: Coelacanths have the lowest values for resting oxygen consumption of any fish and the slowest metabolic rate of any known vertebrate. They usually move very slowly and efficiently, drifting with underwater currents, although they are capable of quick, rapid movement when catching prey or avoiding danger.

Different demography: Coelacanths are large reef-dwelling fishes, reaching at least 179 cm and 98 kg in females and 165 cm and 65 kg in males. Only the giant Napoleon wrasse, potato basses and groupers, among reef fish, match this size. Although they have a low metabolic rate, coelacanths grow quickly, averaging about 6.5 cm per year and reaching about 170 cm (in females) by their 20th year. The latest research suggests that they may live to over 100 years, an age exceeded only by Greenland sharks, some Antarctic fishes and a few others.

Hans Fricke and Karen Hissmann have estimated that there are between 300 and 500 adult coelacanths living off Grande Comore, which may be the world's largest population, although population sizes in the submarine canyons off Tanzania, Madagascar and northern Mozambique are unknown.

Individual identification: The pattern of light spots on the body of individual coelacanths is unique, so it acts as a 'fingerprint' that allows researchers to identify individual animals, offering a rare opportunity. Hans Fricke, Karen Hissmann and Jürgen Schauer have compiled an extraordinary database on coelacanths in the Comoros, spanning 21 years during which they have been able to individually identify and track 145 adult coelacanths from the *Jago*. For example, a 27-kg male coelacanth caught by an 80 year-old Comorian fisherman, Ali Mmadi, and now on display in the Fukushima Aquarium in Japan, was recognized by its pattern of spots as a specimen photographed by Hans Fricke off Grande Comore on 10 different occasions in 1995, over 10 years earlier!

In the iSimangaliso Wetland Park, Kerry Sink, as a team member of ACEP, has so far individually identified and traced 32 coelacanths using a remote-controlled submersible. Few other fishes, and only a handful of birds and mammals on land, have been studied with this intensity.

Social behaviour: Coelacanths are social animals that gather together in caves during the day. As many as 16 coelacanths have been observed in one cave. In November 2008 I visited a cave off the south shore of Grande Comore with Jürgen Schauer in the *Jago*. We spent a serene two hours with five coelacanths in their secret lair.

A group of coelacanths in a cave off Grande Comore

Daily activity pattern: Coelacanths hide in caves from visual predators during the day but venture out to hunt at night, their main hunting method (ambushing) being most effective under the cover of darkness. They are able to hunt at night by using mainly non-visual cues, such as electroreception, to locate their prey. They drift in the ocean currents, like albatrosses in the wind, lazily flapping their 'tassel fins' to and fro, with the large tail fin remaining motionless until they make a lightning-quick pounce on prey. They drift near the bottom, often following gullies into deeper water (250–300 m, occasionally 700 m or more). At dawn they return to their home caves in shallower water, usually at depths from 180 to 210 m.

Illustration by Bennie Krüger

A coelacanth doing a headstand

Headstanding behaviour: During their nocturnal hunting forays, and even while sheltering in caves, coelacanths often stand on their head and bend the middle lobe of their tail fin back and forth. It is possible that these headstands allow them to detect the electrical fields of prey hidden in the sand using their rostral organ, which is similar to the ultra-sensitive electro-receptive organs on the snout of sharks. Hans Fricke and Jürgen Schauer were able to induce coelacanths to do a headstand by emitting weak electrical currents from their submersible!

Although goatfishes do headstands when they blow sand away to reveal prey, and sand-nesting species like some tilapias stand on their heads while digging their nests, the prey-hunting headstands of coelacanths appear to be unique.

Feeding behaviour: Coelacanths feed on cartilaginous and bony fishes as well as squids and cuttlefishes. The coelacanth's position high up in the food chain is illustrated by the fact that they carry significant burdens of toxic pesticides in their tissues. These pesticides are blown from the African mainland onto the sea, where they are absorbed by fishes and accumulate up the food chain, especially in top predators.

Role and value in its ecosystem: As a top predator, the coelacanth plays a pivotal although largely undefined role in the changing ecosystems that it has inhabited over millennia. It is therefore an important component of the marine biota of the Western Indian Ocean and Celebes Sea, and possibly also of the West Indo-Pacific Ocean if they are present there.

Extraordinary breeding strategy: Remarkably, coelacanths, which are among the most ancient fishes, have one of the most advanced breeding strategies – live-bearing, which is shared by about 5% of all fishes as well as by some amphibians, reptiles and all mammals (but no birds). As far as we know, the coelacanth is the only live-bearer in which the male does not have an intromittent organ. The female extrudes her oviduct to receive the sperm.

Coelacanths produce relatively few eggs compared to most fishes, usually between 2 and 67 eggs, although a female caught off Anjouan in March 1955 had an astonishing 197 eggs – by far the most of any coelacanth yet caught. They were in three distinct size groups (numbering 58, 65 and 74 eggs), which indicates

Coelacanth eggs compared to the size of an orange

Eugene Balon

that they produce eggs in consecutive batches, like some sharks and rays. A coelacanth caught off Madagascar in July 2001 contained four small eggs and two embryos. Most fishes produce hundreds or thousands of eggs per spawning; the ocean sunfish is reputed to produce over 30 million minute eggs per spawning!

When fully developed, coelacanth eggs are very large, the size of a grapefruit, making them the largest of any fish. The unborn young consume the yolk in their yolksacs while still inside the mother's body. It is also possible that, after hatching inside the mother, they may feed on unhatched eggs, as in some live-bearing sharks, given that their jaws and teeth are fully formed at birth, and that more eggs are produced than hatch.

Placenta-like organ? John Wourms, James Atz, Dean Stribling and Eugene Balon have suggested that coelacanths have a placenta-like organ, comprising vascular labyrinths that lie between the walls of the yolksac and the uterus; the mother may exchange nutrients and gases through the labyrinth with the unborn embryos. This view is disputed by Phil Heemstra, Len Compagno and others. Coelacanth research has thrived on, and benefited from, controversy!

Tiny brood size: Coelacanth broods (2–26 juveniles per brood are known so far) are among the smallest of any fish, with only the mouth-brooding cichlids,

Dr Norihiro Okada, a professor emeritus, Tokyo Institute of Technology

Coelacanth eggs revealed during a dissection in 2009 at the Tokyo Institute of Technology, observed by the Emperor.

Jerome F Hamlin

Model of a coelacanth yolksac juvenile

surfperches, live-bearing sharks and a few other fishes producing fewer young per brood. Coelacanths give birth to the largest known new-born juveniles of any bony fish (up to 410 mm and 530 grams). The juveniles are well developed at birth and resemble miniature adults, a trait found in only a few fishes such as live-bearing wrasses, swordtails and guppies. Hans Fricke has predicted that coelacanth females may produce only about 140 young during their entire life span, one of the lowest lifetime breeding rates of any fish and more akin to the breeding strategy of a reptile.

Long gestation period: Remarkably, coelacanth young have a gestation period (during which they develop inside the mother's body) of about 36 months (1,095 days), which is by far the longest of any animal and 1.7 times longer than that of the next candidate (African elephant, *Loxodonta africana*: 645 days). The coelacanth is certainly a creature of superlatives!

Hein von Hörsten / Images of Africa

African elephant with young

Parental care? An unsolved mystery is whether the coelacanth protects its young after they have been born. No observations on parental care have yet been made, and there is therefore no direct evidence for it. However, many fishes that give birth to large, well-developed young are known to show parental care,

although this is less common in predators (such as sharks) due to the risk of cannibalism. It does, though, make evolutionary sense for a fish that has made such a large parental investment in a few, large young to optimize their survival.

Based on my studies on the breeding strategies and life-history styles of fishes, I predict that the coelacanth's new-born young are protected by the mother, perhaps in the deep recesses of a cave, during the day. We know that coelacanths are social animals that gather in caves, so it is possible that several females may protect the young together.

Illustration by Bennie Krüger

Artist's impression of a coelacanth guarding its young

Vulnerability to extinction: The coelacanth, in common with species such as the giant panda, great white shark and California condor, has many of the attributes of an animal that is vulnerable to extinction, including rarity, large size, a high position on the food pyramid, low dispersal rates, few offspring, longevity, a tendency to accumulate toxic chemicals, and high levels of specialization. Some of the attributes that we originally thought would also increase the coelacanth's vulnerability to extinction, such as a narrow distribution range, low gene flow and genetic bottlenecking, may, in the light of recent research, not be important. Their susceptibility to deep-set gillnets and blast fishing increases their vulnerability, but the threat posed by traditional hand-line fishing is probably less than originally feared.

Cultural history and impact on humans

The coelacanth has an unusually rich cultural history for a fish. An astounding array of scientists, fishermen, politicians, artists, poets and musicians, and even some practical jokers, have been part of its narrative, and it has had multiple impacts on, and interfaces with, human culture. We should not, however, judge its value purely on the basis of its benefits to humankind, although having socio-historical and socio-cultural importance does allow it to be a unifying force in global campaigns to conserve marine life.

JLB Smith: Smith himself must take some credit for making the coelacanth so famous. Although he was an intensely private person he had a knack for kindling public interest in 'old fourlegs'. The discovery of the first coelacanth in December 1938 created excitement in a dreary world that was about to be engulfed by the Second World War. The human drama and intrigue of the discovery provided an escape from reality and a positive focus of attention for people from all walks of life. By the time the second coelacanth was discovered, in 1952, the world was quite different: it was rapidly recovering from the war, Everest had been 'conquered', the mood was jubilant and people were ready to share in the success of JLB's 14 year-long search. His radio interview in Durban in late December 1952 was one of the most dramatic ever broadcast on South African radio and was translated and broadcast by the BBC in 10 languages. The subsequent publication of *Old Fourlegs*, and its re-publication in 9 languages worldwide, put the icing on the cake.

Promotion of the natural sciences: The discovery of the living coelacanth, and the ongoing surprises that its further study has revealed, has promoted ichthyology, as well as palaeontology and oceanography, worldwide. It has also made people aware that undiscovered large creatures may still lurk in the sea and that we should place a stronger emphasis on 'inner space' research.

Popularization of science: Together with the dinosaurs (especially *Tyrannosaurus rex*), the great whales, the charismatic megaherbivores and some cute species that have been Disneyfied, the coelacanth has played a leading role in popularizing science and making it accessible to lay people. Also, as Keith Thomson (1991, p. 70) has noted, 'At a time when science seemed to have all the answers and threatened to take all the mystery out of life, *Latimeria* made zoology romantic again and science the realm of real people, like fishermen and small-town museum curators'.

Consequences for South African ichthyology: The discovery of the first coelacanth changed the course of South African ichthyology. It persuaded Smith to abandon chemistry in favour of fish studies and to establish a research Department of Ichthyology at Rhodes University. The newly-formed CSIR provided a research grant for Smith in 1946 that enabled him to mount extensive fish-collecting expeditions up the East African coast. This, in turn, led to the publication of the first edition of *The Sea Fishes of Southern Africa* in 1949 and the discovery of the second coelacanth in 1952, described in *Old Fourlegs* (1956). After JLB Smith died, a research institute was established in his name; this institute, now the South African Institute for Aquatic Biodiversity, has a broadened mandate to conduct research on all aquatic organisms. The new Department of Ichthyology & Fisheries Science was established at Rhodes University in 1981 to promote the teaching of ichthyology and fisheries science when the Institute left Rhodes University and became a national museum.

Cacophony of coelacanthophiles: A fascinating array of scientists has been involved in the coelacanth saga. The first player was the Swiss-American palaeontologist Louis Agassiz who described the first coelacanth fossils in Germany in 1839 and coined the name *Coelacanthus*, referring to the hollow spines in the tail fin. He wrongly classified coelacanths with placoderms and other ancient fishes but, in 1861, the British palaeontologist Sir Thomas Huxley ('Darwin's Bulldog') correctly placed them with the lobe-finned fishes.

The next players were the trawler skipper Hendrik Goosen and the remarkable young South African museologist Marjorie Courtenay-Latimer, followed by the passionate crusaders JLB and Margaret Smith, as well as their thoroughly mistaken antagonist, Dr Errol White, from England. Then came another boat skipper, Eric Hunt; an unlikely procession of South African politicians who did (or did not) assist Smith in his passionate pursuit; and the highly competent crew of the famous 'Flying Fishcart'. They were followed by the scholarly Jacques Millot, Jean Anthony and Daniel Robineau from France.

Other colourful characters to join the fray included the mischievous Italian Franco Prosperi, who photographed a fake coelacanth; the legendary Jacques-Yves Cousteau and his *Calypso* crew, who searched in vain for coelacanths; Dirk Thys van den Audenarde and the meticulous Rik Nulens from Belgium; the amazing *Jagonauts* Hans Fricke, Jürgen Schauer, Karen Hissman, Olaf Reinicke and Lutz Kasang from Germany and Rafael Plante from France; and then the modern South African contingent, Phil Heemstra, Peter Timm, Robin Stobbs, Tony Ribbink, Tommy Bornman, Angus Paterson and the ACEP 1, 2 and 3 and ASCLME teams, especially Mike Roberts, Sean Fennessy, Kerry Sink, Rosemary Dorrington, Malcolm Smale, David Vousden, Garth van Heerden, Paul Cowley, Jean Harris, Ryan Palmer and many others.

In England, Adam Locket, George Hughes and Peter Forey; in Canada, Eugene Balon, Christine Flegler-Balon

Coelacanth on display in the President's Palace in Moroni, Grande Comore

Eugene Balon

and Richard Cloutier; and, in the USA, John McCosker, Michael Lagios, Keith Thomson, Len Compagno, James Atz, Robert Griffith, John Wourms, Jack Musick, Hans-Peter Schultze and many others made telling contributions, as did Teruya Uyeno, Toshio Tsutsumi, Naoki Suzuki and Kengi Tanuma from Japan.

Three flamboyant multi-millionaires have even contributed to the coelacanth cause: Herbert Axelrod, erstwhile publisher of *Tropical Fish Hobbyist* magazine, who funded research by John Maisey of the American Museum of Natural History on fossil coelacanths in South America; Paul Allen, co-founder of Microsoft, who made his ship *My Octopus* available to Hans Fricke for research in the Comoros; and 'Shogun Noodle', the diminutive Japanese industrialist who invented instant noodles, who is sponsoring Fricke's latest research in the Pacific Ocean. In this elite company we can also include the mysterious Japanese media mogul who made such substantial cultural contributions to the French and Comorian endeavours that the President of France, Charles de Gaulle, donated a coelacanth to him in 1967!

Has any other fish been blessed with such a diverse assemblage of fans?

Coelacanth diplomacy: Unusually for a fish, dead coelacanths have been donated, with due pomp and ceremony, as goodwill gifts by the Comorian and French governments to a motley assortment of high-level benefactors, businessmen and politicians, a kind of 'coelacanth detente'. Undoubtedly the most

Monaco Aquarium
Jacques-Yves Cousteau, co-inventor of the aqualung

Japanese coelacanth researcher, Teruya Uyeno

unusual gift received by the United Nations on its 40th anniversary in 1985 was a small, 91-cm coelacanth in a beautifully carved mahogany box, which President Ahmed Abdallah Abderemane of the Comoros presented to Secretary-General Javier Perez de Cuellar!

Other examples include:

Specimen	Date donated	Donated by	Donated to
125-cm female	1956	Muséum National d'Histoire Naturelle de Paris	Natural History Museum, London
Length unknown	1967	President Charles de Gaulle, France	Japanese media mogul
11.2 kg	1976	Comoros government	Algeria government
Unknown	1976	Comoros government	Kuwait government
Various	1982, 1983, 1984, 1997	Comoros government	China government
Length unknown	1985	Comoros government	South Korea
176-cm female	1990	Comoros government	President Mitterrand, France
164-cm female	1991	President Djohar, Comoros	South African Foreign Affairs Minister, 'Pik' Botha

Not all coelacanth diplomacy went smoothly. In April 1985 President Ahmed Abdallah of the Comoros sent a specimen to a construction company in Japan but it was confiscated by Japanese customs 'due to an inappropriate export permit' in terms of CITES regulations; it was subsequently appropriated by the Ibaraki Museum. Another coelacanth specimen was hijacked by Tokyo customs officials in 2009 and is now on permanent display in the Customs House in East Kyoto Minato-ku!

Bounty on its head: Although the coelacanth is now listed on Schedule I of CITES and may not be traded for financial gain, generous rewards were initially offered for its capture, first by JLB Smith (£10, then £100) and then by the French (£200) and the Americans ($5,000). Furthermore, according to Hans Fricke, the worldwide museum trade offered between US$400 and $2,000 per

coelacanth for several years, but specimens on the black market fetched far more. At one stage the black market in the Comoros offered formalin-fixed specimens for about US$1,000 each. In 1992 two live coelacanths were offered for sale in Germany for US$130,000 each on the price list of the illegal international aquarium trade, but they could not be traced. Traditional fishermen who catch a *gombessa* usually receive a fraction of the sale price but it is still worth more than an average year's earnings, the equivalent of winning the lottery.

Since its CITES listing, and because it is a large fish that cannot easily be transported without being noticed, the illegal trade is probably minimal. As a result, a number of specimens have accumulated in the freezers of hotels, commercial fishing companies and research facilities in the Comoros and Tanzania, and some have been lost due to power failures.

Live coelacanths on display in a public aquarium would have enormous public appeal and must represent a very tempting economic opportunity to the Comoros, which is still wracked by an unstable economic and political climate; at one stage it had 18 different governments in 20 years! In 1989 the Toba Aquarium in Japan mounted a US$2 million expedition, sponsored by Mitsubishi, to catch live coelacanths in the Comoros, but was unsuccessful. Other aquaria have expressed interest in displaying the fish but they are restricted by the CITES protection. Hans Fricke (1997) has suggested that, 'In future, science will probably be used as an excuse for the first public display and this will be followed by an exponential increase in the market price'.

Capture by humans: Until August 1991, when a coelacanth was caught in a trawl net off Mozambique, all coelacanths (except the first) had been caught by traditional fishermen using hand lines. Following the deployment of deep-set gillnets targeting sharks, many coelacanths have been caught in their deadly meshes in recent decades, especially off Tanzania and Madagascar. Notwithstanding this recent development, nearly 80% of all coelacanths caught to date have been landed by traditional fishermen, an unusual circumstance for such a rare and important fish.

Socio-history: Before 1938 the coelacanth was regarded as a useless fish in the Comoros, without value to traditional fishermen. Since then it has become the symbol of the new national identity of the Federal Islamic Republic of the Comoros, and a source of intense civic pride. Furthermore, fishermen, who are generally of low social rank, gain social prestige when they catch a *gombessa*.

Its importance has been honoured in many ways, including 'coelacanth diplomacy', the issue of postage stamps, postcards and money depicting the fish, art- and craftworks, the naming of the main street in the capital, Moroni, as 'Boulevard du Coelacanthe' and the 'Hotel Coelacanthe', as well as its display in a magnificent mahogany cabinet in the presidential palace.

Coelacanths on display: Coelacanths are archived in museum collections and are on display in museums, universities, research institutes, hotels and private homes in at least 32 countries, probably more, and fibreglass models of juveniles and adults are on display worldwide.

Hans Fricke, the author and Jürgen Schauer with the first coelacanth in the East London Museum in 1987

Educational value: Coelacanths have been of immense educational value because of their important contribution to our understanding of key evolutionary processes. The local *Phuhlisa* programme and many other educational projects around the world have produced a plethora of educational aids on 'old fourlegs'.

Vanhamel, M., Museum voor Dierkunde, KU Leuven, Belgium

Coelacanth in the Museum of Zoology, Leuven, Belgium

Emotional impact: Few fishes have had the emotional impact of the coelacanth. In June 1987 Hans Fricke made the first public presentation of his unedited film on the living coelacanth at a conference in Grahamstown. He described his impression afterwards, 'I did not realize what emotional importance the coelacanth had for Grahamstown and the whole of South Africa … More than one thousand people were absolutely quiet … Never before had I sensed such a tension in an audience … When the first moving images appeared on the screen the audience broke into frenetic applause … Eugene Balon told me later that some people had tears in their eyes.' Humphry Greenwood from the Natural History Museum in London exclaimed, 'This is the most extraordinary biological film of the century'.

Coelacanth inventory: The coelacanth inventory has regularly been updated by various scientists and archivists since 1972 and is now compiled by Rik Nulens and collaborators. Each coelacanth catch in the inventory is allocated a CCC number and full details of its site, time, date, depth, distance from shore and method of capture, name of fisherman, weight and length, condition, sex, method of preservation and transport, tissues available for study, how and where it was studied, where it is archived or displayed, and the first literature reference (if known). So far (to July 2016) over 314 specimens have been documented.

If we combine the inventory of dead specimens with the catalogue of live individuals that have been identified through their unique patterns of white dots by the teams from the *Jago* (145 individuals) and ACEP (32), then we must surely have the most comprehensive catalogue of any animal in the world, dead and alive? Not even very rare and critically endangered (yet more accessible) species, such as the California condor, Przewalski's horse, black and Sumatran rhinos, leatherback turtle, giant panda, mountain gorilla, pangolin or Yangtze finless porpoise have been inventoried to this level of detail.

An inspirational fish

General symbolism: In art, poetry, literature, performing arts and crafts and politics the coelacanth has been represented variously as a symbol of surprise, rarity, longevity, tenacity, immortality, survival, a phoenix-like resurrection from the past, links to ancient roots, primitiveness, a living fossil, and even the concept of 'fossil fuel'. Like other fishes, it has also come to symbolize virility, fecundity, promiscuity and elusiveness. In everyday language the term 'coelacanth' is used, somewhat unfairly, to refer to people who are old-fashioned, conservative, unable to cope with modern life but still alive due to their tenacity.

Inspiration for artists: Some visual artworks featuring the coelacanth predate the discovery of the living fish. Two silver artefacts from Bilbao and Toledo in Spain, which clearly depict coelacanths, have been seen by Hans Fricke and dated by expert silversmiths to the 17th or 18th century. The unusual shape of the fish probably inspired the artist, but where did he obtain his models?

A coelacanth sculpture atop a tower in Ulaan Baatar, Mongolia

Eugene Balon

Mike Bruton

'Feline Fantasy' by B Quirk

Mike Bruton

Soapstone sculpture of a coelacanth on display in Moroni, Grande Comore, in November 2008

'Coelacanth to Boku' by Sakanaction

Mike Bruton

Stainless steel sculpture of a man inside a coelacanth by Uwe Pfaff

Mike Bruton

Tinka-tinka painting of the coelacanth by Tanzanian artist Suleiman Awaz

Rebecca Ranney

Artwork by Rebecca Ranney

Woody Blackwell

Metal sculpture of a coelacanth by Woody Blackwell

Fish prints using the traditional Japanese method of fish printing (known as *gyotaku*) have been produced in the Comoros.

Coelacanths have frequently been depicted in modern artworks. Perhaps the most remarkable is the linocut by Hylton Mann of South Africa, called *Oncanthusphere*, which uses coelacanths to symbolize our excessive dependence on fossil fuels. It depicts three businessmen assembled around a coelacanth egg, the source of life, while the fish spirals off into a desolate landscape that symbolizes its impending extinction and that of fossil fuels.

Other artists who have depicted coelacanths have included Stephen Rautenbach, Stephen Colegate, Uwe Pfaff, Ravenari, Roy Reynolds, B Duhem, B Quirk, Charlotte Firbank-King, Erik la Gattuta, Doug Katagiri, Khimera, Ellen Marcus, Mark Shultz, Woody Blackwell, Rebecca Ranney, Hemlock, Christopher Manuel, K Holland, Angela

Ellen Marcus

'Coelacanth on Bicycle' (2009) by Ellen Marcus

Amber Ansley, Beesocks

Marty Bobroskie

A coelacanth-inspired charm by Marty Magic Jewelry (www. martymagic.com)

Emily Smith, cryptovolans

Coelacanths have even inspired fabric designs.

Mike Bruton

A coelacanth made from used tea bags by Grahamstown crafters

Oliver, Tiffany Bozic, Anya Stasenko, Peter Boehme and Salva Leontyev. Outdoor sculptures of coelacanths from metal and soapstone are common in the Comoros and South Africa and even appear in such unlikely places as atop the clock tower in Ulaan Baatar, Mongolia!

Inspiration for designers: Many designers have adopted the coelacanth motif as the basis for their compositions.

Inspiration for playwrights: The famous South African actor and producer, Nicholas Ellenbogen, produced and performed a play called *The Coelacanth Project* in 1991 based on our research and, in 2003, another well-known South African playwright, Andrew Buckland, produced and performed in a play entitled *Plunge*, also about 'old fourlegs'.

Inspiration for craftsmen and -women: Coelacanth craftworks are very common. They probably first originated in the Comoros where exquisite filigree

Mike Bruton

A wooden coelacanth puzzle

Mike Bruton

Coelacanth lamp sculpture

Mike Bruton

Coelacanth made from coloured fabrics, Grahamstown

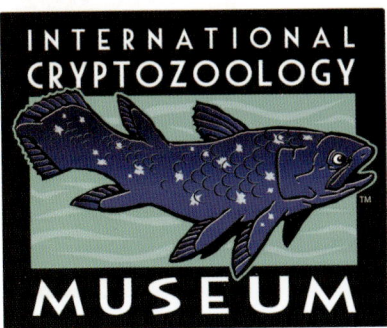

Logo of the International Cryptozoology Museum

gold-plated pendants were made in the 1980s and 1990s. In the Albany District around Grahamstown in South Africa a plethora of coelacanth craftworks and jewellery has been made using materials as varied as wood, cardboard, soapstone, ceramics, steel wire, recycled aluminium cans, ring-can openers, colourful fabrics and even used tea bags and elephant dung!

Coelacanth robots: Japanese technologists have even made mechanical and robotic coelacanths which may, in future, be displayed in aquaria instead of live fishes.

Inspiration to poets, authors, songwriters and movie makers: The coelacanth has featured in a song by C Rand, poems by FM Storey (who wrote about its 'no progression' and 'no regression'), Horace Shipp (why it wants to stay as it is) and Ogden Nash (adoration for its timeless existence). Coelacanths appeared in their first movie in 1954, where one is portrayed as the hero in the *Monster of the Black Lagoon* directed by J Arnold; and, in 1992, LJ Langley created a tribe of vicious submarine vampires represented as coelacanths. They have also been widely mentioned in children's books, including the popular *The Big Fish Tin Lin* (Klaus Kordon, Simon & Schuster, 1992) and the author's *The Four-Legged Fish* (Cambridge University Press, 2000).

Coelacanths on postage stamps: Coelacanths have featured on the postage stamps of at least 22 countries, including six where they are known to occur (Comoros, Indonesia, Madagascar, Mozambique, South Africa, Tanzania) and 16 where they are not known to occur (Benin, Guinée, Guyana, Ivory Coast, Japan, Kuwait, Liberia, Mali, Mauretania, North Korea, Palau, Russia, Sao Tomé and Principé, The Gambia, Togo, and Turks and Caicos). The Comoros has issued 12 different sets of coelacanth stamps. In 1989, at the instigation of the JLB Smith Institute of Ichthyology, South Africa issued four postage stamps depicting the coelacanth, JLB Smith and Marjorie Courtenay-Latimer, the JLB Smith Institute of Ichthyology, and the *Jago* submersible.

First day cover from South Africa, 1989 (above) and a close-up of the stamps (below)

First coelacanth stamp issued by the Comoros, signed by JLB Smith

The coelacanth has been adopted as a symbol of the new national identity of the Comoros.

East London Museum 1988 commemorative cover

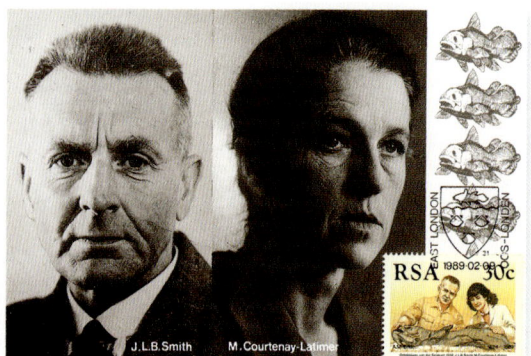

Miniature sheet issued in error by the South African Post Office, showing Margaret Smith instead of Marjorie Courtenay-Latimer

ELPEX 1968 commemorative cover

The 50th anniversary of the capture of the first coelacanth in 1988 was celebrated in Grahamstown and around the world. A set of four postage stamps was issued by the postal authorities in South Africa to commemorate this occasion.

In 1989 the South African Post Office even issued a rare and now very collectible error on a miniature sheet that depicted Margaret Smith instead of Marjorie Courtenay-Latimer.

Coelacanths on money: The 1,000 CF bank note and 55 CF coin in the Comoros depict the coelacanth, as do a €10 coin issued by France, a 2 Maticais coin from Mozambique, 1,000 and 4,000 Kwacha coins from Zambia, and a silver 10c coin from South Africa. In addition, South Africa minted a

Comorian coin depicting the coelacanth

24-carat gold commemorative coin depicting the coelacanth in 1998 to celebrate the 60th anniversary of the first capture and, in 2013, a 10-cent silver coin to commemorate the establishment of the iSimangaliso Wetland Park.

Coelacanths on a 1,000 CF note from the Comoros

Harold Macmillan

http://www.gac.culture.gov.uk

Political metaphors: Winston Churchill once referred to a back bencher in the British House of Lords, who had remained silent for nearly 20 years and then got up to make a great speech, as 'that coelacanth of a man'. In 1973, eight years after Churchill's death, George Lichtheim described him as follows, 'In one of his aspects he is a political coelacanth, a prehistoric monster fished up from the depths of the past (thus he appears to British left-wingers, who are nonetheless secretly proud of him)'.

Harold Macmillan was described by Ben Pimlott as, 'The coelacanth of British politicians'; and his biographer, Peter Catterall, wrote, 'In retrospect, he can be seen as the coelacanth of modern British politics – the last of the old and the first of the new – who contained in his crafted persona both elements of a foregone era, and quicksilver signs of adapting to the modern one.'

Time magazine once referred to Richard Nixon as the 'coelacanth of American anti-communism' and the *Washington Times* described the political position of the African National Congress, fossilized but still alive, as the 'Coelacanth face of the ANC'.

Inspiration for cartoonists: Coelacanth cartoons and caricatures are common because of the creature's rich symbolism. In Britain a coelacanth was drawn in 1940 with the head of Neville Chamberlain and a swastika as a tail fin to illustrate the influence of the Nazis on the backboneless politician; and even President Robert Mugabe of Zimbabwe has been depicted as a 'living fossil'. Some coelacanth cartoons have bemoaned nuclear testing, industrial pollution or the retrograde evolution of humans, whereas others have celebrated the evolutionary transition from water to land and the evolution of four-legged animals, or deplored our meagre knowledge of the ocean depths.

Symbol of marine conservation: The coelacanth has, for many, become the symbol for marine conservation, the 'panda of the seas'.

The coelacanth and the panda have both become symbols of conservation.

Inspiration: More broadly, the coelacanth has inspired extensive multinational collaborative research, epitomized by ACEP, among many others. It has also inspired many careers in ichthyology, fisheries science and marine science, in Africa and internationally, and its discovery effectively led to the establishment of the world-famous JLB Smith Institute of Ichthyology, now the South African Institute for Aquatic Biodiversity (SAIAB), in Grahamstown. JLB's book *Old Fourlegs* itself inspired many ichthyologists to pursue a career in this field, including Hans Fricke and the author.

Commodity value: Oil from the coelacanth is used as a laxative, similar to the castor oil of the oilfish (*Ruvettus*). *Ruvettus* oil is also used as a mosquito repellent and anti-malarial drug; coelacanth oil probably has the same properties. The rumour that fluid from the notochord of the coelacanth is a life-prolonging elixir has, of course, no basis in fact.

Promotional value: The symbol of the coelacanth has been widely used to promote entrepreneurial companies, such as the dive group 'Gombessa plongée' in the Comoros, as well as to sell utensils, magazines, newspapers, books and even cigarettes. The coelacanth has also received worldwide media coverage, arguably more than that of any other fish.

'Le Coelacanthe' cigarettes

Future recreational use: It would be difficult and expensive to facilitate visits by tourists in submersibles to coelacanths in their natural habitat, but the installation of low-light digital cameras that beam images up to a tourist facility on land would create an opportunity for visitors to the Comoros to be enthralled and inspired by real-time sightings of the fish.

Right to exist: The survival of the coelacanth is enshrined in the conservation ethic that all organisms on Earth have the right to exist, *or at least to struggle to exist,* and that this right is not dependent on their actual or potential use to humans.

Coelacanths are an important part of our heritage, not because they are of nutritional or commercial value, as are many fishes, but because of their evolutionary, scientific, ethical and cultural significance and the way in which they symbolize longevity and tenacity. They also have psychological, inspirational and intellectual value and occupy a unique place in the human consciousness. Their rich symbolism should be used to promote their long-term conservation.

I challenge the reader to name any other animal that has such an impressive suite of natural and cultural history attributes!

POSTSCRIPT

I trust that this annotated version of *Old Fourlegs* has shed light on the character of JLB Smith and placed his research on the coelacanth in a modern context. Despite the multiple advances in science and technology in this Information Age, the simple story of 'old fourlegs' still has enormous appeal. Furthermore, the saga is not over: it continues to unfold with dramatic new discoveries on extinct and living coelacanths and enlightening insights into the crucial step in evolution when animals first ventured onto land.

Ultimately, the coelacanth story is about survival – survival of the ancient lineage of the coelacanth but also about the continuance of the much more recent, and fragile, human experiment. The coelacanth story has raised awareness of the importance of biodiversity conservation, especially in the oceans. This is very important as we are essentially terrestrial animals with an earthbound point of view. In many of our research and conservation endeavours we have ignored the oceans (and freshwaters) relative to the land and sky, but we do so at our peril.

All life-support systems on 'Earth' begin and are mediated in the oceans, yet we have damaged aquatic systems in many ways. For instance, since the start of the Industrial Revolution, we have dumped over 500 billion tonnes of CO_2 from the atmosphere into the oceans, increasing their acidity by about 30%. Ocean temperatures have also increased by 1°C in the last 30 years. The oceans do not need us but we need them because our quality of life, and that of all the other inhabitants of the planet, depends on their welfare.

Although we arose from and survive through a biological process, we are no longer part of wild Nature. The majority of us are servants to our machines, trapped in unsustainable urban environments, the first animal to domesticate itself and to lose its ecological place. But how wrong we are to consider ourselves to be outside of Nature! The only tenable way forward is to understand and work in harmony with Nature, not to conquer it.

The coelacanth, as a symbol of marine conservation, is a beacon of hope in a sometimes bleak world. By allowing us to look into the past, it has empowered us to glimpse the future.

This perspective has made us realize that we need to undergo a fundamental change in our mind-set and behaviour if we are to become more mature and responsible custodians of the biosphere.

70 years of 'Old Fourlegs'

Latest Comoros sighting marks the anniversary of the discovery of the 'extinct' coelacanth, writes **James Clarke**

The Sunday Independent, 2008

GLOSSARY OF TERMS

Brood size Number of young produced at a single breeding event.

Coelacanth Rare order of fishes that includes two living species in the genus *Latimeria*: the African coelacanth, *Latimeria chalumnae*, and the Indonesian coelacanth, *L. menadoensis*. They belong to the oldest known living lineage of Sarcopterygii (lobe-finned fishes and tetrapods), which means that they are more closely related to lungfishes, reptiles and mammals than they are to the common ray-finned fishes. Since both coelacanth species are threatened, it is the most endangered order of animals in the world.

Cretaceous Geological epoch extending from about 145 to 65 million years before present.

Cretaceous extinction Massive extinction event that took place 80–65 million years ago and caused about three-quarters of the plant and animal species on Earth, including all non-flying dinosaurs, to go extinct. With the exception of some cold-blooded species in marine and freshwaters, such as sea turtles and crocodiles, no tetrapods weighing more than about 25 kg survived.

Crossopterygii Group of lobe-finned fishes, also known as the Sarcopterygii, whose living representatives include the lungfishes and coelacanths, and the tetrapods.

Demography Statistical study of animal populations and the individuals comprising them, including their size, structure, distribution and growth rates and changes in relation to birth, migration, ageing and death.

Electroreception Ability of an animal to detect natural electrical stimuli, found almost exclusively in aquatic or amphibious animals since water is a much better electrical conductor than air.

Fishapod Informal name for an animal that is transitional between fishes and tetrapods, such as the tiktaalik.

Genetic bottlenecking Sharp reduction in the size and the genetic diversity of a population due to environmental events or human activities. Thereafter, a smaller population with a smaller genetic diversity remains to pass on genes to future generations. Also known as the Founder Effect.

Gestation period Period of time during which unborn eggs and young develop inside the mother.

Gombessa Comorian name for the coelacanth.

Haemoglobin Iron-containing protein in red blood cells that carries oxygen around the body.

Ichthyology Scientific study of fishes.

Intracranial joint Joint between the two main parts of the brain case in coelacanths that helps them to open their jaws wide and capture and eat large prey.

Intromittent organ Organ of a male animal that delivers sperm into the female during copulation.

iSimangaliso Wetland Park World Heritage Site, established in December 1999, derived from the Greater St Lucia Wetland Reserve that was first formed in northern Zululand, South Africa, in 1895.

Lateral line Sensory system consisting of cells within tubular canals extending over the head and along the sides of a fish, enabling it to detect pressure changes and vibrations.

Lazarus species Species that disappears from the fossil record only to appear again later. The term refers to the story in the Gospel of John in which Jesus Christ raises Lazarus from the dead.

Live-bearing Giving birth to live young (juveniles) that resemble the adult. Also called viviparous.

Living fossil Living species that appear to be similar to species otherwise known only from fossils, typically with no close living relatives. Coined by Charles Darwin when discussing the platypus and South American lungfish; examples include the coelacanth and the maidenhair tree, *Ginkgo biloba*.

Medial In the midline of the body (not paired).

Metabolic rate Rate of energy expenditure per unit of time by warm-blooded animals at rest.

Missing link Term coined by Charles Darwin to describe a species that has traits common to both its ancestors and descendants.

Nessa Oilfish, *Ruvettus pretiosus*, which is targeted by traditional Comorian fishermen, who occasionally catch coelacanths as a bycatch.

Notochord Flexible, rod-shaped tube, found in the embryos of all chordates, that forms a primitive axis. In some chordates, such as hagfishes, coelacanths, lampreys, sturgeons, lungfishes and axolotls, it persists throughout life as the main axial support of the body.

Palaeontology Scientific study of fossils and other evidence of ancient life.

Parental investment Any parental expenditure of time, energy or food that benefits the offspring at a cost to the parent's ability to invest in its own survival, its future ability to breed and its ability to care for other young.

Placenta Organ that connects the developing foetus to the uterine wall to allow food uptake, waste elimination and gas exchange via the mother's blood supply. Placentas are a defining characteristic of placental mammals but are also found in some fishes.

Platypus *Ornithorhynchus anatinus,* a semi-aquatic, egg-laying mammal that is endemic to eastern Australia, including Tasmania. Together with four species of echidnas it is the only living representative of the monotremes, the only mammals that lay eggs instead of giving birth.

Quastenflosse German name of the coelacanth. Literally 'tassel fins'.

Rhipidistia Group of extinct lobe-finned fishes that gave rise to the coelacanths and lungfishes and include the ancestors of the tetrapods.

Rostral organ An organ unique to the coelacanth that is used to detect the electrical fields of prey and comprises a large, gel-filled cavity in the snout with three pairs of canals leading to the outside.

Roudi Gempylid fish, *Promethichthys prometheus*, which is used as bait by traditional Comorian fishermen.

Spiral valve Corkscrew-shaped lower portion of the intestine of some sharks, sturgeons, paddlefishes, rays, skates, bichirs, lungfishes and coelacanths. It is internally twisted to increase the surface area of the intestine so as to increase nutrient absorption.

Tapetum lucidum Layer of tissue behind the retina in the eye that reflects visible light back through the retina, thus increasing the light available to the photoreceptors. It contributes to the superior night vision of nocturnal and deep-sea animals.

Tetrapods Four-legged animals and their descendants, including living and extinct amphibians, reptiles, birds and mammals. They evolved from lungfishes about 395 million years ago.

Tiktaalik Extinct lobe-finned fish from the late Devonian period with many features similar to those of four-legged animals and close to the evolutionary transition from fish to amphibians.

Tuatara Three species of reptiles endemic to New Zealand that resemble lizards but are the only living representatives of a distinct group of ancient reptiles that flourished about 200 million years ago. They are of great interest in the study of the evolution of lizards and snakes and for the reconstruction of the appearance and habits of dinosaurs and crocodiles.

Yolksac Membranous sac attached to an embryo that provides early nourishment in the form of yolk in bony fishes (including the coelacanth), sharks, reptiles, birds and mammals.

ACRONYMS USED IN THE TEXT

ACEP African Coelacanth Ecosystem Programme, a multi-disciplinary research programme, led by SAIAB, on the coelacanth and its habitats in the Western Indian Ocean. ACEP was divided into three phases: ACEP 1, 2 and 3.

APG Association for the Protection of Gombessa/Association pour la Preservation du Gombessa

ASCLME Agulhas and Somali Current Large Marine Ecosystem project, a GEF-funded multi-disciplinary initiative that arose from ACEP Phase 1

CCC Coelacanth Conservation Council/Conseil pour la Conservation du Coelacanthe

CITES Convention on International Trade in Endangered Species

CNDRS Centre National de Documentation et de la Recherche Scientifique des Comoros

CSIR Council for Scientific and Industrial Research in Pretoria, South Africa, established in 1946

DDT Dichlorodiphenyltrichloroethane, a colourless and almost odourless organochlorine insecticide

DNA Deoxyribonucleic acid, a molecule that encodes the genetic instructions used in the development and functioning of all living organisms

GEF Global Environmental Facility, an interagency project initiated in 1992 that co-ordinates technical assistance and funding for programmes on biodiversity, climate change, water and the ozone. GEF has provided $14.5 billion in grants and mobilized $75.4 billion in additional financing for about 4,000 projects.

GIS Geographic Information System

ICT International Coelacanth Trust

IUCN International Union for the Conservation of Nature and Natural Resources, the world's main authority on the conservation status of species

JASEC Japanese Scientific Expedition of the Coelacanth

MNHM Muséum National d'Histoire Naturelle de Paris

MPA Marine Protected Area

NEPAD New Partnership for Africa's Development

NRF National Research Foundation, Pretoria, South Africa

PCB Polychlorinated biphenyl, an organic chlorine compound

RNA Polymeric molecule involved in coding, decoding, regulation and expression of genes

ROV Remote-operated vehicle

SAIAB South African Institute for Aquatic Biodiversity in Grahamstown, previously the JLB Smith Institute of Ichthyology

SCUBA Self-contained underwater breathing apparatus, invented by Émile Gagnan and Jacques Cousteau in France in 1943

UNDP United Nations Development Programme

REFERENCES AND FURTHER READING

Amemiya, CT & 98 co-authors. 2013. The African coelacanth genome provides insights into tetrapod evolution. *Nature, London* 496: 311–316.

Anon. 1955. Observations on the first living coelacanth. *Science* 121(3146): 544–545.

Anon. 2013. *Building on the South African Coelacanth Legacy.* South African Institute for Aquatic Biodiversity, Grahamstown. 45 pp.

Anthony, J. 1976. *Opération Coelacanthe.* Arthaud, Paris. 200 pp.

Anthony, J. 1980. Évocation des travaux français sur *Latimeria* notamment depuis 1972 (Review of French research on *Latimeria* particularly since 1972). *Proceedings of the Royal Society of London Series B* 208: 349–367.

Anthony, J & Millot, J. 1972. Première capture d'une femelle de coelacanthe en état de maturité sexuelle. *Academie des Sciences, Paris Séries D* 274: 1925–1926.

Atz, JW. 1976. *Latimeria* babies are born, not hatched. *Underwater Naturalist* 9 (4): 4–7.

Balon, EK. 1991. Probable evolution of the coelacanth's reproductive style: lecithotrophy and orally feeding embryos in cichlid fishes and in *Latimeria chalumnae. Environmental Biology of Fishes* 32: 249–265.

Balon, EK. 1990. The living coelacanth endangered: a personalized tale. *Tropical Fish Hobbyist* 38 (February): 117–129.

Balon, EK. 1990. Tracking the coelacanth: a follow-up tale. *Tropical Fish Hobbyist* 38 (March): 122–131.

Balon, EK, Bruton, MN & Fricke, H. 1988. A fiftieth anniversary reflection on the living coelacanth, *Latimeria chalumnae:* some new interpretations of its natural history and conservation status. *Environmental Biology of Fishes* 23: 241–280.

Barnett, P. 1953. *Sea Safari with Professor Smith.* Business Services, Durban. 158 pp.

Baxter, M. 1994. Mike Bruton – a distinguished career. *ICHTHOS* 44: 7–8.

Bell, S. 1969. *Old Man Coelacanth.* Voortrekkerpers, Johannesburg. 141 pp.

Bemis, WE & Simon, AM. 1997. Coelacanth catches. *Science* 278: 370.

Benno, B, Verheij, E, Stapely, J, Rumisha, C, Ngatunga, B, Abdallah, A & Kalombo, H. 2006. Coelacanth, *Latimeria chalumnae* (Smith, 1939) discoveries and conservation in Tanzania. *South African Journal of Science* 102: 486–490.

Bogart, JP, Balon, EK & Bruton, MN. 1994. The chromosomes of the living coelacanth and their remarkable similarity to those of one of the most ancient frogs. *Journal of Heredity* 8(4): 322–325.

Boy, G. 2001. Kenya's first coelacanth. *Swara* 24: 24–26.

Bruton, MN. 1982. *The life and work of Margaret M Smith.* JLB Smith Institute of Ichthyology. 12 pp.

Bruton, MN. 1986. A passion for fishes. *South African Journal of Science* 82: 622–623.

Bruton, MN. 1989. The coelacanth – can we save it from extinction? *WWF Reports* Oct/Nov, 10–12.

Bruton, MN. 1989. Does the coelacanth occur in the Eastern Cape? *The Naturalist* 33(3): 5–13.

Bruton, MN. 1989. Fifty years of coelacanths. *South African Journal of Science* 85: 205.

Bruton, MN. 1989. The living coelacanth fifty years later. *Transactions of the Royal Society of South Africa* 47: 19–28.

Bruton, MN (ed.). 1989. *Alternative Life-History Styles of Animals.* Perspectives in Vertebrate Science 6: 1–616. Kluwer Academic Publishers, Dordrecht.

Bruton, MN. 1989. The ecological significance of alternative life-history styles, pp. 503–533. In: *Alternative Life-History Styles of Animals.* Perspectives in Vertebrate Science 6. Kluwer Academic Publishers, Dordrecht.

Bruton, MN (ed.). 1990. *Alternative Life-history Styles of Fishes.* Developments in Environmental Biology of Fishes 10: 1–327. Kluwer Academic Publishers, Dordrecht.

Bruton, MN. 1990. Trends in the life-history styles of vertebrates: an introduction to the second ALHS volume. *Environmental Biology of Fishes* 28: 7–16.

Bruton, MN. 1990. The conservation of alternative life-history styles: a conclusion to the second ALHS volume. *Environmental Biology of Fishes* 28: 309–313.

Bruton, MN. 1991. The meaning of the JAGO expedition. *ICHTHOS* 30:1–3.

Bruton, MN. 1992. The mingled destinies of coelacanth and men. *ICHTHOS* 33: 415.

Bruton, MN. 1993. Additions and corrections to the inventory of *Latimeria chalumnae*: II. *Environmental Biology of Fishes* 36: 398–405.

Bruton, MN. 1993. Alterations and additions to the coelacanth inventory: III. *Environmental Biology of Fishes* 38: 400–401.

Bruton, MN. 1994. Lungfishes and Coelacanth. pp. 70–74. In: Paxton, JR & WN Eschmeyer (ed.) 1994. *Encyclopedia of Animals: Fishes.* Academic Press, San Diego.

Bruton, MN. 1994. The epigenesis of an epigeneticist: an interview with Eugene Balon. *South African Journal of Science* 90: 270–275.

Bruton, MN. 1995. Have fishes had their chips? The dilemma of threatened fishes. *Environmental Biology of Fishes* 43: 1–27.

Bruton, MN. 1995. Threatened fishes of the world: *Latimeria chalumnae* Smith, 1939. *Environmental Biology of Fishes* 43: 104.

Bruton, MN. 2000. *The Four-Legged Fish.* Cambridge University Press, Cambridge. 24 pp.

Bruton, MN. 2015. *When I was a Fish. Tales of an Ichthyologist.* Jacana Media, Cape Town. 310 pp.

Bruton, MN. 2016. *Traditional Fishing Methods of Africa.* Cambridge University Press, Cape Town. 96 pp.

Bruton, MN & Armstrong, MJ. 1991. The demography of the coelacanth *Latimeria chalumnae*. *Environmental Biology of Fishes* 32: 301–311.

Bruton, MN, Cabral, AJP & Fricke, H. 1992. First capture of a coelacanth, *Latimeria chalumnae* (Pisces, Latimeriidae) off Mozambique. *South African Journal of Science* 88: 225–227.

Bruton, MN & Coutouvidis, SE. 1991. An inventory of all known specimens of the coelacanth *Latimeria chalumnae*, with comments on trends in the catches. *Environmental Biology of Fishes* 32: 371–390.

Bruton, MN, Coutouvidis, SE & Pote, J. 1991. Bibliography of the living coelacanth *Latimeria chalumnae*, with comments on publication trends. *Environmental Biology of Fishes* 32: 403–433.

Bruton, MN & Matthews, S. 2000. *Wonders of the Ocean.* Struik, Cape Town. 64 pp.

Bruton, MN & Stobbs, RE. 1991. The ecology and conservation of the coelacanth *Latimeria chalumnae*. *Environmental Biology of Fishes* 32: 313–339.

Buxton, CS, Bruton, MN, Hughes, GR & Stobbs, RE. 1988. Recommendations on the proclamation of marine conservation legislation and the establishment of marine reserves in the Federal Islamic Republic of the Comoros. *Investigational Report of the JLB Smith Institute of Ichthyology* 28: 38 pp.

Cloutier, R & Forey, PL. 1991. Diversity of extinct and living actinistian fishes (Sarcopterygii). *Environmental Biology of Fishes* 32: 59–74.

Compagno, LJV. 1979. Coelacanth: shark relatives or bony fishes? *Occasional Papers of the California Academy of Science* 134: 45–55.

Courtenay-Latimer, M. 1979. My story of the first coelacanth. *Occasional Papers of the California Academy of Science* 134: 6–10.

Daeschler, EB, Shubin, NH & Jenkins, FA. 2006. A Devonian tetrapod-like fish and the evolution of the tetrapod body plan. *Nature, London* 440: 757–763.

De Vos, L & Oyugi, D. 2002. First capture of a coelacanth *Latimeria chalumnae* Smith, 1939 (Pisces: Latimeriidae), off Kenya. *South African Journal of Science* 98 : 345–347.

Desylva, DP. 1966. Mystery of the silver coelacanth. *Sea Frontiers* 12: 172–175.

Diamond, JM. 1965. The biology of coelacanths. *Nature, London* 315: 18.

Du Toit, CA. 1953. Some problems of the coelacanth restated. *South African Journal of Science* 49: 332–333.

Erdmann, MV. 1999. An account of the first living coelacanth known to scientists from Indonesian waters. *Environmental Biology of Fishes* 54: 439–443.

Erdmann, MV. 2000. New home for 'old forelegs'. How the coelacanth was discovered on the other side of the Indian Ocean. *California Wild* 53: 8–13.

Erdmann, MV & Caldwell, RL. 2000. How new technology put a coelacanth among the heirs of Piltdown man. *Nature, London* 406: 343.

Erdmann, MV, Caldwell, RL, Jewett, SL & Tjakrawidjaja, A. 1999. The second recorded living coelacanth from north Sulawesi. *Environmental Biology of Fishes* 54: 445–451.

Erdmann, MV, Caldwell, RL & Moosa, MK. 1998. Indonesian 'King of the Sea' discovered. *Nature, London* 396(6700): 336.

Forey, PL. 1980. *Latimeria*: a paradoxical fish. *Proceedings of the Royal Society of London* Series B 208: 369–384.

Forey, PL. 1988. Golden jubilee for the coelacanth *Latimeria chalumnae*. *Nature, London* 336: 727–732.

Forey, PL. 1990. The coelacanth fish: progress and prospects. *Science Progress* 74: 53–67.

Forey, PL. 1991. *Latimeria chalumnae* and its pedigree. *Environmental Biology of Fishes* 32: 75–97.

Forey, PL. 1998. *History of the coelacanth fishes.* Chapman & Hall, London. 419 pp.

Forey, PL. 1998. A home from home for coelacanths. *Nature, London* 395: 319–320.

Forster, GR. 1974. The ecology of *Latimeria chalumnae* Smith: results of field studies from Grande Comore. *Proceedings of the Royal Society of London* Series B 186: 291–296.

Fricke, H. 1987. Im Reich der lebenden Fossilien. *Geo,* 10 October 1987: 14–34.

Fricke, H. 1988. Coelacanths: the fish that time forgot. *National Geographic* 173: 824–838.

Fricke, H. 1992. Coelacanth tissue bank. *Nature, London* 357: 105.

Fricke, H. 1997. Living coelacanths: values, eco-ethics and human responsibility. *Marine Ecology Progress Series* 161: 1–15.

Fricke, H. 2001. Coelacanths: a human responsibility. *Journal of Fish Biology* 59: 332–338.

Fricke, H. 2007. *Die Jagd nach dem Quastenflosser, Der Fisch, der aus der Urzeit Kam.* Verlag C.H. Beck, Münich. 302 pp.

Fricke, H & Hissmann, K. 2000. Feeding ecology and evolutionary survival of the living coelacanth *Latimeria chalumnae*. *Marine Biology* 136: 379–386.

Fricke, H & Hissmann, K. 1990. Natural habitat of coelacanths. *Nature, London* 346: 323–324.

Fricke, H & Hissmann, K. 1992. Locomotion, fin coordination and body form of the living coelacanth *Latimeria chalumnae*. *Environmental Biology of Fishes* 34: 329–356.

Fricke, H & Hissmann, K. 1994. Home-range and migrations of the living coelacanth *Latimeria chalumnae*. *Marine Biology* 120: 171–180.

Fricke, H & Hissmann, K. 2000. Feeding ecology and evolutionary survival of the living coelacanth *Latimeria chalumnae*. *Marine Biology* 136: 379–386.

Fricke, H, Hissmann, K, Schauer, J, Erdmann, M, Moosa, MK & Plante, R. 2000. Biogeography of the Indonesian coelacanth. *Nature, London* 403: 38.

Fricke, H, Hissmann, K, Schauer, J & Plante, R. 1995. Yet more danger for coelacanths. *Nature, London* 374: 314.

Fricke, H, Hissmann, K, Schauer, J, Reinicke, O, Kasang, L & Plante, R. 1991. Habitat and population size of the coelacanth *Latimeria chalumnae* at Grande Comore. *Environmental Biology of Fishes* 32: 287–300.

Fricke, H & Plante, R. 1988. Habitat requirements of the living coelacanth *Latimeria chalumnae* at Grande Comore, Indian Ocean. *Die Naturwissenschaften* 75: 149–151.

Fricke, H & Plante, R. 2001. Silver coelacanths from Spain are not proofs of a pre-scientific discovery. *Environmental Biology of Fishes* 61: 461–463.

Fricke, H, Reinicke, O, Hofer, H & Nachtigall, W. 1987. Locomotion of the coelacanth *Latimeria chalumnae* in its natural environment. *Nature, London* 329: 331–333.

Fricke, H, Schauer, J, Hissmann, K, Kasang, L & Plante, R. 1991. Coelacanth *Latimeria chalumnae* aggregates in caves: first observations on their resting habitat and social behavior. *Environmental Biology of Fishes* 30: 281–285.

Froese, R & Palomares, MLD. 2000. Growth, natural mortality, length-weight relationship, maximum length and length-at-first-maturity of the coelacanth *Latimeria chalumnae*. *Environmental Biology of Fishes* 58: 45–52.

Gon, O. 1996. Fifty years of marine fish systematics at the JLB Smith Institute of Ichthyology. *Transactions of the Royal Society of South Africa* 51: 45–78.

Gorr, T & Kleinschmidt, T. 1993. Evolutionary relationships of the coelacanth. *American Scientist* 81: 72–82.

Green, AN, Uken, R, Ramsay, P, Leuci, R & Perrit, S. 2009. Potential sites for suitable coelacanth habitat using bathymetric data from the western Indian Ocean. *South African Journal of Science* 105: 151–154.

Greenwood, PH. 1968. Professor JLB Smith; obituary. *Nature, London* 217: 690–691.

Greenwood, PH. 1989. Fifty years a 'living fossil' – the coelacanth fish *Latimeria chalumnae*. *Biologist* 36: 15–19.

Greenwood, PH. 1993. *Latimeria chalumnae* – the living coelacanth. *ICHTHOS Special Publication* 1: 1–18.

Griffith, RW. 1973. A living coelacanth in the Comoro Islands. *Discovery* 9(1): 27–33.

Griffith, RW. 1991. Guppies, toadfish, lungfish, coelacanths and frogs: a scenario for the evolution of urea retention in fishes. *Environmental Biology of Fishes* 32: 199–218.

Griffith, RW & Thompson, KS. 1973. *Latimeria chalumnae*: reproduction and conservation. *Nature, London* 242: 617–618.

Griffith, RW & Thompson, KS. 1973. Observations on a dying coelacanth. *American Zoologist* 12: 730.

Heemstra, PC, Freeman, ALJ, Yan Wong, H, Hensley, DA & Rabesandratana, HD. 1996. First authentic capture of a coelacanth, *Latimeria chalumnae* (Pisces: Latimeriidae), off Madagascar. *South African Journal of Science* 92: 150–151.

Heemstra, PC, Fricke, H, Hissmann, K, Schauer, J, Smale, M & Sink, K. 2006. Interactions of fishes with particular reference to coelacanths in the canyons at Sodwana Bay and the St Lucia Marine Protected Area of South Africa. *South African Journal of Science* 102: 461–465.

Hess, A. 2004. Quastenflosse in Tanzania (first Tanzanian coelacanth discovery). *Habari*, June 2004 19(2): 2–4.

Hissmann, K & Fricke, H. 1996. Movements of the epicaudal fin in coelacanths. *Copeia* 1996: 606–615.

Hissmann, K, Fricke, H & Schauer, J. 1998. Population monitoring of the coelacanth (*Latimeria chalumnae*). *Conservation Biology* 12: 758–765.

Hissmann, K, Fricke, H & Schauer, J. 2000. Pattern of time and space utilisation in coelacanths (*Latimeria chalumnae*), determined by ultrasonic telemetry. *Marine Biology* 136: 943–952.

Hissmann, K, Fricke, H, Schauer, J, Ribbink, AJ, Roberts, MJ, Sink, K & Heemstra, PC. 2006. The South African coelacanths – an account of what is known after three submersible expeditions. *South African Journal of Science* 102: 491–500.

Hughes, GM & Itazawa, Y. 1972. The effect of temperature on the respiratory function of coelacanth blood. *Experientia* 28: 1,247.

Jackson, PBN. 1987. Margaret Smith the humanitarian: a time to keep. *ICHTHOS* 17: 17–19.

Jackson, PBN. 1996. Variations on a theme: the three directors of the first fifty years of the JLB Smith Institute of Ichthyology. *Transactions of the Royal Society of South Africa* 51: 33–43.

Lampert, KP, Fricke, H, Hissmann, K, Schauer, J, Blassmann, K, Ngatunga, BP & Schartl, M. 2012. Population divergence in East African coelacanths. *Current Biology* 22(11): 439–440.

Locket, NA. 1976. A future for the coelacanth? *New Scientist* 70: 456–458.

Locket, NA. 1980. Some advances in coelacanth biology. *Proceedings of the Royal Society of London* Series B 208: 265–307.

Locket, NA & Griffith, RW. 1972. Observations on a living coelacanth. *Nature, London* 237(5351): 175.

Malherbe, EH. 1981. *Never a Dull Moment.* Howard Timmins, Cape Town. 419 pp.

McCabe, H & Wright, J. 2000. Tangled tale of a lost, stolen and disputed coelacanth. *Nature, London* 406: 114.

McCosker, JE. 1979. Inferred natural history of the living coelacanth. pp. 17–24. In: JE McCosker & MD Lagios (eds) 1979. The Biology and Physiology of the Living Coelacanth. *Occasional Papers of the California Academy of Sciences* 134, San Francisco.

McCosker, JE & Lagios, MD (eds). 1979. The Biology and Physiology of the Living Coelacanth. *Occasional Papers of the California Academy of Sciences* 134: 1–175.

McKenzie, D. 1995. Can the coelacanth be saved? *New Scientist* 146: 6.

Meyer, A. 1993. Coelacanth controversy. *American Scientist* 81: 209–210.

Millot, J. 1953. Notre coelacanthe. *Revue Madagascar, Tananarive* 1: 18–20.

Millot, J. 1954. New facts about coelacanths. *Nature, London* 174: 426–427.

Millot, J. 1954. Les nouveaux coelacanthes. *Nature, Paris* 3228: 1210–124.

Millot, J. 1954. Les troisième coelacanthe. *Naturaliste malgache* suppl. 1: 1–26.

Millot, J. 1955. À propos des coelacanthes. *Nature, Paris* 3241: 202–203.

Millot, J. 1955. First observations on a living coelacanth. *Nature, London* 175: 362–363.

Millot, J & Anthony, J. 1958. Anatomie de *Latimeria chalumnae* I: Squelette et muscles. CNRS, Paris.

Millot, J & Anthony, J. 1965. Anatomie de *Latimeria chalumnae* II: Système nerveux et organes des sens. CNRS, Paris.

Millot, J & Anthony, J. 1974. Les oeufs du coelacanthe. *Science at Nature, Paris* 121: 1–4.

Minshull, J. 2009. Memories of the past – the Zimbabwe coelacanth. *African Fisherman* 20(3): 8–11.

Musick, JA, Bruton, MN & Balon, EK (eds). 1991. *The Biology of* Latimeria *and Evolution of Coelacanths.* Kluwer Academic Publishers, Dordrecht. 446 pp.

Myking, LM. 1977. Old Four Legs: the living fossil. *Sea Frontiers* 23: 334–341.

Nulens, R, Scott, L & Herbin, M. 2011. An updated inventory of all known specimens of the coelacanth, *Latimeria* spp. *Smithiana Special Publication* 3: 1–52.

Owens, HL, Bentley, AC & Townsend Peterson, A. 2012. Predicting suitable environments and potential occurrences for coelacanths (*Latimeria* spp.). *Biodiversity and Conservation* 21: 577–587.

Plante, R, Fricke, H & Hissmann, K. 1998. Coelacanth population, conservation and fishery activity at Grande Comore, Western Indian Ocean. *Marine Ecology Progress Series* 166: 231–236.

Pote, J. 1996. Historical highlights: the JLB Smith Institute of Ichthyology and the Department of Ichthyology & Fisheries Science 1946–1996. *Transactions of the Royal Society of South Africa* 51: 5–31.

Pouyaud, L, Wirjoatmodjo, S, Rachmatika, I, Tjakrawidjaja, A, Hadiaty, RK & Hadie, W. 1999. Une nouvelle espèce de coelacanthe. Preuves génétiques et morphologiques. *Comptes rendus de l'Académie des Sciences – Sciences de la vie* 322(3): 261–267.

Ribbink, AJ & Roberts, MJ. 2006. African Coelacanth Ecosystem Programme: an overview of the conference contributions. *South African Journal of Science* 102: 409–415.

Roberts, MJ, Ribbink, AJ, Morris, T, van den Berg, MA, Engelbrecht, DC & Harding, RT. 2006. Oceanographic environment of the Sodwana Bay coelacanths (*Latimeria chalumnae*), South Africa. *South African Journal of Science* 102: 435–443.

Schaeffer, B. 1952. *Latimeria* and the history of the coelacanth fishes. *Transactions of the New York Academy of Sciences* 15(2): 170–178.

Schartl, M, Hornung, U, Hissmann, K, Schauer, J & Fricke, H. 2005. Relatedness among East African coelacanths. *Nature, London* 435: 901.

Shubin, N. 2009. *Your Inner Fish*. Penguin, London. 237 pp.

Skelton, PH & Lutjeharms, JRE (eds). 1996. The JLB Smith Institute of Ichthyology – 50 Years. *Transactions of the Royal Society of South Africa* 51: 1–320.

Smith, CL, Rand, CS, Schaeffer, B & Atz, JW. 1975. *Latimeria*, the living coelacanth, is ovoviviparous. *Science* 190(4219): 1,105–1,106.

Smith, JLB. 1931. New and little known fishes from the south and east coasts of South Africa. *Records of the Albany Museum* 4(1): 145–160.

Smith, JLB. 1939. A living fish of Mesozoic type. *Nature, London* 143(3620): 455–456.

Smith, JLB. 1939. The living coelacanthid fish from South Africa. *Nature, London* 143(3627): 748–750.

Smith, JLB. 1939. A surviving fish of the order Actinistia. *Transactions of the Royal Society of South Africa* 27(1): 47–50.

Smith, JLB. 1940. *A Simplified System of Organic Identification*. Nasionale Pers, Cape Town. 48 pp.

Smith, JLB. 1940. A living coelacanthid fish from South Africa. *Transactions of the Royal Society of South Africa* 28: 1–106.

Smith, JLB. 1940. A living fossil. *Cape Naturalist* 1: 321–328.

Smith, JLB. 1941. *A System of Qualitative Organic Analysis*. Nasionale Pers, Cape Town. 63 pp.

Smith, JLB. 1949. *The Sea Fishes of Southern Africa*. Central News Agency, Cape Town. 550 pp.

Smith, JLB. 1953. The second coelacanth. *Nature, London* 171: 99–101.

Smith, JLB. 1953. The second coelacanth. *Copeia* (1): 72.

Smith, JLB. 1953. The two coelacanths. *South African Museum Association Bulletin* 5(8): 206–210.

Smith, JLB. 1953. Problems of the coelacanth. *South African Journal of Science* 49(9): 279–281.

Smith, JLB. 1955. Live coelacanths. *Nature, London* 176: 473.

Smith, JLB. 1956. *Old Fourlegs. The Story of the Coelacanth*. Longmans Green & Co., London. 260 pp.

Smith, JLB. 1965. *The Sea Fishes of Southern Africa*. Central News Agency, Cape Town. 580 pp.

Smith, JLB. 1968 (posthumous). *Our Fishes*. Voortrekkerpers, Johannesburg. 263 pp.

Smith, JLB. 1968 (posthumous). *High Tide*. Books of Africa. Cape Town. 165 pp.

Smith, JLB. 1977 (posthumous). *Smith's Sea Fishes*. Valiant Press, Sandton. 580 pp.

Smith, JLB & Smith, MM. 1969. *The Fishes of the Seychelles*. JLB Smith Institute of Ichthyology, Grahamstown. 223 pp.

Smith, MM. 1969. JLB Smith, his life, work, bibliography and list of new species. *Rhodes University Department of Ichthyology Occasional Paper* 16: 185–215.

Smith, MM. 1979. The influence of the coelacanth on African ichthyology. *Occasional Papers of the California Academy of Sciences* 134: 11–16.

Smith, MM. 1980. The search for the world's oldest fish. *Oceans* 3(6): 26–36.

Smith, MM. 1987. It all started this way. *ICHTHOS* 15: 1–2.

Smith, MM & Heemstra, PC (eds). 1986. *Smiths' Sea Fishes*. Macmillan, Johannesburg. 1,047 pp.

Stobbs, RE. 1989. Laxative lipids and the survival of the living coelacanth. *South African Journal of Science* 85(9): 557–558.

Stobbs, RE. 1996. The changing face of *Latimeria* – and more mythology? *ICHTHOS* 50: 3–6.

Stobbs, RE & Bruton, MN. 1991. The fishery of the Comoros, with comments on its possible impact on coelacanth survival. *Environmental Biology of Fishes* 32: 341–359.

Suzuki, N, Suyehiro, Y & Hamada, T. 1985. Initial report of expeditions for coelacanth – Part 1 – Field studies in 1981 and 1983. *Scientific Papers of the College of Arts and Sciences, the University of Tokyo* 35: 37–79.

Thomson, KS. 1966. The history of the coelacanth. *Discovery* 2(1): 27–32.

Thomson, KS. 1969. The biology of the lobe-finned fishes. *Biological Reviews* 44: 91–154.

Thomson, KS. 1973. New observations on the coelacanth fish, *Latimeria chalumnae*. *Copeia* 1973(4): 813–814.

Thomson, KS. 1973. Secrets of the coelacanth. *Natural History, New York* 82: 58–65.

Thomson, KS. 1991. The story of the coelacanth. *Oceanus* 34: 38–43.

Thomson, KS. 1991. *Living Fossil. The Story of the Coelacanth*. Norton, New York. 252 pp.

Thornycroft, RE & Booth, AJ. 2012. Computer-aided identification of coelacanths, *Latimeria chalumnae*, using scale patterns. *Marine Biology Research* 8: 300–306.

Uyeno, T. 1991. Observations on locomotion and feeding of released coelacanths, *Latimeria chalumnae*. *Environmental Biology of Fishes* 32: 267–273.

Uyeno, T & Tsutsumi, T. 1991. Stomach contents of *Latimeria chalumnae* and further notes on its feeding habits. *Environmental Biology of Fishes* 32: 275–279.

Watson, DMS. 1921. On the coelacanth fish. *Annals and Magazine of Natural History* 8: 320–337.

Weinberg, S. 1999. *A Fish Caught in Time*. Fourth Estate Limited, London. 239 pp.

White, E. 1953. The coelacanth fishes. *Discovery, London* 14: 113–117.

White, E. 1953. More about the coelacanths. *Discovery, London* 14: 332–335.

Coelacanth necklace

www.paleopanthera.com